Proceedings in Technology Transfer

The book series "Proceedings in Technology Transfer" provides a valuable platform for researchers, practitioners, and policymakers to share their experiences and insights related to technology transfer. Our aim is to facilitate knowledge exchange between academic researchers and industry professionals, accelerating the introduction of new technologies and driving economic growth.

Our series publishes books and conference proceedings that focus on original research findings, best practice recommendations, emerging trends, and technology evaluations. The series covers all topics which explore the transfer of technology in various industry sectors that require innovations and economic growth.

Join us in advancing the technology transfer and its impact on innovation. Publish your proceedings with us and offer insights and expertise to professionals in your field.

Andreas Öchsner
Editor

Collaborative Research Advancing Engineering Solutions for Real-World Challenges 2

The 2024 Postgraduate Seminar in Esslingen

Springer

Editor
Andreas Öchsner
Faculty of Mechanical and Systems Engineering
Esslingen University of Applied Sciences
Esslingen, Germany

ISSN 2948-2321 ISSN 2948-233X (electronic)
Proceedings in Technology Transfer
ISBN 978-3-031-79092-8 ISBN 978-3-031-79093-5 (eBook)
https://doi.org/10.1007/978-3-031-79093-5

This Springer imprint is published by the registered company Springer Nature Switzerland AG
The registered company address is: Gewerbestrasse 11, 6330 Cham, Switzerland

Preface

This monograph contains the revised manuscripts of the 4th Ph.D. seminar, which was held on 21st June 2024 at Esslingen University, Germany. Universities of applied sciences in Germany are nowadays also involved in Ph.D. programs, whereas the academic title is finally granted by these applied universities (sometimes in some kind of consortium) or in cooperation with classical universities. This fact results in an increasing number of Ph.D. students at universities of applied sciences. However, these students are many times scattered around different laboratories and research directions and a binding organizational structure, as in the case of classical universities, might be missing. In order to strengthen the integration of these students, Esslingen University offers an annual Ph.D. seminar. This fourth seminar was again open to other universities of applied sciences and even classical universities. It is hoped that this format is beneficial for the Ph.D. students and the scientific cooperation among universities.

Esslingen, Germany
September 2024

Andreas Öchsner

Preface

[illegible]

Contents

Postgraduate Lectures

Postgraduate Lectures

Rods with Long-Range Force Interactions: Governing Equations, Boundary Conditions and Benchmark Problems

Alvina Oksanchenko(✉) and Konstantin Naumenko

Institut für Mechanik, Otto-von-Guericke-Universität Magdeburg, Universitätsplatz 2, Magdeburg 39106, Germany
alvina.oksanchenko@ovgu.de, konstantin.naumenko@ovgu.de

Abstract. In this study a theory of rods is developed and analyzed by taking into account long-range force interactions between cross sections. The balance of momentum and principle of virtual work are formulated. Different variants of the displacement and force boundary conditions are introduced by the specifying force and virtual power fluxes through a cross section of the rod. With a special constitutive assumption for the interaction force a peridynamic type theory is derived. Based on the analytical solutions under both static and dynamic loadings and for several types of boundary conditions including fixed-fixed and fixed-free edges, benchmark problems are formulated. The numerical mesh free type method is applied to solve the governing equations and numerical results are compared with analytical solutions.

Keywords: Long-range forces · Peridynamics · Uni-axial rod · Virtual work · Analytical solution

1 Introduction

Over past decades significant progress is observed in development of generalized non-local theories in continuum mechanics. Motivated by new applications, such as micro- and nanometer scale components, metamaterials, composites with high contrast in properties and laminates with extremely thin layers, several extensions of the classical continuum mechanics (CCM) are proposed in the literature.

Many weakly non-local theories are derived by introducing higher order deformation gradients as arguments of the strain energy density [1,2]. With a variational principle, strain gradient theories of different order can be formulated. In the case of elasticity, Nth-order gradient terms result to a in a system of partial differential equations of the $2N$th order [3]. Such theories include higher order stress tensors and require additional boundary conditions corresponding to the order of included gradients.

On the other hand strongly-nonlocal theories are based on integral type strain measures and long-range forces. Peridynamics (PD) is an example of such

A. Öchsner (ed.), *Collaborative Research Advancing Engineering Solutions for Real-World Challenges 2*, Proceedings in Technology Transfer,
https://doi.org/10.1007/978-3-031-79093-5_1

extended continuum theory. Integro-differential equations of motion are derived within PD [4]. Special numerical solution procedures are proposed, usually applying a mesh free methods, [5–7]. Results of many numerical studies illustrate the ability of PD to capture complex phenomena such as initiation of crack patterns [8–10], crack branching [11], crack kinking [12], crack interaction with heterogeneities [13,14], among others. The recent review on applications of PD theory is presented in [15].

By analogy with CCM, robust PD theories for elementary structures can be developed by formulating governing equations directly for deformable lines for beams, planes in the case of plates and surfaces for shells. PD equations of motion for straight rods are derive in [16,17]. Discretized PD equations of motion for the axial displacement and the deflection for straight beams are presented in [18]. PD integro-differential equations of motion are derived in [19,20] for straight beams and in [21,22] for initially curved beams. Discretized PD equations for plates and shells are presented in [23]. A linear PD shell theory with respect to the displacements and small rotations of the base surface is developed in [24]. A non-linear PD plate theory with finite rotations is presented in [25] a non-linear PD plate theory considering finite rotations is presented.

An issue of any numerical method to solve PD equations is the analysis of results as they behave by reducing the horizon size. For elastic solids the solutions must converge to those of CCM as the horizon size approaches zero. Another point is the effect of PD constraints and boundary conditions. To address these issues closed-form analytical solutions are required. Unlike CCM, where many analytical solutions for rods beams, plates, shells and three-dimensional solids are derived, only several explicit analytical solutions to PD equations are available in the literature. For initially straight rods analytical solutions to PD equations are presented in [17,26–28]. For initially straight beams with various boundary constraints series solutions to PD equations are derived in [20,29]. Governing PD equations and explicit solutions for buckling of straight beams are given in [19]. Analytical solutions to PD equations for three-dimensional infinite media are obtained in [30,31].

The aim of this study is to present governing equations of mechanics with long range force interactions based on the example of the uni-axial rod. The balance of momentum and the principle of virtual work will be derived. With a constitutive assumption for the elastic rod, the bond force density can be related to the bond strain as discussed within the peridynamic theory. By specifying the peridynamic fluxes through the cross sections of the rod the a class of boundary conditions will be presented. The derived equations will be solved by the mesh free method discussed in the literature. To verify the accuracy of the numerical methods the results will be compared with closed-form analytical solutions for several benchmark problems.

2 Peridynamic Theory of Rods

In this section the governing equations for the peridynamic theory of rods is discussed. The balance of momentum is The special attention is given to the principle of virtual work, where the new concepts of absorbed and supplied virtual works are given. Based on this the peridynamic boundary conditions will be introduced.

2.1 Balance of Momentum

Consider a straight rod, as shown in Fig. 1. For a part P of the rod the momentum is computed as follows

$$I(\mathrm{P}) = \int\limits_{(\mathrm{P})} A\rho v \mathrm{d}x, \tag{1}$$

where A is the cross sectional area, v is the velocity and x is the axial coordinate.

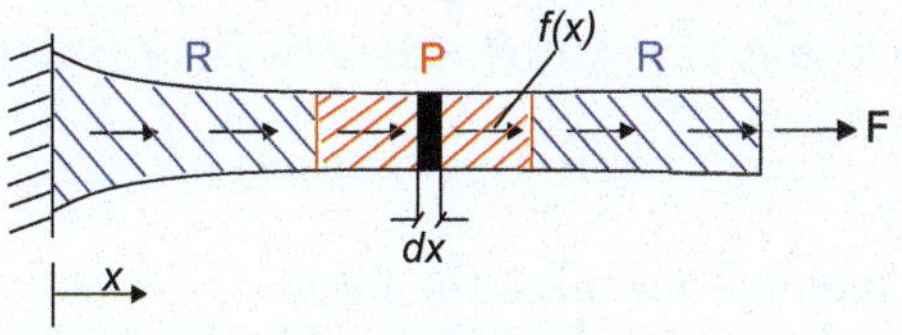

Fig. 1. Rod with a variable cross section. Geometry and loading

The balance of momentum for a part of the rod can be formulated as follows

$$\frac{\mathrm{d}I(\mathrm{P})}{\mathrm{d}t} = F(\mathrm{R},\mathrm{P}) + F\left(\mathrm{Env},\mathrm{P}\right), \tag{2}$$

where $F(\mathrm{R},\mathrm{P})$ is the force acting on the part P from the environment R, Fig. 1. The long-range force acting from an infinitesimal part $\mathrm{d}x'$ on an infinitesimal part $\mathrm{d}x$, see Fig. 2 can be expressed by the force density $n\left(x',x\right)$ as follows

$$F\left(\mathrm{d}x',\mathrm{d}x\right) = n\left(x',x\right)\mathrm{d}x'\mathrm{d}x \tag{3}$$

Consequently the force acting from the environment R on the element $\mathrm{d}x$ is expressed by

$$F(\mathrm{R},\mathrm{d}x) = \int\limits_{(\mathrm{R})} n\left(x',x\right)\mathrm{d}x'\mathrm{d}x \tag{4}$$

The resultant force from the environment R on a part of the rod P is computed as follows

$$F(\mathrm{R},\mathrm{P}) = \int\limits_{(\mathrm{P})}\int\limits_{(\mathrm{R})} n\left(x',x\right)\mathrm{d}x'\mathrm{d}x \tag{5}$$

Applying the balance of momentum to two disjoint finite parts P_I and P_{II} and the part $\mathrm{P}=\mathrm{P}_I \bigcup \mathrm{P}_{II}$ it can be shown that

$$F\left(\mathrm{P}_{II},\mathrm{P}_I\right) + F\left(\mathrm{P}_I,\mathrm{P}_{II}\right) = 0, \quad F(\mathrm{P},\mathrm{P}) = 0 \tag{6}$$

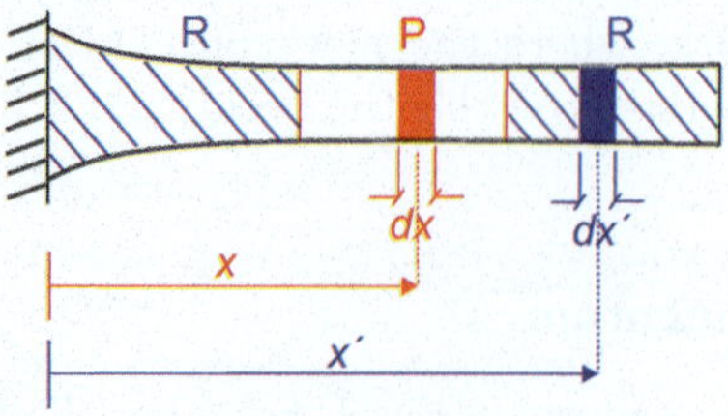

Fig. 2. Infinitesimal parts of the rod with coordinates x and x'

Furthermore, for the bond force density we have

$$n(x', x) = -n(x, x'), \quad n(x, x) = 0 \tag{7}$$

For the region B such that $\mathrm{B} = \mathrm{P} \cup R$ and with Eq. 6 it follows

$$F(\mathrm{R}, \mathrm{P}) = F(\mathrm{B}, \mathrm{P}) \tag{8}$$

Now the balance of momentum takes the form

$$\int_{(\mathrm{P})} \rho A \dot{v} \mathrm{d}x = \int_{(\mathrm{P})} \int_{(\mathrm{B})} n(x', x) \, \mathrm{d}x' \mathrm{d}x + \int_{(\mathrm{P})} f(x) \mathrm{d}x \tag{9}$$

Equation 9 is valid for any part P of the rod. Therefore, the localization yields

$$\rho A \dot{v} = \int_{(\mathrm{B})} n(x', x) \, \mathrm{d}x' + f(x) \tag{10}$$

2.2 Principle of Virtual Work

In the case of statics Eq. 10 takes the form

$$\int_{(\mathrm{B})} n(x', x) \, \mathrm{d}x' + f(x) = 0 \tag{11}$$

Multiplying Eq. 11 by the virtual displacement $\delta u(x)$ and integrating over the part P provides

$$\int_{(\mathrm{P})} \int_{(\mathrm{B})} n(x', x) \, \delta u(x) \mathrm{d}x' \mathrm{d}x + \int_{(\mathrm{P})} f(x) \delta u(x) \mathrm{d}x = 0 \tag{12}$$

With the following decomposition

$$n \delta u = n \delta \bar{u} + \frac{1}{2} n \delta u_{\Delta}, \quad \bar{u} = \frac{u + u'}{2}, \quad u_{\Delta} = u' - u \tag{13}$$

the principle of virtual work takes the form

$$L_a(B,P) = L_s(R,P) + L_f(P), \tag{14}$$

where $L_a(B,P)$ is the work absorbed by the part P, $L_s(R,P)$ is the work supplied from the environment R and $L_f(P)$ is the work of the body force f defined as follows

$$L_s(R,P) = \int_{(P)}\int_{(R)} n\delta\bar{u}dx'dx, \quad L_f(P) = \int_{(P)} f\delta u dx, \quad L_a(B,P) = \int_{(P)}\int_{(B)} \frac{1}{2}n\delta u_\Delta dx'dx \tag{15}$$

2.3 Constitutive Law and Strain Energy Density

Figure 3a shows a finite line element (bond) $\xi = x' - x$ in the reference configuration. The corresponding line element in the actual configuration is $u' - u + x' - x = u_\Delta + \xi$, see Fig. 3b. The strain for the line element (bond strain) can be defined as follows

$$\varepsilon_b = \frac{|u_\Delta + \xi| - |\xi|}{|\xi|} = \frac{|u_\Delta + \xi|}{|\xi|} - 1 \tag{16}$$

For deformations with $u_\Delta + \xi > 0$, i.e. by excluding penetration of points under compression, Eq. 16 is simplified to

$$\varepsilon_b = \frac{u_\Delta}{\xi} \tag{17}$$

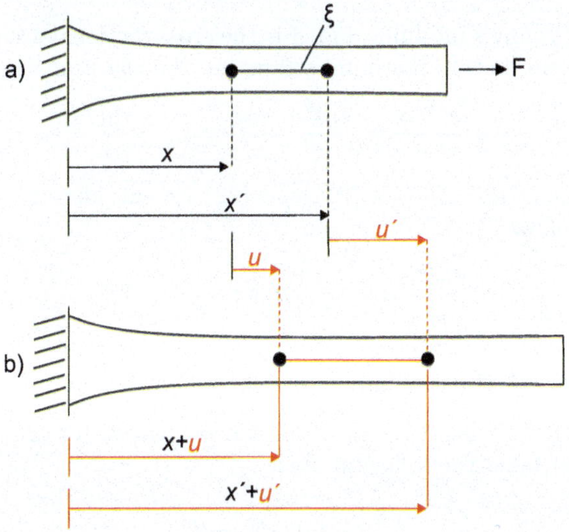

Fig. 3. Axial deformation of the rod. **a** Reference configuration, **b** actual configuration

For the linear-elastic rods the following constitutive equation can be applied

$$n(x', x) = C\varepsilon_b(x', x)\,\omega(|x' - x|)\,\mathrm{sgn}(x' - x), \tag{18}$$

where C is a stiffness parameter and ω is defined as follows

$$\omega(x', x) = \begin{cases} 1, |x' - x| \leq \delta_{\mathrm{H}}, \\ 0, \text{otherwise} \end{cases} \tag{19}$$

where δ_{H} is the horizon size. The constitutive Eq. 18 takes the form

$$n(x', x) = \begin{cases} C\dfrac{u' - u}{|x' - x|}, 0 < |x' - x| \leq \delta_{\mathrm{H}} \\ 0, \qquad \text{otherwise} \end{cases} \tag{20}$$

Following classifications in the literature on PD theories, [4], the constitutive assumptions (20) corresponds to the bond-based theory. In this case the force interaction in a bond $x' - x$ depends only on its own deformation, and deformations of other bonds connected to x do not influence its mechanical state.

For elastic materials the absorbed virtual work $L_{\mathrm{a}}(\mathrm{B}, \mathrm{P})$ is the variation of the strain energy density. Indeed from Eqs. 27 and 20 we obtain

$$L_{\mathrm{a}}(\mathrm{B}, \mathrm{P}) = \delta \int\limits_{(\mathrm{P})} \bar{e}_s(x)\mathrm{d}x, \quad \bar{e}_s(x) = \int\limits_{x-\delta_H}^{x+\delta_H} \frac{C}{4}\frac{u_\Delta^2}{|x' - x|}\mathrm{d}x' \tag{21}$$

For the homogeneous deformation $u(x) = \varepsilon_0 x$ the stored PD strain energy density $\bar{e}_s$ must coincide the strain energy density of CCM, which is e_s defined as follow

$$e_s(x) = \frac{1}{2}EA\left(\frac{\mathrm{d}u}{\mathrm{d}x}\right)^2, \tag{22}$$

where E is the Young's modulus and A is the cross sectional area of the rod. For homogeneous deformation the bond strain and the classical strain coincide, i.e.

$$\varepsilon_b = \frac{u' - u}{x' - x} = \frac{\varepsilon_0 x' - \varepsilon_0 x}{x' - x} = \varepsilon_0, \quad \frac{\mathrm{d}u}{\mathrm{d}x} = \varepsilon_0 \tag{23}$$

By equating the strain energies (21) and (22) we can identify the PD stiffness parameter as follows

$$\bar{e}_s = C\frac{\varepsilon_0^2\delta_{\mathrm{H}}^2}{4} = e_s = \frac{1}{2}EA\varepsilon_0^2, \quad \Rightarrow \quad C = \frac{2EA}{\delta_H^2} \tag{24}$$

2.4 Equation of Motion and Boundary Conditions

With the constitutive law (20) and the stiffness parameter (24), the balance of momentum (10) takes the following form

$$\rho A\dot{v} = \int\limits_{x-\delta_H}^{x+\delta_H} \frac{2EA}{\delta_H^2}\frac{u(x') - u(x)}{|x' - x|}\mathrm{d}x' + f(x) \tag{25}$$

In order to solve PD equation of motion, PD boundary conditions are required. In the classical theory of rods the boundary conditions for the displacement and/or its first derivative are defined for the end point. In PD the boundary conditions must be defined for the displacement ans/or bond force density within the region around the end point. To discuss examples of boundary conditions consider the PD fluxes through a cross section of the rod including the force flux (5) and the supplied virtual work (27). Let the coordinate of the cross section be x_c Since the range of the bond force density is limited by the horizon size δ_H the force supplied from the environment into the rod through the cross section x_C is computed as follows

$$F(\mathrm{R_c}, \mathrm{P_c}) = \int\limits_{(\mathrm{P_c})}\int\limits_{(\mathrm{R_c})} n\,(x', x)\,\mathrm{d}x'\mathrm{d}x = \int\limits_{x_\mathrm{c}-\delta_\mathrm{H}}^{x_\mathrm{c}}\int\limits_{x_\mathrm{c}}^{x+\delta_\mathrm{H}} n\,(x', x)\,\mathrm{d}x'\mathrm{d}x \tag{26}$$

For the supplied virtual power we obtain

$$L_\mathrm{s}(\mathrm{R_c}, \mathrm{P_c}) = \int\limits_{(\mathrm{P_c})}\int\limits_{(\mathrm{R_c})} n\delta\bar{u}\mathrm{d}x'\mathrm{d}x = \int\limits_{x_\mathrm{c}-\delta_\mathrm{H}}^{x_\mathrm{c}}\int\limits_{x_\mathrm{c}}^{x+\delta_\mathrm{H}} n\delta\bar{u}\mathrm{d}x'\mathrm{d}x \tag{27}$$

As an example consider a free boundary at x_c. In this case we have to set $F(\mathrm{R_c}, \mathrm{P_c}) = 0$ and $L_\mathrm{s}(\mathrm{R_c}, \mathrm{P_c}) = 0$. By taking into account Eq. 20 we obtain the following boundary condition

$$u(x_\mathrm{c} + \xi) = u(x_\mathrm{c} - \xi), \qquad \xi \in [0, \delta] \tag{28}$$

Another example is the fixed edge such that $u(x_\mathrm{c}) = 0$. To assure that the virtual power is not supplied we have to set

$$u(x_\mathrm{c} + \xi) = -u(x_\mathrm{c} - \xi), \qquad \xi \in [0, \delta] \tag{29}$$

3 Numerical Solutions

In this section, different numerical solutions for static and dynamic conditions for the rod are presented. In each case, various boundary conditions are considered, such as fixed boundary conditions and fixed-free boundary conditions.

3.1 Static Loading

For the numerical solution the discretized equation of motion (25) is formulated. To this end the rod is discretized by the node lattice with N nodes such that the distance between the nodes is $\Delta x = L/N$ with L being the length of the rod. The horizon size is chosen $\delta_\mathrm{H} = 3.015\Delta x$, as recommended in [5]. For the case of static loading we obtain

$$C\sum_k \frac{U_k - U_i}{|\xi_{ki}|} V_k + f_i = 0, \tag{30}$$

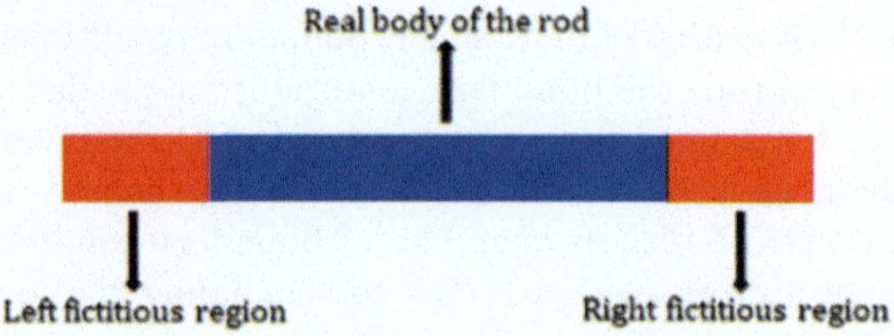

Fig. 4. Rod model with fictitious regions at the ends

where U_i is the displacement value at the node i, $\xi_{ki} = x_k - x_i$, with x_i being the coordinate of the node i, V_i is the size of the node i and f_i is the body force value at the node i.

3.1.1 Fixed-Fixed Boundaries A rod with a length $L = 1$ m is considered. The Young's modulus is specified as $E = 200$ GPa. The axial body force is applied in the form of $f(x) = 1000x(1-x)$. Boundary conditions (BCs) in PD are applied to the fictitious regions with the length δ_{H}, with (Fig. 4). For the fixed-fixed ends the BSc (29) are applied. Therefore we have to set

$$\begin{cases} U(-\eta) = -U(\eta) \text{ for } 0 \leq \eta \leq \delta_{\mathrm{H}} \\ U(L-\eta) = -U(L+\eta) \text{ for } 0 \leq \eta \leq \delta_{\mathrm{H}} \end{cases} \tag{31}$$

Equations 30 and BCs (31) are put in the matrix form as follows

$$[K][U] = [F], \tag{32}$$

where $[K]$ is the PD stiffness matrix, $[U]$ is the nodal displacement vector and $[F]$ is the loading vector. Solving the linear system (29) the nodal displacements are obtained. Figure 5 illustrates the displacement versus location of the rod for $N = 100$.

3.1.2 Fixed-Free Boundary Conditions The same properties of the rod as in the previous case are considered, except that the boundary conditions are fixed on the left side and free on the right side. Taking into account Eqs. 28 and 29 we obtain

$$\begin{cases} U(-\eta) = -U(\eta) \text{ for } 0 \leq \eta \leq \delta_{\mathrm{H}} \\ U(L-\eta) = U(L+\eta) \text{ for } 0 \leq \eta \leq \delta_{\mathrm{H}} \end{cases} \tag{33}$$

The axial displacement versus axial coordinate are presented in Fig. 6.

3.2 Dynamic Loadings

For the dynamic case the following discretized form of Eq. 25 is obtained

$$\rho \ddot{U}_k = C \sum_k \frac{U_i - U_k}{|\xi_{ik}|} V_k + f_i \tag{34}$$

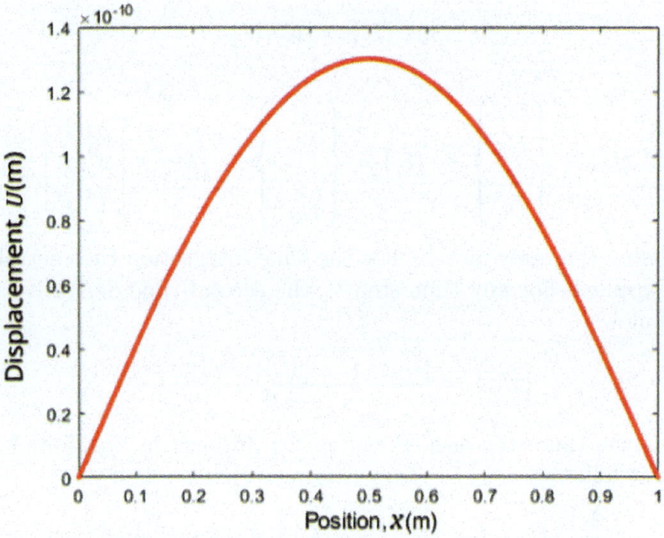

Fig. 5. Numerical result for the displacement versus axial coordinate of the rod with fixed-fixed BCs

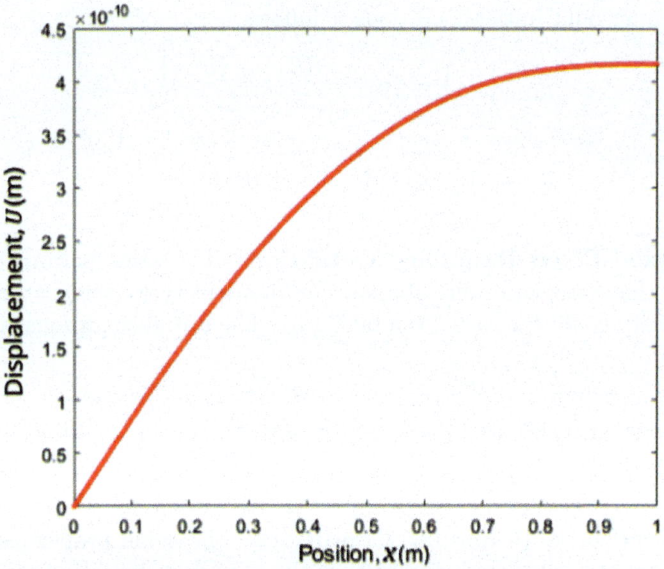

Fig. 6. Numerical result for displacement versus axial coordinate of the rod with fixed-free BCs

Equation 34 can be put into the following matrix form

$$[M]\{\ddot{U}\} = [K]\{U\}, \tag{35}$$

where

$$\{U\} = \begin{Bmatrix} U_1 \\ U_2 \\ \cdots \\ U_n \end{Bmatrix}, \quad \{\ddot{U}\} = \begin{Bmatrix} \ddot{U}_1 \\ \ddot{U}_2 \\ \cdots\cdot \\ \ddot{U}_n \end{Bmatrix}, \quad [M] = \begin{bmatrix} \rho\,0\,0 \\ 0\,\rho\,0 \\ 0\,0\,\rho \end{bmatrix},$$

with $[M]$ being the mass matrix. For the time integration the central difference method is applied. For any time step t_n the second time derivative is approximated as follows

$$\left\{\ddot{U}_n\right\} = \frac{\{U_{n-1}\} - 2\{U_n\} + \{U_{n+1}\}}{\Delta t^2}, \tag{36}$$

where Δt is the time step size. Plugging Eq. 36 back in Eq. 35 and after some calculation it follows

$$\frac{[M]}{\Delta t^2}\{U_{n+1}\} = \left(2\frac{[M]}{\Delta t^2} + [K]\right)\{U_n\} + \frac{[M]}{\Delta t^2}\{U_{n-1}\} \tag{37}$$

For the initial time step t_0 we obtain

$$\frac{[M]}{\Delta t^2}\{U_1\} = \left(2\frac{[M]}{\Delta t^2} + [K]\right)\{U_0\} + \frac{[M]}{\Delta t^2}\{U_{-1}\} \tag{38}$$

Taking into account initial conditions it follows

$$\left\{\dot{U}_0\right\} = \{v_0\} = \frac{\{U_1\} - \{U_{-1}\}}{2\Delta t}$$

and

$$\{U_{-1}\} = \{U_1\} - 2\Delta t\left\{\dot{U}_0\right\} \tag{39}$$

3.2.1 Fixed-Fixed Boundary Conditions The same parameters for the rod as for the static case were chosen. For the density we set $\rho = 7850$ kg/m^3. The boundary conditions are given by Eq. 31. The initial conditions are assumed as follows

$$u_0(x) = 0.01x(x-1)^2, \quad v_0(x) = -100x(x-1)^2$$

Figure 7 illustrates the axial displacement versus time for three positions of the rod.

3.2.2 Fixed-Free Boundary Conditions The same properties as in the previous case are considered, except that the boundary conditions are fixed-free boundary conditions, defined by Eqs. 33. Initial conditions are specified as

$$u_0(x) = 0.001x, \quad v_0(x) = 0$$

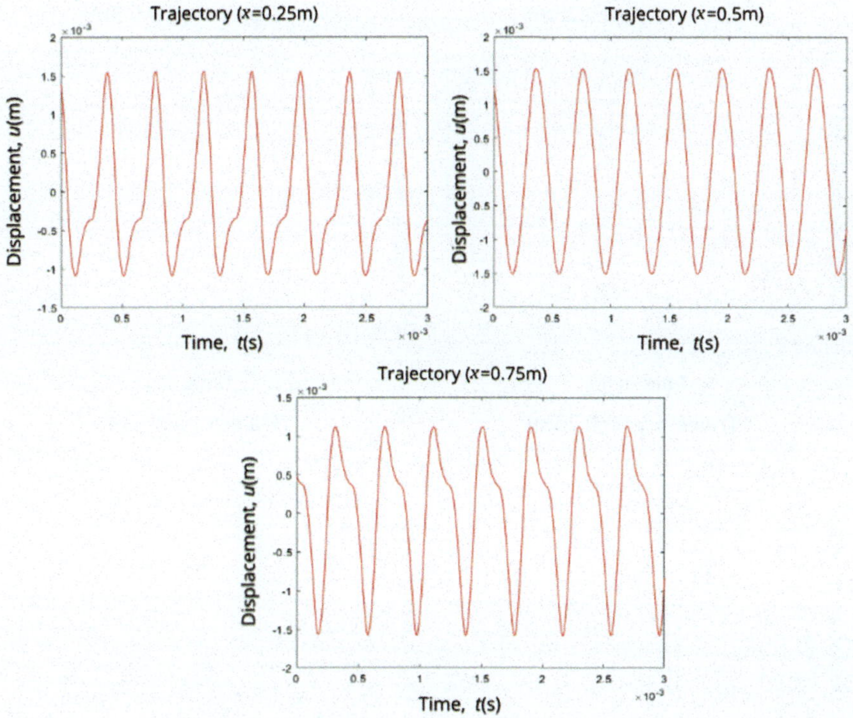

Fig. 7. Axial displacement versus time for the rod with fixed-fixed boundary conditions at three positions $x = 0.25$ m, $x = 0.5$ m and $x = 0.75$ m

4 Verification of Numerical Solutions

To verify the reliability of the numerical methods, we compare the results with closed form analytical solutions of PD equations for the rod derived in [16]. For the static loading case and fixed-fixed boundary conditions the solution has the following form

$$u(x) = \frac{2}{CL} \sum_{n=1}^{\infty} \frac{\int_0^L f(x) \sin\left(\frac{n\pi x}{L}\right) \mathrm{d}x}{\int_{-\delta_{\mathrm{H}}}^{\delta_{\mathrm{H}}} \frac{1}{|\xi|} \left[1 - \cos\left(\frac{n\pi\xi}{L}\right)\right] \mathrm{d}\xi} \sin\left(\frac{n\pi x}{L}\right) \tag{40}$$

Figure 9 illustrates the plot of Eq. 40 with the parameters presented in Sect. 3.1.1. For the considered horizon size the numerical results presented in Fig. 5 agree well with the graph of analytical solution in Fig. 11. For the fixed-free BCs the closed-form analytical solution reads

$$u(x) = \frac{2}{CL} \sum_{n=1}^{\infty} \frac{\int_0^L f(x) \sin\left(\frac{(2n-1)\pi x}{2L}\right) \mathrm{d}\,x}{\int_{-\delta_{\mathrm{H}}}^{\delta_{\mathrm{H}}} \left[1 - \cos\left(\frac{(2n-1)\pi\xi}{2L}\right)\right] \mathrm{d}\xi} \sin\left(\frac{(2n-1)\pi x}{2L}\right) \tag{41}$$

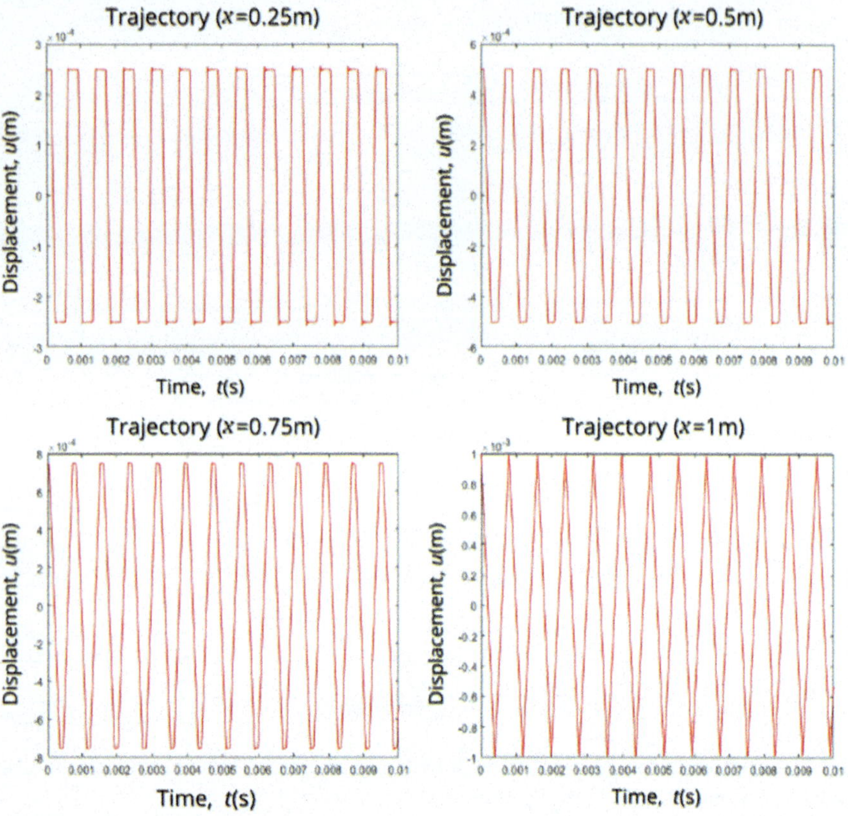

Fig. 8. Axial displacement versus time for the rod with fixed-free boundary conditions at four positions $x = 0.25$ m, $x = 0.5$ m, $x = 0.75$ m and $x = 1$ m

Figure 10 shows the axial displacement versus axial coordinate according to Eq. 41 with the parameters presented in Sect. 3.1.2. Again, for the considered horizon size the numerical results presented in Fig. 6 agree well with the analytical solution. In the case of dynamic loading with fixed-fixed boundary condition the closed form analytical solution reads [16]

$$u(x,t) = \sum_{n=1}^{\infty} \left[A_n \cos\left(\sqrt{\omega_n \frac{C}{\rho} t}\right) + B_n \sin\left(\sqrt{\omega_n \frac{C}{\rho} t}\right) \right] \sin\left(\frac{n\pi x}{L}\right), \tag{42}$$

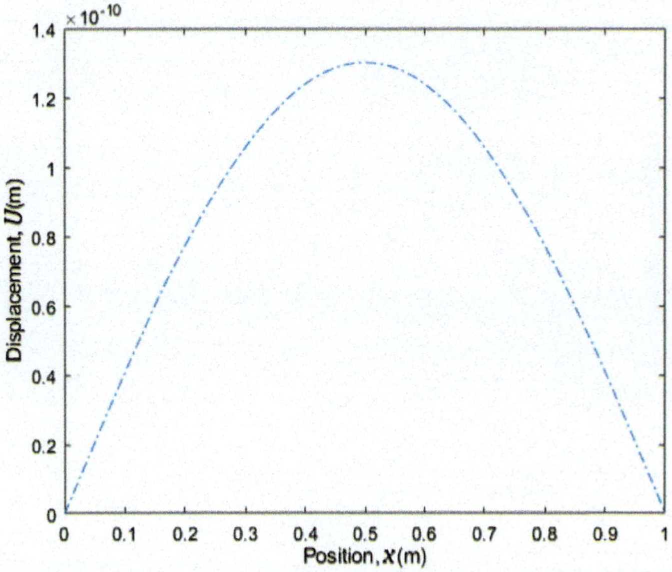

Fig. 9. Plot of analytical solution for displacement of the rod with fixed-fixed boundaries

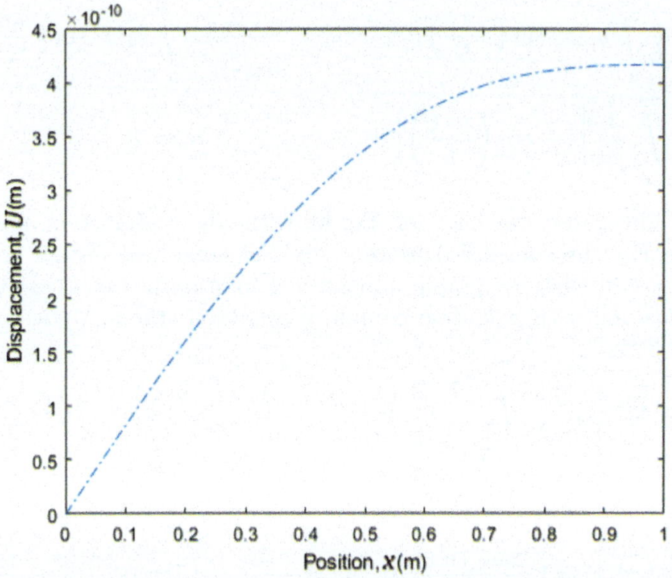

Fig. 10. Plot of analytical solution for displacement of the rod with fixed-fixed boundaries

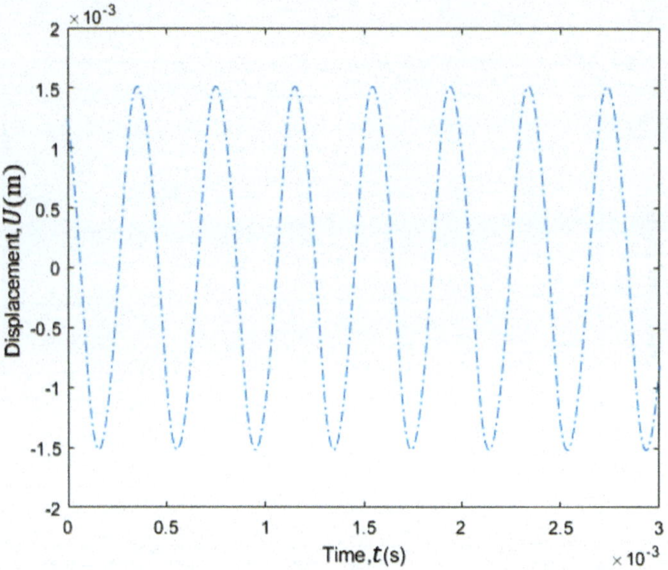

Fig. 11. Axial displacement versus time for the rod with fixed-fixed boundaries at the position $x = 0.5$ m (analytical solution)

where

$$A_n = \frac{2}{L}\int_0^L u_0(x)\sin\left(\frac{n\pi x}{L}\right)\mathrm{d}x, \quad B_n = \frac{2}{L\sqrt{\omega_n \frac{C}{\rho}}}\int_0^L v_0(x)\sin\left(\frac{n\pi a}{L}\right)\mathrm{d}x,$$

$$\omega_n = \int_{-\delta_{\mathrm{H}}}^{\delta_{\mathrm{H}}} \frac{1}{|\xi|}\left[1-\cos\left(\frac{n\pi\xi}{L}\right)\right]\mathrm{d}\xi$$

Figure 11 illustrates the plot of Eq. 40 with the parameters presented in Sect. 3.2.1. For the considered horizon size the numerical results presented in Fig. 7 agree well with the graph of analytical solution in Fig. 11. In the case of dynamic loading with fixed-free boundary conditions the analytical solution is given as follows

$$u(x,t) = \sum_{n=1}^{\infty}\left[A_n\cos\left(\sqrt{\omega_n\frac{C}{\rho}t}\right) + B_n\sin\left(\sqrt{\omega_n\frac{C}{\rho}t}\right)\right]\sin\left(\frac{(2n-1)\pi x}{2L}\right), \tag{43}$$

where

$$A_n = \frac{2}{L}\int_0^L u_0(x)\sin\left(\frac{(2n-1)\pi x}{2L}\right)\mathrm{d}x, \; B_n = \frac{2}{L\sqrt{\omega_n \frac{C}{\rho}}}\int_0^L v_0(x)\sin\left(\frac{(2n-1)\pi x}{2L}\right)\mathrm{d}x,$$

$$\omega_n = \int_{-\delta_{\mathrm{H}}}^{\delta_{\mathrm{H}}} \frac{1}{|\xi|}\left[1-\cos\left(\frac{(2n-1)\pi\xi}{2L}\right)\right]\mathrm{d}\xi$$

Figure 12 shows the axial displacement versus time at the position $x = 0.5$ m according to Eq. 41 with the parameters presented in Sect. 3.2.2. The numerical results presented in Fig. 8 agree well with the graph of analytical solution in Fig. 12.

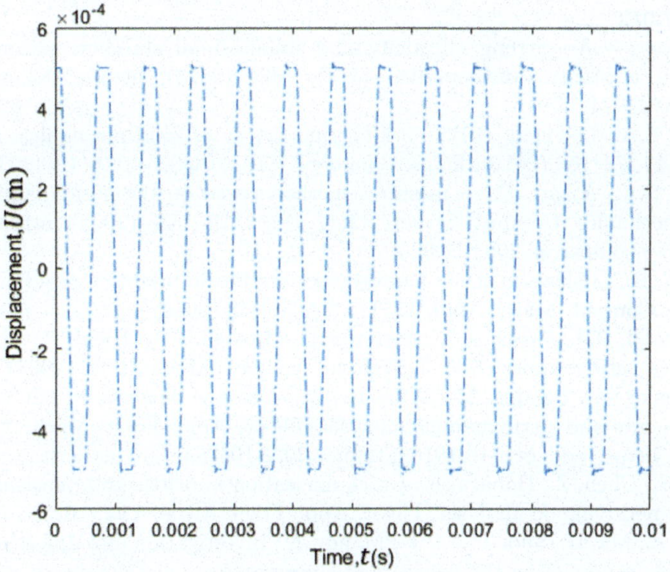

Fig. 12. Axial displacement versus time for the rod with fixed-free boundaries at the position $x = 0.5$ m (analytical solution)

5 Conclusions

The aim of this study is to analyze governing equations of mechanics with long range force interactions. To this end the uni-axial rod is considered. The general forms of the balance of momentum and the principle of virtual work are presented. For a special constitutive assumption for the bond force density the peridynamic equations for the rod discussed in the literature are obtained. By specifying the peridynamic fluxes through the cross sections of the rod the boundary conditions can be specified. Two examples including the fixed edge and the force-free edge are considered. In both examples the virtual power is not supplied through the cross section.

Numerical solutions based on a mesh-free method discussed in the literature are presented illustrating the behavior of the rod in static and dynamic cases. To validate the developed numerical codes, the results were compared with available closed-form analytical solutions to PD equations. The very good agreement between the solutions is observed for different cases of boundary conditions and loadings. This provides a basis for further investigations.

References

1. Bertram, A.: Compendium on Gradient Materials. Springer, Berlin (2023). https://doi.org/10.1007/978-3-031-04500-4
2. Cordero, N.M., Forest, S., Busso, E.P.: Second strain gradient elasticity of nanoobjects. J. Mech. Phys. Solids **97**, 92–124 (2016). https://doi.org/10.1016/j.jmps.2015.07.012
3. Eremeyev, V.A.: Strong ellipticity and infinitesimal stability within nth-order gradient elasticity. Mathematics **11**(4), 1024 (2023). https://doi.org/10.3390/math11041024
4. Silling, S.A., Lehoucq, R.B.: Peridynamic theory of solid mechanics. Adv. Appl. Mech. **44**, 73–168 (2010). https://doi.org/10.1016/S0065-2156(10)44002-8
5. Silling, S.A., Askari, E.: A meshfree method based on the peridynamic model of solid mechanics. Comput. Struct. **83**(17–18), 1526–1535 (2005). https://doi.org/10.1016/j.compstruc.2004.11.026
6. Seleson, P., Littlewood, D.J.: Convergence studies in meshfree peridynamic simulations. Comput. Math. Appl. **71**(11), 2432–2448 (2016)
7. Littlewood, D.J., Parks, M.L., Foster, J.T., Mitchell, J.A., Diehl, P.: The peridigm meshfree peridynamics code. J. Peridyn. Nonlocal Model. 1–31 (2023)
8. Naumenko, K., Pander, M., Würkner, M.: Damage patterns in oat glass plates: experiments and peridynamics analysis. Theor. Appl. Fract. Mech. **118**, 103264 (2022). https://doi.org/10.1016/j.tafmec.2022.103264
9. Niazi, S., Chen, Z., Bobaru, F.: Crack nucleation in brittle and quasi-brittle materials: a peridynamic analysis. Theor. Appl. Fract. Mech. **112**, 102855 (2021)
10. Wang, Z., Ma, D., Suo, T., Li, Y., Manes, A.: Investigation into different numerical methods in predicting the response of aluminosilicate glass under quasi-static and impact loading conditions. Int. J. Mech. Sci. **196**, 106286 (2021)
11. Mehrmashhadi, J., Bahadori, M., Bobaru, F.: On validating peridynamic models and a phase-field model for dynamic brittle fracture in glass. Eng. Fract. Mech. **240**, 107355 (2020). https://doi.org/10.1016/j.engfracmech.2020.107355
12. Diana, V., Ballarini, R.: Crack kinking in isotropic and orthotropic micropolar peridynamic solids. Int. J. Solids Struct. **196**, 76–98 (2020). https://doi.org/10.1016/j.ijsolstr.2020.03.025
13. Rahimi, M.N., Kefal, A., Yildiz, M., Oterkus, E.: An ordinary state-based peridynamic model for toughness enhancement of brittle materials through drilling stop-holes. Int. J. Mech. Sci. **182**, 105773 (2020)
14. Zhang, Y., Deng, H., Deng, J., Liu, C., Yu, S.: Peridynamic simulation of crack propagation of non-homogeneous brittle rock-like materials. Theor. Appl. Fract. Mech. **106**, 102438 (2020)
15. Oterkus, E., Oterkus, S.: Recent advances in peridynamic theory: a review. AIMS Mater. Sci. **11**(3), 515–546 (2024)
16. Yang, Z., Ma, C.-C., Oterkus, E., Oterkus, S., Naumenko, K.: Analytical solution of 1-dimensional peridynamic equation of motion. J. Peridyn. Nonlocal Model. 1—9 (2022). https://doi.org/10.1007/s42102-022-00086-1
17. Mikata, Y.: Analytical solutions of peristatic and peridynamic problems for a 1D infinite rod. Int. J. Solids Struct. **49**(21), 2887–2897 (2012). https://doi.org/10.1016/j.ijsolstr.2012.02.012
18. Shen, G., Xia, Y., Hu, P., Zheng, G.: Construction of peridynamic beam and shell models on the basis of the micro-beam bond obtained via interpolation method. Eur. J. Mech. A Solids **86**, 104174 (2021)

19. Yang, Z., Naumenko, K., Altenbach, H., Ma, C.-C., Oterkus, E., Oterkus, S.: Beam buckling analysis in peridynamic framework. Arch. Appl. Mech. **92**(12), 3503–3514 (2022). https://doi.org/10.1007/s00419-022-02245-8
20. Yang, Z., Naumenko, K., Ma, C.-C., Chen, Y.: Closed-form analytical solutions for the defection of elastic beams in a peridynamic framework. Appl. Sci. **13**(18), 10025 (2023)
21. Aung, Z.Y., Vo, D., Suttakul, P., Atroshchenko, E., Bui, T.Q., Rungamornrat, J.: Peridynamic formulations for planar arbitrarily curved beams with Euler-Bernoulli beam model. Thin-Walled Struct. 112278 (2024)
22. Yang, Z., Naumenko, K., Ma, C.-C., Oterkus, E., Oterkus, S.: Peridynamic analysis of curved elastic beams. Eur. J. Mech. A Solids **101**, 105075 (2023)
23. Vazic, B., Oterkus, E., Oterkus, S.: Peridynamic model for a Mindlin plate resting on a Winkler elastic foundation. J. Peridyn. Nonlocal Model. 1–10 (2020)
24. Chowdhury, S.R., Roy, P., Roy, D., Reddy, J.: A peridynamic theory for linear elastic shells. Int. J. Solids Struct. **84**, 110–132 (2016). https://doi.org/10.1016/j.ijsolstr.2016.01.019
25. Naumenko, K., Eremeyev, V.A.: A non-linear direct peridynamics plate theory. Compos. Struct. **279**, 114728 (2022)
26. Nishawala, V.V., Ostoja-Starzewski, M.: Peristatic solutions for finite one-and two-dimensional systems. Math. Mech. Solids **22**(8), 1639–1653 (2017). https://doi.org/10.1177/1081286516641180
27. Silling, S.A., Zimmermann, M., Abeyaratne, R.: Deformation of a peridynamic bar. J. Elast. **73**(1), 173–190 (2003). https://doi.org/10.1023/B:ELAS.0000029931.03844.4f
28. Weckner, O., Abeyaratne, R.: The effect of long-range forces on the dynamics of a bar. J. Mech. Phys. Solids **53**(3), 705–728 (2005). https://doi.org/10.1016/j.jmps.2004.08.006
29. Yang, Z., Naumenko, K., Altenbach, H., Ma, C.-C., Oterkus, E., Oterkus, S.: Some analytical solutions to peridynamic beam equations. ZAMM Z. Angew. Math. Mech. **102**(10), 202200132 (2022). https://doi.org/10.1002/zamm.202200132
30. Mikata, Y.: Linear peridynamics for isotropic and anisotropic materials. Int. J. Solids Struct. **158**, 116–127 (2019)
31. Mikata, Y.: Analytical solutions of peristatics and peridynamics for 3D isotropic materials. Eur. J. Mech. A Solids 104978 (2023). https://doi.org/10.1016/j.euromechsol.2023.104978

Investigation of Interlayer Influence on the Dynamic Impact Response of the Laminated Glass Plate

Olha Sukhanova[1,2(✉)], Oleksiy Larin[2], and Konstantin Naumenko[1]

[1] Institute of Mechanics, Otto-von-Guericke-University Magdeburg, Universitätsplatz 2, Magdeburg 39106, Germany
konstantin.naumenko@ovgu.de

[2] Department of Mathematical Modeling and Intelligent Computing in Engineering, National Technical University "Kharkiv Polytechnic Institute", 2 Kyrpychova str., Kharkiv 61002, Ukraine
olha.sukhanova@ovgu.de, oleksiy.larin@khpi.edu.ua

Abstract. The aim of the work is to investigate the behavior of laminated glass plates subjected to ball drop impact. Laminated glasses are known for their safety and impact resistance. One of the important features of laminates is a sufficiently thin polymer interlayer with a pliable structure. This study evaluates the mechanical properties of a laminated glass plate configuration consisting of two glass layers and one layer of polymeric interlayer. The finite element method was used to perform the computational model, that focuses on the dependencies of displacements and first principal stresses on transverse shear strain and the interlayer plate thickness. Results indicate significant insights into the deformation characteristics and stress distribution within the laminated glass structure. The study's outcomes provide a foundation for further research and development, particularly in optimizing the interlayer properties and thickness for better impact resistance. This work contributes to advancing the knowledge in the field of laminated glass structures, promoting the development of more resilient and safer glass products for various applications.

Keywords: Laminated glass · Polymer interlayer · Elasticity modulus · Impact · Shear stiffness · Finite element analysis

1 Introduction

Laminated glass panels are important materials that are widely used in the construction field, such as automotive industry, civil engineering, glazing and photovoltaic applications [1–3]. These panels are subjected to different quasi-static,

Oleksiy Lari and Konstantin Naumenko contributed equally to this work.

A. Öchsner (ed.), *Collaborative Research Advancing Engineering Solutions for Real-World Challenges 2*, Proceedings in Technology Transfer,
https://doi.org/10.1007/978-3-031-79093-5_2

dynamic and non-stationary thermal loadings. As a result of external loadings, glass layers and encapsulant materials may experience deformation, damage and ageing. One of the purposes of the laminates is protection from environmental influences. In addition, laminated glass modules have superior rigidity and strength compared to modules with monolithic glass. The load values can be very high to cause microcrack failures and module functionality loss. The parameters affecting the bearing capacity are the thickness and the material properties of the layers. Since the polymer interlayer is organic, exposure to the environment can lead to a change in mechanical properties over time [4]. Several parameters can influence interlayer properties. For example, a temperature increase leads to a change in the shear modulus. Therefore, a reliable assessment of the stiffness properties is needed by the testing of a prototype, e.g. by bending tests of a beam or a plate.

To evaluate the test results, robust structural analysis methods are required to relate the global deformation state to the applied external load. To analyse deformed states in laminated glass elements, several approaches to structural mechanics can be applied. Classical and first-order shear deformation theories of beams and plates are used for laminated glasses with shear-stiff interlayers [5,6]. The key point in applying first-order shear deformation theory is to establish the material properties of the laminate, in particular the transverse shear stiffness. Although the closed-form relationships are developed to find effective elastic stiffness of a laminate from the properties of layers, e.g. [7], numerical methods are necessary to determine the transverse shear strain when the material behaves inelastically [8–10].

Laminated glass can be examined using solid finite elements combined with three-dimensional elasticity theory for the numerical analysis. However, due to substantial differences in the material properties between the layers and the small thickness of the interlayer, achieving accurate results requires significant numerical effort [4]. Special attention must be given to the finite element meshing of the interlayer to accurately calculate transverse shear strains and associated stresses. Various finite element formulations, including those with user-defined elements for laminates, have been developed [11–13]. Additionally, three-dimensional solid mechanics approaches are utilized to address issues such as transverse normal deformation, stress concentrations, and impact from ball drop [14–17].

Based on previous studies, it is clear that the interlayer influences the dynamics of laminated glass panels. Its thickness and elasticity are affected and from a practical point of view, it is important to see the regularity of the influence of both, as well as how the stresses are distributed in the interlayer to correctly distribute in terms of boundary conditions and relative influences and to understand the mechanisms of energy absorption during impacts.

Thus, this paper aims to investigate the laminated glass under impact loading. The particular interest is the influence of the interlayer properties on the laminate behaviour. An emphasis is placed on discretization including finite element mesh and time stepping, as well as boundary and contact conditions. Displacements, the first principal stresses of the glass layers and interlayer after the ball drop will be presented in detail as functions of time. The dependence of elastic modulus and thickness of interlayer on displacements and first principal stresses will be shown and analysed.

2 Ball Impact Simulation Model

In the current study, a hit on a composite plate with a solid steel ball of mass 1 kg with an initial velocity of 8.85 m/s is considered. This is the speed at which the ball falls during the free ball drop test from a height of 4 m, which corresponds to the majority of experimental data in the literature. The square plate of size 500 mm consisted of two layers of glass with a thickness of 2.9 mm and one interlayer with a thickness of 0.36 mm, as shown in Fig. 1. The impact of the ball was simulated in the centre of the plate. As boundary conditions, the glass is simply supported on all sides. The material properties of steel and glass are pointed out in Table 1. The elasticity modulus of the interlayer is varied in the range between 7 and 70,000 MPa.

Table 1. Material properties of the simulation model

Material	Density, ρ $(\mathrm{kg/m^3})$	Young's modulus, E (GPa)	Poisson ratio, ν
Steel	7850	210	0.3
Glass	2500	70	0.22
Interlayer	960	Varied	0.45

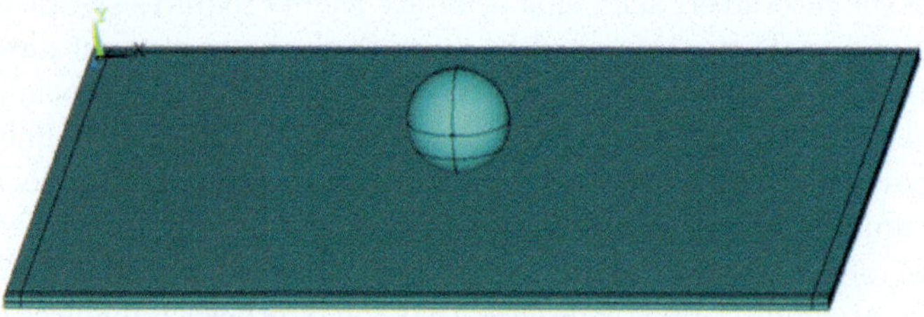

Fig. 1. Simulation model

The impact of the ball on the glass plate is analysed using the explicit approach in the framework of three-dimensional modelling by the finite element method (FEM). Linear finite elements (FE) with 8 nodes and 3 degrees of freedom in each are used. The interaction of the ball and the plate is considered as a one-way contact by the surface-to-surface algorithm. The resistance caused by air during impact is neglected. The time step size is chosen as 3×10^{-5} s. For the simulation model, the FE mesh with a 2 mm element size is chosen. The mesh is presented in Fig. 2.

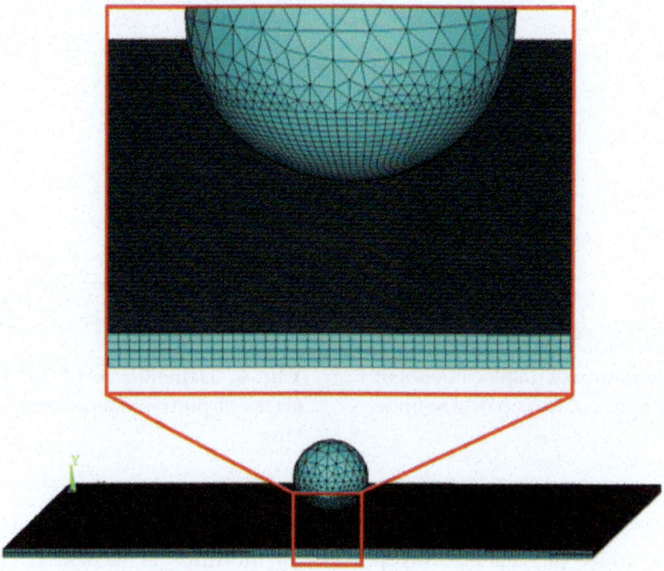

Fig. 2. FE mesh of the model

To model realistic free support in the model, on the bottom surface the frame with the size of 1 cm by the plate perimeter edges are created. The nodes on the frame were selected and constrained according to their displacements in the y-plane.

3 Analysis Results

Transient analysis was done for the steel ball hit on a composite plate in the CAE software Ansys. The type of nonlinear statement was chosen for the current study. The results of the maximum displacement of the ball, plate and first principal stress of the plate were calculated.

Figures 3 and 4 illustrate maximum displacements (w) of the ball bottom node and the plate top centre node and the dependence of plate first principal stress ($s1$) on time at the plate bottom centre node with Young's modulus E = 7 MPa. According to the results within the first interaction of the ball and plate from the initial time moment till 0.002 s the first stress peak occurs with the value around 300 MPa. Then the ball comes off the plate for a short time. Within the second interaction of the ball and plate from 0.003 s till 0.0045 s the second stress peak occurs with a value around 475 MPa.

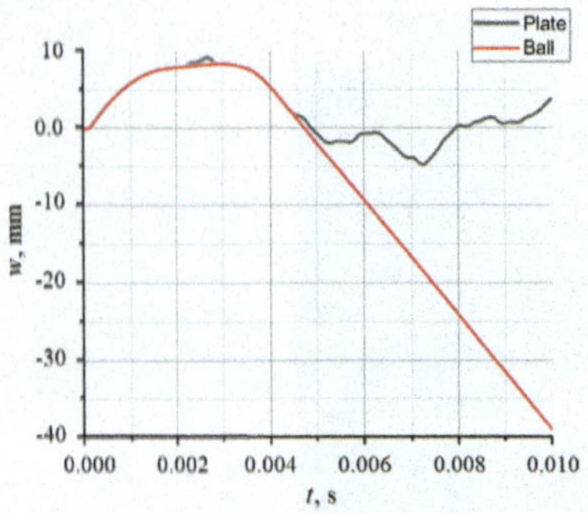

Fig. 3. Dependence of displacements of ball bottom node, plate top centre node on time

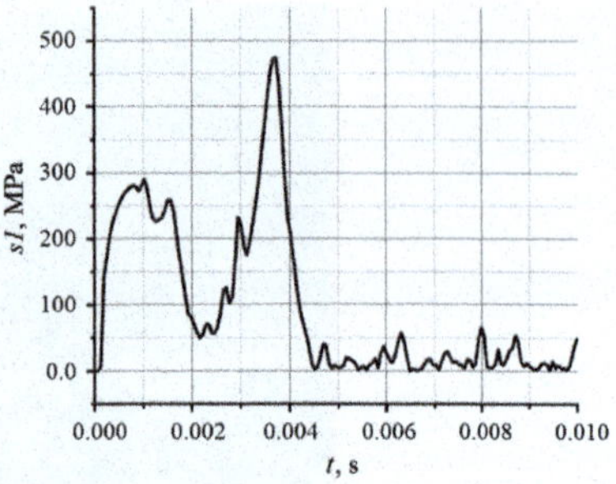

Fig. 4. Dependence of 1st principal stress of plate bottom centre node on time

In Fig. 5 the dependence of the maximum displacements at the bottom middle node of the plate on the interlayer Young's modulus is shown. It can be seen that displacements decrease with increasing the modulus value. In Fig. 6 the dependence of the first and second peaks of maximum first principal stresses at the bottom middle node of the plate on Young's modulus of interlayer is shown. It can be seen that stresses increase with increasing the modulus and then stabilize.

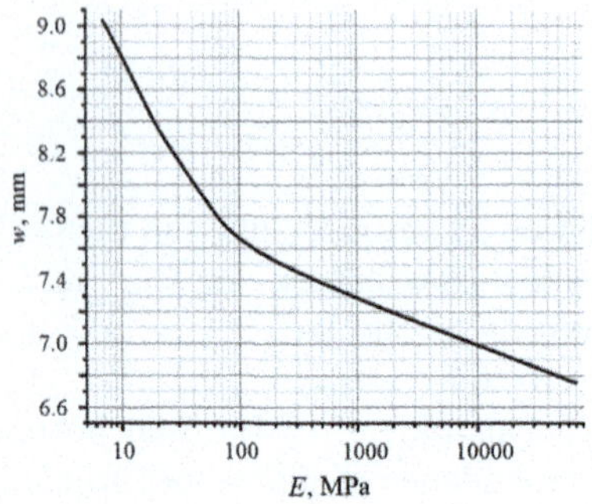

Fig. 5. Dependence of maximum displacement of centre bottom plate node on Young's modulus of interlayer

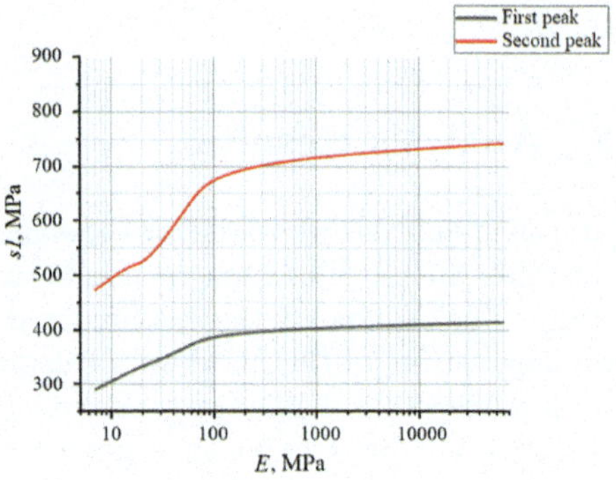

Fig. 6. Dependence of first and second peak of 1st principal stress of centre bottom plate node on Young's modulus of interlayer

Table 2 shows displacements, first principal stresses and strains at different time points. It can be seen that elastic waves spread through the composite, and the maximum displacements occur in the centre of the plate. The stresses are shown at the bottom layer of the plate. It takes on cruciform shapes and spreads through the plate. The strains are shown at the interlayer. It can be seen that at the time moment when the ball is in contact with the plate the maximum deformations occurred at the edges of the plate.

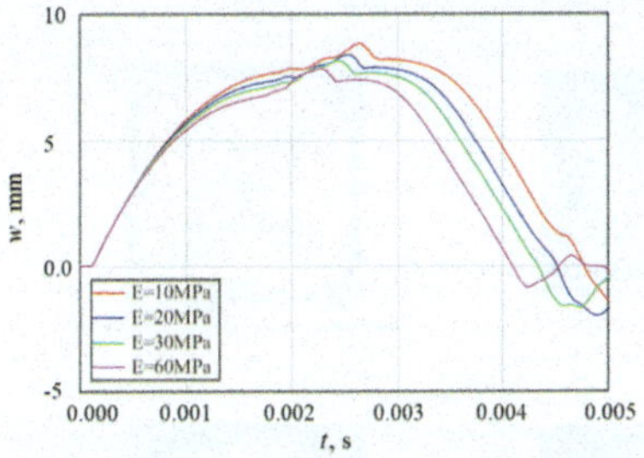

Fig. 7. Dependence of displacements of bottom centre plate node on time

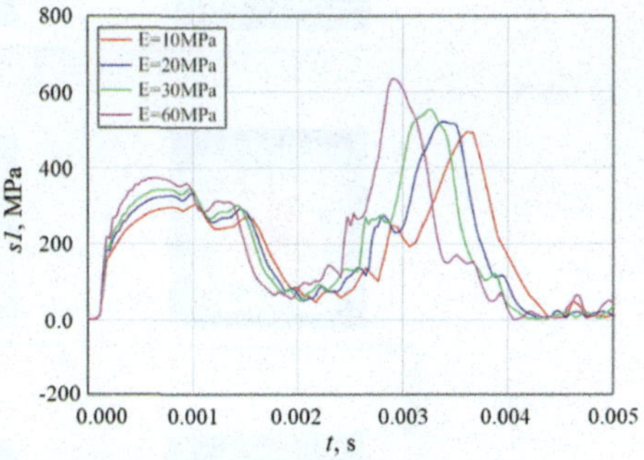

Fig. 8. Dependence of 1st principal stress of bottom centre plate node on time

In Figs. 7 and 8 the dependences of displacements and 1st principal stresses of bottom centre plate node on time with different Young's modulus of interlayer are shown. It can be seen that with an increase of the module, the displacement peaks decrease by 10% with each step, and the contact time of the ball with the plate is reduced. With an increase of the module, the stress peaks increase by 10% with each step, and the contact time of the ball with the plate is reduced.

Then Young's modulus was set to 15 MPa. The interlayer with varied thicknesses of 0.36, 0.72, 1.08 and 1.44 mm was used.

Figures 9 and 10 show the dependences of displacements and 1st principal stresses of bottom centre plate node on time with different thickness of interlayer. It can be seen that with an increase in the interlayer thickness, the displacements and the contact time of the ball with the plate are increased. With an increase in the interlayer thickness, the stress peaks decrease by 15% with each step.

Table 2. Displacements, first principal stresses and strains

Time (s)	w (mm)	$s1$ (MPa)	$\varepsilon 1$
0.75e-3	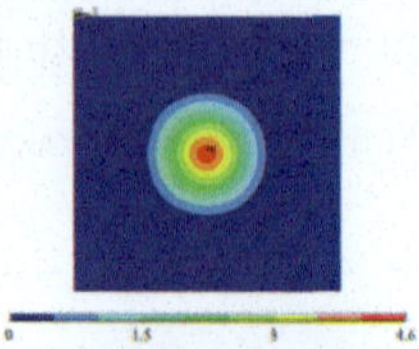	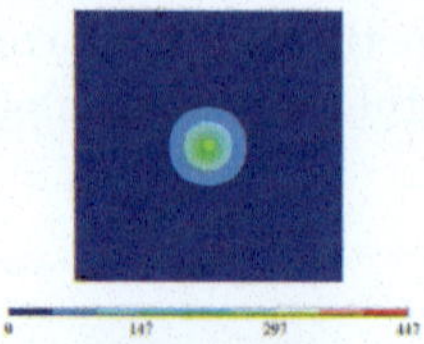	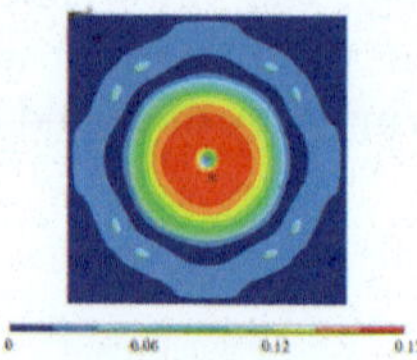
2e-3	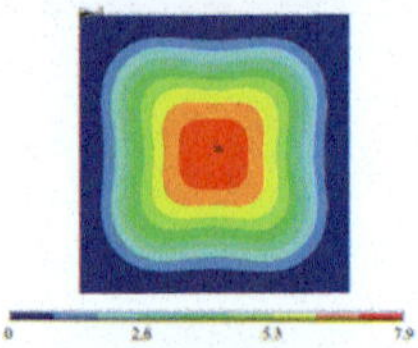	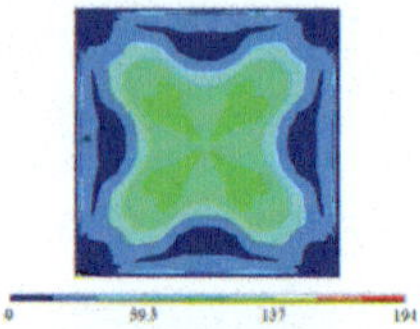	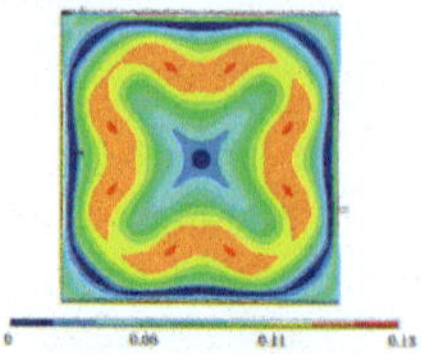
2.7e-3	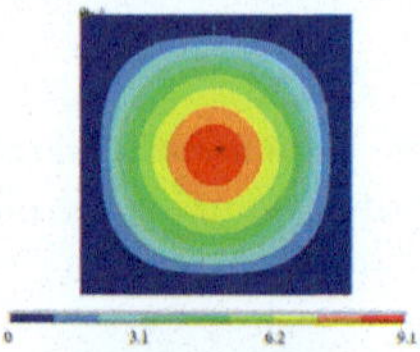	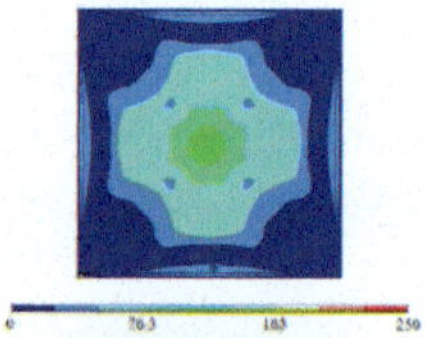	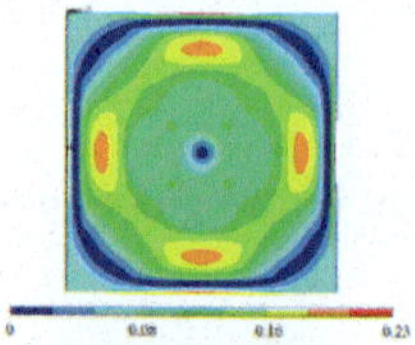
3.3e-3	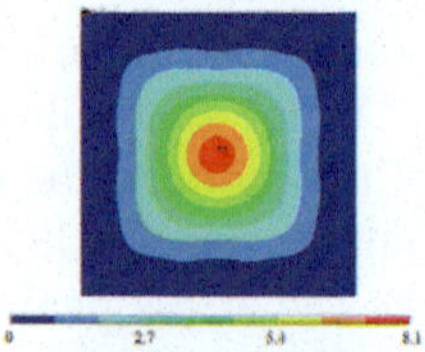	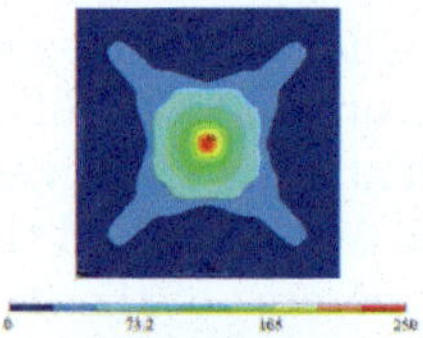	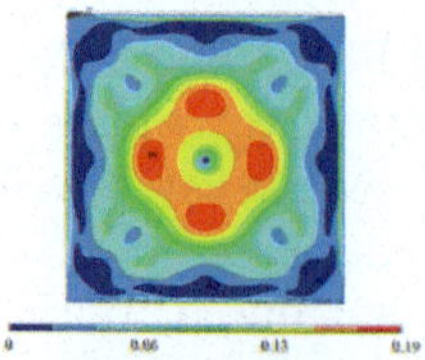
3.7e-3	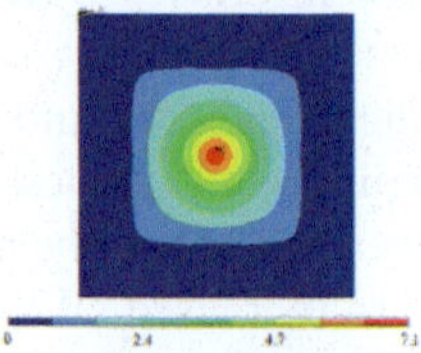	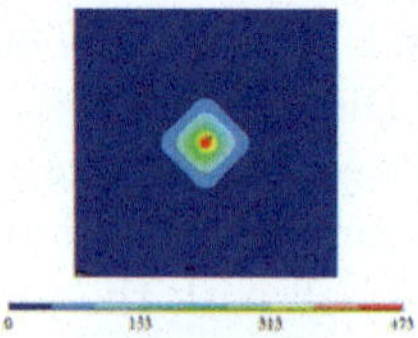	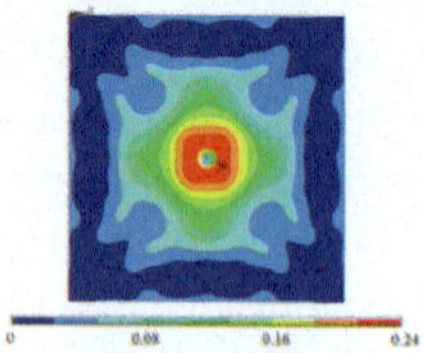
4e-3	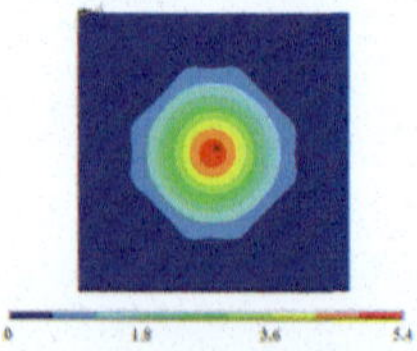	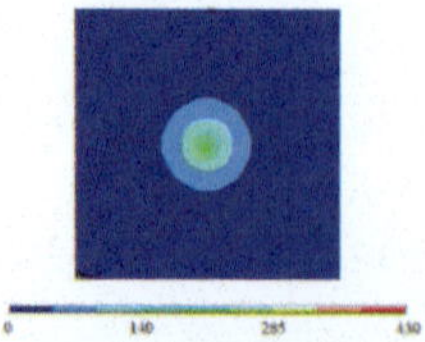	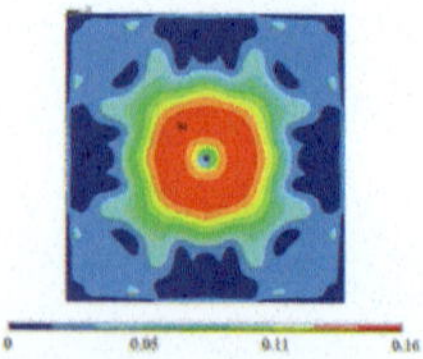

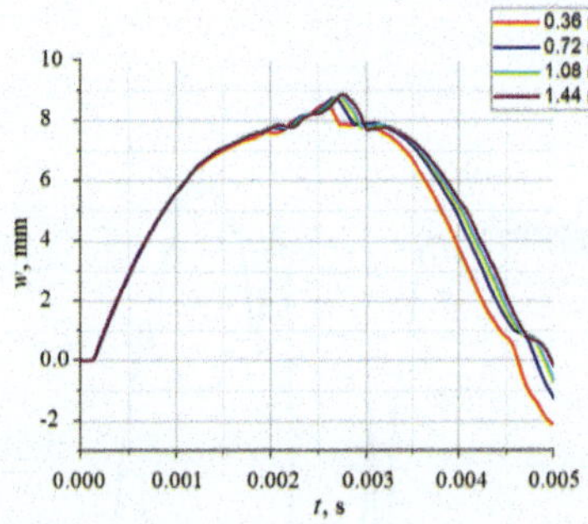

Fig. 9. Dependence of displacements of bottom centre plate node on time

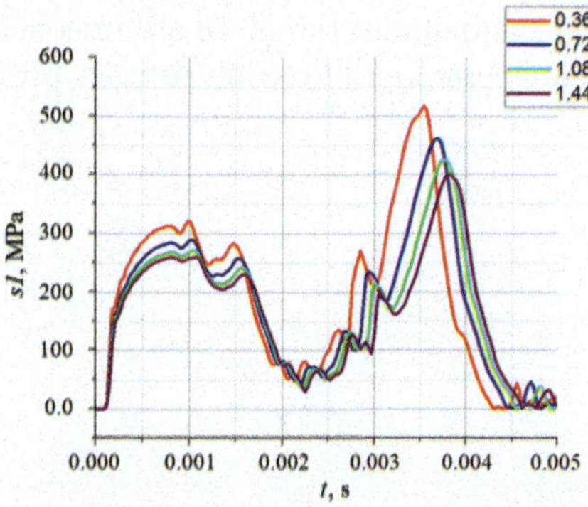

Fig. 10. Dependence of 1st principal stress of bottom centre plate node on time

In Fig. 11 the dependence of the maximum displacements at bottom middle node of the plate on interlayer thickness is shown. It can be seen that displacement peaks increase by 5% with each step with increasing the interlayer thickness. In Fig. 12 the dependence of the first and second peaks of maximum first principal stresses at bottom middle node of the plate on interlayer thickness is shown. It can be seen that stresses decrease with increasing the interlayer thickness.

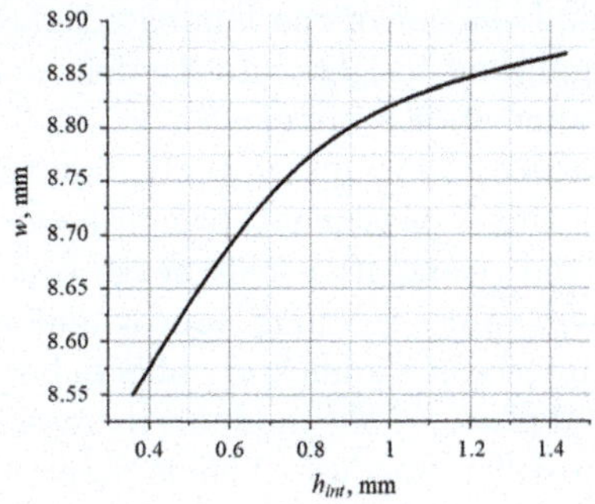

Fig. 11. Dependence of maximum displacement of centre bottom plate node on thickness of interlayer

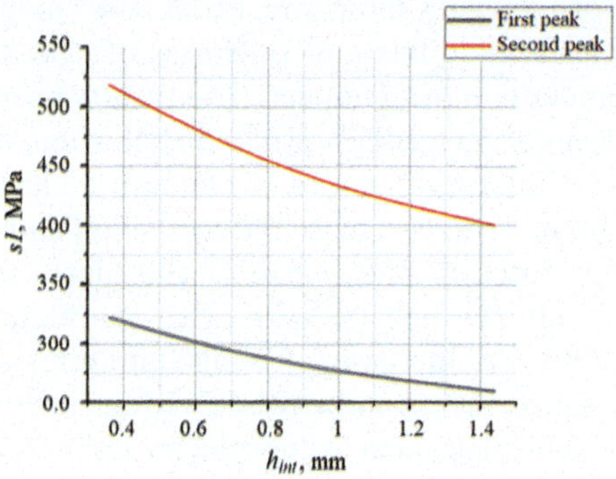

Fig. 12. Dependence of first and second peak of 1st principal stress of centre bottom plate node on thickness of interlayer

In Figs. 13 and 14 the displacements of the bottom node of the ball and the bottom middle node of the plate and the contact pressure of the bottom node of the ball are shown. It can be seen that contact pressure occurs during moments when the ball and the plate are in contact. And when the ball and the plate are not in contact - there is no contact pressure.

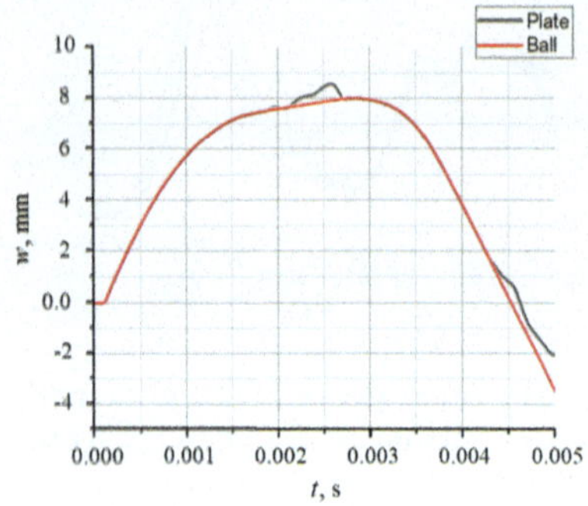

Fig. 13. Dependence of displacements of plate bottom middle node on time

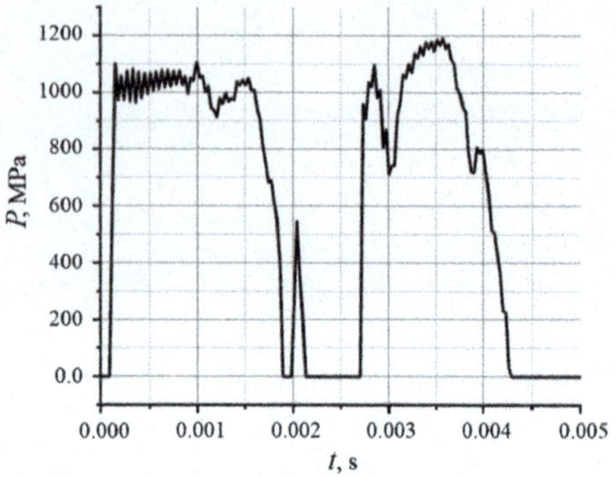

Fig. 14. Dependence of contact pressure of plate bottom middle node on time

4 Conclusions

This study investigates the dynamic impact response of laminated glass plates with varying interlayer properties, focusing on the influence of the interlayer's elastic modulus and thickness on the dynamic behavior of the laminated glass under ball drop impact. The results provide significant insights into the deformation characteristics, stress distribution, and optimization strategies for enhancing the impact resistance of laminated glass structures. The impact of a ball on the laminated glass plate induces significant stress and displacement, primarily concentrated at the centre of the plate. Initial and secondary stress peaks occur within the first 0.0045 s, with the highest stress values reaching approximately 475 MPa. Increasing Young's modulus of the interlayer significantly affects both displacements and stresses within the laminated glass. As the modulus increases, the displacement values decrease, indicating a stiffer response. This is observed with a 10% reduction in displacement peaks for each step increase in modulus. Conversely, the first principal stress values increase with a higher modulus, stabilizing after a certain point. Stress peaks increase by 10% with each step increase in modulus. Varying the thickness of the interlayer (from 0.36 mm to 1.44 mm) also influences the impact response. Thicker interlayers result in increased displacements and longer contact times between the ball and plate. Displacement peaks increase by 5% with each step increase in thickness. The stress peaks decrease with thicker interlayers, showing a 15% reduction for each increase in thickness. This indicates better stress distribution and energy absorption with thicker interlayers.

References

1. Brokmann, C., Alter, C., Kolling, S.: Experimental determination of failure strength in automotive windscreens using acoustic emission and fractography. Glass Struct. Eng. **4**(2), 229–241 (2019)
2. Naumenko, K., Eremeyev, V.A.: A layer-wise theory of shallow shells with thin soft core for laminated glass and photovoltaic applications. Compos. Struct. **178**, 434–446 (2017)
3. Zemanová, A., Zeman, J., Šejnoha, M.: Comparison of viscoelastic finite element models for laminated glass beams. Int. J. Mech. Sci. **131**, 380–395 (2017)
4. Smetankina, N., Semenets, O., Merkulova, A., Merkulov, D., Misura, S.: Two-stage optimization of laminated composite elements with minimal mass. In: Arsenyeva, O., Romanova, T., Sukhonos, M., Tsegelnyk, Y. (eds.) Smart Technologies in Urban Engineering STUE 2022. Lecturer Notes in Network System, vol. 536, pp. 473–485. Springer, Cham (2023). https://doi.org/10.1007/978-3-031-20141-7_42
5. Adámek, V.: The limits of Timoshenko beam theory applied to impact problems of layered beams. Int. J. Mech. Sci. **145**, 128–137 (2018)
6. Smetankina, N., Merkulova, A., Merkulov, D., Postnyi, O.: Dynamic response of laminate composite shells with complex shape under low-velocity impact. In: Nechyporuk, M., Pavlikov, V., Kritskiy, D. (eds.) Integrated Computer Technologies in Mechanical Engineering—2020. ICTM 2020. Lecture Notes in Network System, vol 188, pp. 280–291. Springer, Cham (2021). https://doi.org/10.1007/978-3-030-66717-7_22
7. Altenbach, H., Eremeyev, V.A., Naumenko, K.: On the use of the first order shear deformation plate theory for the analysis of three-layer plates with thin soft core layer. ZAMM-J. Appl. Math. Mech./Z Angew. Math. Mech. **95**(10), 1004–1011 (2015). https://doi.org/10.1002/zamm.201400206
8. Sukhanova, O., Larin, O., Naumenko, K., Altenbach, H.: Dynamics of curved laminated glass composite panels under impact loading. In: Altenbach, H., Amabili, M., Mikhlin, Y.V. (eds.) Nonlinear Mechanics of Complex Structures. Advanced Structured Materials, vol. 157, pp. 67-85. Springer, Cham (2021). https://doi.org/10.1007/978-3-030-75890-5_6
9. Altenbach, H., Eremeyev, V.A.: On the bending of viscoelastic plates made of polymer foams. Acta Mech. **204**(3–4), 137–154 (2009). https://doi.org/10.1007/s00707-009-0118-5
10. Larin, O.O.: Probabilistic model of fatigue damage accumulation in rubber like materials. Strength Mater. **47**, 849–858 (2015). https://doi.org/10.1007/s11223-015-9722-3
11. Ugrimov, S., Smetankina, N., Kravchenko, O., Yareshchenko, V.: Analysis of laminated composites subjected to impact. In: Nechyporuk, M., Pavlikov, V., Kritskiy, D. (eds.) Integrated Computer Technologies in Mechanical Engineering—2020. ICTM 2020. Lecture Notes in Network Systems, vol. 188, pp. 248-258. Springer, Cham (2021). https://doi.org/10.1007/978-3-030-66717-7_19
12. Aßmus, M., Naumenko, K., Altenbach, H.: A multiscale projection approach for the coupled global-local structural analysis of photovoltaic modules. Compos. Struct. **158**, 340–358 (2016). https://doi.org/10.1016/j.compstruct.2016.09.065
13. Li, G., Cinefra, M., Carrera, E.: Coupled thermo-mechanical finite element models with node-dependent kinematics for multi-layered shell structures. Int. J. Mech. Sci. **171**, 105379 (2020). https://doi.org/10.1016/j.ijmecsci.2019.105379

14. Janda, T., Schmidt, J., Hála, P., Konrád, P., Zemanová, A., Sovják, R., Zeman, J., Šejnoha, M.: Reduced order models of elastic glass plate under low velocity impact. Comput. Struct. **244**, 106430 (2021). https://doi.org/10.1016/j.compstruc.2021.106430
15. Pelfrene, J., Kuntsche, J., Van Dam, S., Van Paepegem, W., Schneider, J.: Critical assessment of the post-breakage performance of blast loaded laminated glazing: experiments and simulations. Int. J. Impact Eng. **88**, 61–71 (2016). https://doi.org/10.1016/j.ijimpeng.2015.10.008
16. Vedrtnam, A., Pawar, S.: Laminated plate theories and fracture of laminated glass plate—a review. Eng. Fract. Mech. **186**, 316–330 (2017). https://doi.org/10.1016/j.engfracmech.2017.10.014
17. Vedrtnam, A., Pawar, S.: Experimental and simulation studies on fracture and adhesion test of laminated glass. Eng. Fract. Mech. **190**, 461–470 (2018). https://doi.org/10.1016/j.engfracmech.2017.12.015

Flame Propagation in Piston Lands of a Spark-Ignited Hydrogen Internal Combustion Engine

Saraschandran Kottakalam(✉), Dennis Maier, Robin Beck, and Gregor Rottenkolber

Faculty of Mobility and Technology, Esslingen University of Applied Sciences, Kanalstrasse, Esslingn am Neckar 73728, Baden-Württemburg, Germany
{saraschandran.kottakalam,gregor.rottenkolber}@hs-esslingen.de, beck_robin@gmx.de

Abstract. Hydrogen is a carbon-free alternative to fossil fuels that can be used in internal combustion engines allowing the transfer of know-how gained over the last 100 years of internal combustion engine development. Mass market adoption of these powertrains depend on making them zero-impact vehicles. This means restricting their exhaust emissions even though they are, in most cases, lower than the gasoline counterparts. With this in mind, experiments were carried out in an optically accessible hydrogen-fuelled engine to understand flame propagation through the piston top and ring lands, and to understand oil combustion in these regions. The results were then compared to similar phenomena with gasoline operation in the same engine. A correlation between the observed optical phenomena in the piston lands and the exhaust measurement was also derived which lays the path for improvement of hydrogen internal combustion engines towards zero-impact-emission powertrains. Moreover, indications for the cause of abnormal combustion phenomena could be observed.

Keywords: Top land flame propagation · Ring land flame propagation · Hydrogen internal combustion engines

1 Introduction

Hydrogen is considered a viable alternative to fossil fuels in internal combustion engines (ICEs) [1]. This is because, when produced with green energy, hydrogen internal combustion engines (H_2-ICEs) reduce global as well as local emissions in comparison to fossil fuels. As a result, it remains an actively researched topic [2].

Compared to traditional gasoline fuels, hydrogen presents several advantages and challenges as a fuel for ICEs. This comparison is necessary because this is ultimately what dictates the transfer of information from the development of

A. Öchsner (ed.), *Collaborative Research Advancing Engineering Solutions for Real-World Challenges 2*, Proceedings in Technology Transfer,
https://doi.org/10.1007/978-3-031-79093-5_3

Table 1. Properties of hydrogen compared to gasoline [3]

Property	Unit	Gasoline	Hydrogen
Stoichiometric air requirement	kg_{air}/kg_{fuel}	14.0	34.3
Ignition limits	λ-Range	0.4–1.4	0.13–10
Self-ignition temperature	°C	230–450	585
Minimum ignition energy	mJ	0.24	0.017
Diffusion coefficient	cm^2/s	0.05	0.61
Laminar flame speed	cm/s	40	230

gasoline engines into H_2-ICEs. Table 1 provides an overview of the properties of hydrogen compared to gasoline.

In comparison to gasoline, hydrogen has a higher laminar flame speed, very low ignition energy, and a wider range of ignition limits even though the self-ignition temperature is higher. Because of these properties, hydrogen mixtures (near stoichiometric) can be easily ignited, flame propagation is easier and leads to efficient combustion. However, the same properties make it extremely sensitive to abnormal ignitions (preignition or detonation) [3,4]. This means, the temperatures of the components inside the combustion chamber has to remain below the self-ignition temperature to avoid these issues which is generally not a concern in gasoline operations. This can be overcome by lean operation due to colder combustion. However, the energy density of hydrogen engines would be impacted by this since hydrogen requires more air for combustion. Another way to do this is by water injection, which is a strategy used in motorsport engines that ensures cooling of the combustion chamber while maintaining the energy density since they operate at near stoichiometric levels [5].

The main property that makes hydrogen attractive as an alternative fuel for a zero-impact internal combustion engine is the lack of carbon in the molecule. This significantly reduces CO, CO_2 and HC emissions. These small amounts of emissions come from the combustion of lubricating oil. Therefore, oil combustion is critical for H_2-ICEs [6]. Due to a combination of properties shown in the Table 1, hydrogen has a smaller quenching distance than gasoline. This means it burns closer to the walls and also burns through the smallest of crevices under engine conditions. Not only does this lead to evaporation and ignition of the thin oil film on the walls, but also negatively influences the efficiency and longevity of the engine. A higher amount of oil combustion leads to higher exhaust emissions. Research has been performed on exhaust emissions from H_2-ICEs showing higher than expected emissions due to oil combustion [6]. Several studies also provide a detailed analysis of emissions under different operational conditions, as well as possible sources of emissions. This includes the interaction of the hydrogen jet with the thin oil film at the wall and the accumulation of ignitable mixtures in the piston rings [7–9]. Flame quenching in piston top lands could be a huge contributor to these emissions as in the case of gasoline operation.

The availability of a research engine with optical access to the piston ring region along with a hydrogen infrastructure at the Automotive Powertrain Laboratory provides a unique opportunity to see the hydrogen flame propagation through piston ring lands and compare it to the emission measurements to gain a deeper understanding of this phenomena to develop powertrains with zero-impact-emission.

As already mentioned, mitigation of abnormal combustion phenomena such as pre-ignition, backfire, or knock due to the high ignitability of hydrogen is another major open subject that requires solution. The piston top lands are suspected to be the cause of some of these phenomena [10]. The results described in this paper provide a similar indication.

2 Top Land Flame Propagation

Top land flame propagation (TLFP), which is the propagation of the flame through the top of the piston, is known to cause HC emissions in gasoline-ICEs [11]. Since emission reduction is also important in gasoline engines, this is a field that has been studied. However, the different properties of hydrogen compared to gasoline, listed in Table 1, lead to different conditions in the combustion chamber and the piston land due to the different behavior of hydrogen flames. To better understand this phenomenon in H_2-ICEs, similar research was performed.

From research, it was found that the quenching distance of the mixture influences this phenomenon significantly, i.e. if the mixture propagates through the gap between the piston and the cylinder wall or not.

2.1 Quenching Distance

The quenching distance of a flame depends on several properties of a homogeneous ignitable mixture, such as diffusivity, laminar flame speed, Lewis number, pressure, temperature and testing process (like parallel plate method, direct quenching method). In engine operations, this means that the larger the quenching distance is for a mixture, the less likely that TLFP occurs.

Equation (1) was introduced to calculate the quenching distance of a mixture in engine [12].

$$W_c = K_1 \cdot p_{\max}^{-0.9} \cdot T_w^{-0.5} \tag{1}$$

Here, the quenching distance, W_c in mm, depends on the pressure, $p_{\max}$ in MPa, the wall temperature, T_w in K, and the empirical constant, K_1. The value of K_1 depends on various parameters such as the engine-design, fuel used, the stoichiometry of the mixture, and the engine-operating parameters (load and rpm). This equation was validated by subsequent research and a K_1 of 14.8 was obtained for gasoline, $\lambda = 1$ mixture based on experiments [13]. This value of K_1 was further utilised for the study of TLFP in the same research engine for gasoline operation [14]. The engine specifications are detailed in Sect. 2.

However, K_1 for hydrogen mixtures were not found in the literature. Therefore, from the experimentally obtained quenching distances for hydrogen mixtures under standard conditions ($p_{max} = 0.1$ MPa and $T_w = 273$ K) [15], K_1 for different hydrogen mixtures were calculated. According to this study, for stoichiometric hydrogen mixtures, W_c was measured to be 0.6 mm. Another independent study performed under similar conditions obtained a similar value of 0.64 mm [16]. Using Eq. (1) and the quenching distance of 0.64 mm and standard temperature and pressure, a K_1 value of 1.2 was calculated for $\lambda = 1$ hydrogen mixture under standard conditions. This forms a good foundation for a rough estimation of the K_1 value for hydrogen mixtures which contribute to the understanding of TLFP in hydrogen engines. Further, K_1 value was calculated for $\lambda = 2$ to be 3.8 from a quenching distance of approximately 2 mm at standard conditions [15].

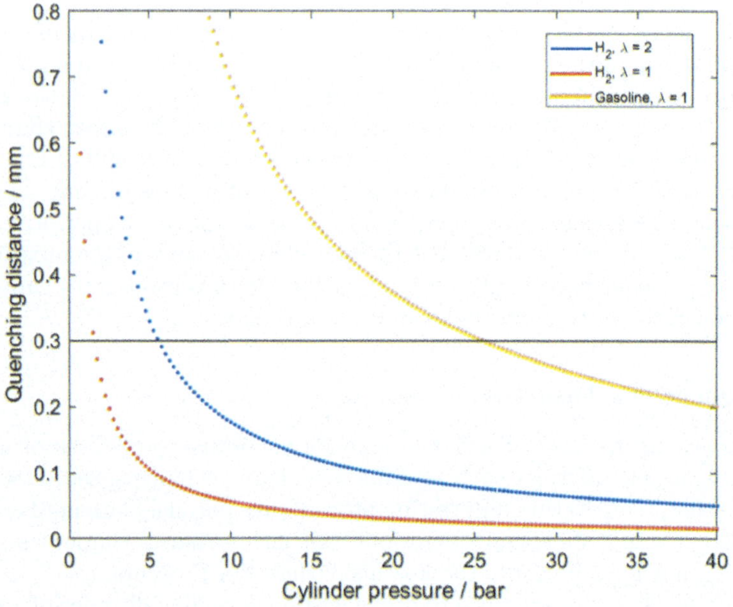

Fig. 1. Quenching distances as a function of cylinder pressure for different mixtures calculated using Eq. (1)

Using the K_1 value obtained, Fig. 1 shows the quenching distances calculated for different pressures keeping the temperature of the walls constant at 433 K at the piston rings. This was obtained as an average temperature of the measurements performed in the same engine [14]. Considering the gap at the top land of the pistons to be 0.3 mm, the minimum pressure that needs to be present in the combustion chamber for the flame to propagate through this gap can be seen to be ~ 25 bar for gasoline, ~ 5.75 bar for hydrogen at $\lambda = 2$ and ~ 1.75

bar for hydrogen at $\lambda = 1$. Under the same conditions, the pressure required for stoichiometric hydrogen mixtures to propagate through the piston top lands is less than 1/10th of that for the gasoline mixtures. This means that according to Eq. (1), the criteria for TLFP is easily met for the hydrogen mixtures and there should be a homogeneous premixed flame propagating through the piston top land consistently.

From the research performed previously in the same engine for gasoline operation, it was found that when these conditions are met, i.e. when the pressure is above the required pressure, homogeneous premixed flame propagated into the piston top land and was quenched at the first ring gap. This propagation is aided by the pressure difference in the combustion chamber and the first ring land and the flow due to this occurring through the ring gap. The same trend was seen for the diffusion flames which was quenched at the ring gap due to the cooling effect occurring at the gap region [14].

2.2 Optical Detection of TLFP

The light emitted during hydrogen combustion is documented for both premixed and for diffusion hydrogen flames. For the studies described in this article, the emissions for homogeneous premixed hydrogen flames were considered relevant because hydrogen is considered to mix well with air because it is injected very early. The highest emission intensity should be expected in the ultraviolet spectrum with emission due to hydroxyl radical formation with a peak of around 300 nm. These emissions are not detectable by the human eye and the high-speed camera used. However, the vibrationally excited H_2O-molecules, a product of hydrogen combustion, emits light in the infrared spectrum and the near-infrared region. Some of these light emissions can be detected by the high-speed camera (starting at 600 nm) [17]. In contrast, diffusion flames, which are the combustion of lubricating oil, emit strong signals throughout the visible wavelength range [18].

For the studies conducted as part of this paper, optical observations were performed with a black and white camera (details in Sect. 3.2). So, light emissions sources must be differentiated based on the emission intensity. In this case, since the emissions from hydrogen-premixed flames would be very weak signals, the oil combustion, which is manifested as diffusion flames, is immediately noticeable because they are extremely bright in comparison. Hydrogen has a notable lack of emissions in the visible light range from 400 to 600 nm and appropriate optical filters could also be used to observe only that spectrum of light, separating oil combustion from the hydrogen flames.

In the case of gasoline operation, premixed flames emit blueish light, and the diffusion flame, as in the case of hydrogen combustion, would be extremely bright broad emission [14].

However, observing phenomena with different emission intensities simultaneously poses some challenges and requires post-processing which is explained in Sect. 4.

3 Experimental Setup

This section provides a detailed explanation of the engine test-bed where the experiments were performed along with the instruments used for optical and exhaust measurements.

3.1 Optically Accessible Research Engine

The tests were performed in a single-cylinder research engine, the laser-optical transparent tribology engine (LOTTE), based on Mercedes-Benz M274 engine. This engine has two sapphire glass windows on the thrust and anti-thrust sides, providing view of the piston ring regions close to the top dead centre (TDC). The combustion chamber becomes visible at 24 °C after TDC, while the first compression ring stays visible at TDC. The optical cylinder used permit operation, similar to a normal engine without a window. Figure 2 shows the engine setup and camera arrangement used for the experiments along with the exhaust measurement setup.

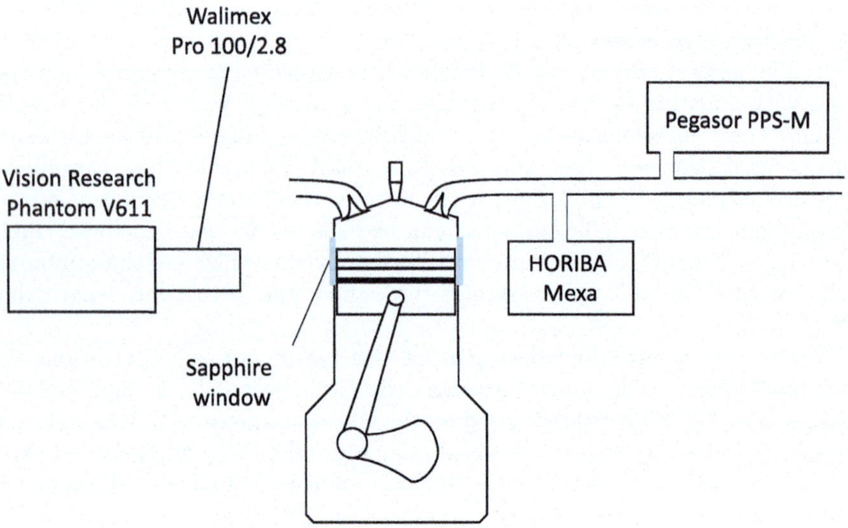

Fig. 2. Experimental setup of the testbed

Table 2 provides the detailed specifications of the engine. The engine was originally developed to investigate tribological phenomena in the piston ring region during gasoline operations. To facilitate hydrogen operation, the standard ignition system previously used was replaced by a cold spark plug placed between the exhaust valves with a Bosch ignition system that prevents residual discharges. Both gasoline and hydrogen injections were performed with a piezo injector at pressures of 200 bar and 50 bar respectively. The first compression ring is fixed on the piston so that the ring gap remains in the middle of the glass window on the anti-thrust side. Boost pressure up to 10 bar is made available externally through the testbed. The lubricating oil used is without additives as required by other optical measurement techniques such as laser-induced fluorescence (LIF).

Table 2. Technical details of the single cylinder research engine [19]

Combustion chamber	Pent-roof, 4 valve
Compression ratio	9.8
Displacement [cm^3]	498
Bore/Stroke [mm]	83/92
Max. rpm [min^{-1}]	6000
Max. IMEP [bar]	25
Fuel injection	Direct injection with a piezo-injector

3.2 Optical Analysis

The optical measurements were performed with a Vision Research Phantom V611 high-speed camera with a detection wavelength range of 400–1000 nm. The full resolution of 1280 × 800 pixels can be recorded at 6.242 fps. It is paired with a Walimex Pro 100mm f2.8 macro lens. The images were recorded with LaVision Davis software. Figure 2 shows the camera positioned towards the anti-thrust side of the engine, which is the configuration used for all the results mentioned in this paper.

3.3 Exhaust Analysis

The primary exhaust gas analysis was performed using HORIBA Mexa 7170DEGR. This includes the CO, CO_2, HC and NO_x emission measurements details of which can be seen in Table 3.

The exhaust particle measurements were carried out with MS4 Pegasor PPS-M particle sensor and the detailed specifications are provided in Table 4. However, it does not provide a particle size distribution. Since it is calibrated with hydrocarbon fuels, measurement results may be considered to be qualitatively accurate.

The exhaust analysis was performed after a steady state was reached with the results averaged over a period of one minute.

Table 3. HORIBA Mexa sensor specifications

Sensor	Type	Measured Gas	Range	Resolution
AIA-721	NDIR	CO_{Low}	0–2.500 ppm	50 ppm
			0–5.000 ppm	100 ppm
AIA-722	NDIR	CO_{High}	0–12% vol.	0.5% vol.
FIA-726D	H.FID	HC	0–500 ppmC	10 ppmC
			0–50.000 ppmC	1.000 ppmC
CLA-756	H.CLD	NO/NO_x	0–500 ppm	10 ppm
			0–50.000 ppm	1.000 ppm

Table 4. Pegasor PPS-M specifications

Particle size range	23 nm– 2.5 μm
Particle size resolution	300 cm^{-3} - 10^9 cm^{-3}
Particle mass range	1 $\mu g/m^3$ - 250 mg/m^3
Inlet sample temperature	Up to 200 °C
Trap voltage	400 V

3.4 Operational Conditions

A fair comparison of gasoline and hydrogen operations is possible only if the conditions are similar. Keeping that in mind, an operation point of 8 bar IMEP and 1500 rpm, with a constant mass fuel burned (MFB50) of 15 °C aTDC was chosen for the comparison of exhaust emissions and for comparison of optical flame propagation in the piston lands.

As mentioned earlier, the hydrogen was injected at 50 bar and the gasoline was injected at 200 bar pressure. For good mixing, the gasoline was injected at −285 °C after top dead centre (aTDC). Since the injector used was not designed for hydrogen injection, long injection durations were required to have enough hydrogen inside the engine at these 'low' pressures. Therefore, the injection started from −210 °C aTDC. An earlier start of injection was not possible due to the increased risk of backfire. The measurements were performed with stoichiometric ($\lambda = 1$) and lean ($\lambda = 2$) hydrogen mixtures. The same operation point with stoichiometric gasoline mixture was taken as a reference.

4 Image Processing

As mentioned in the previous Sect. 2.2, the emission signal from flame propagation in the ring lands is weak and difficult to detect from the native 12-bit grayscale image from the camera. So the raw images are exported using LaVision DaVis software and processed with a Matlab script for better visualization. This section details the image processing steps used in the Matlab script used to process the raw images.

Once exported from the DaVis software, the following steps are performed for visualization:

1. An average of the first few images in a cycle before the flame is visible in the view field was taken to obtain a reference image and subtracted from the images with flame propagation to remove the background noise.
2. The average value of the signal of a small section of each image in the bottom left corner of each image is taken and subtracted from each image to remove image to image zero signal.
3. A false colormap 'Turbo' was applied on the captured images to increase the contrast of low signals.
4. Finally, for understanding the location of the piston rings, an overlay of piston ring locations was made based on the crank angle when the image is captured and the location of piston calculated for each crank angle.

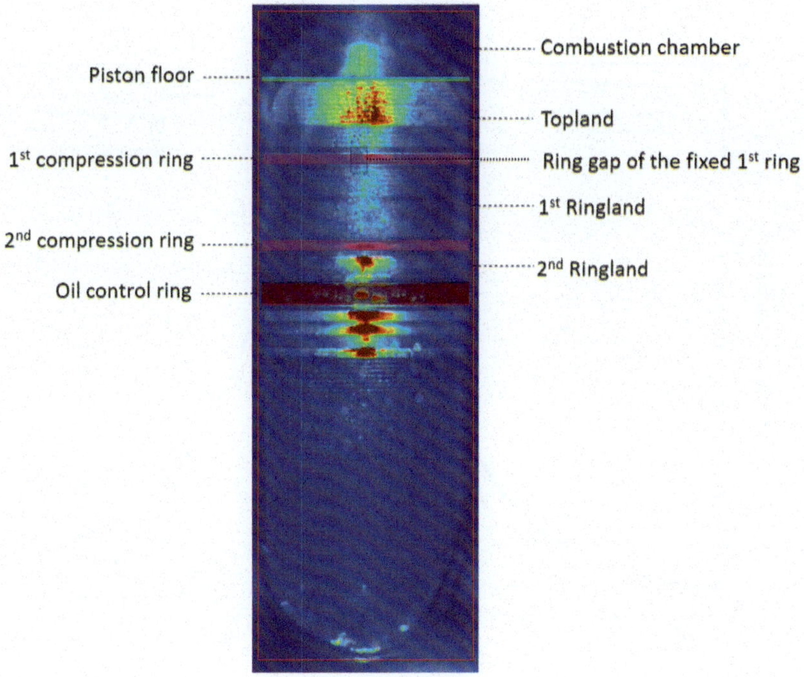

Fig. 3. Labeled image with overlayed locations of the piston rings and lands

Figure 3 shows an example of the image resulting from the post-processing. Here, the piston rings and the regions are labeled. The figure shows the piston ring region observed through the glass window by the visible high-speed camera when illuminated by a high-powered LED.

5 Results

Once the experiments were carried out, both the optical and exhaust measurements were analysed separately to understand the difference between hydrogen and gasoline operation. Further, the results were correlated to obtain better understanding of the TLFP in hydrogen operation.

5.1 H_2-Gasoline Exhaust Emission Comparison

Using the experimental setup mentioned in the previous section, tests were performed with both hydrogen and gasoline fuels. The operation point was fixed as mentioned in Sect. 3.4. Figure 4 shows the mean pressure curve (100 cycles) obtained for this operation conditions achieved with hydrogen operated at $\lambda = 1$ (H_2 stoich), $\lambda = 2$ (H_2 lean) and gasoline $\lambda = 1$ operation (Gasoline). These mean pressure traces for H_2 stoich, H_2 lean, and gasoline are represented in red, blue and orange respectively.

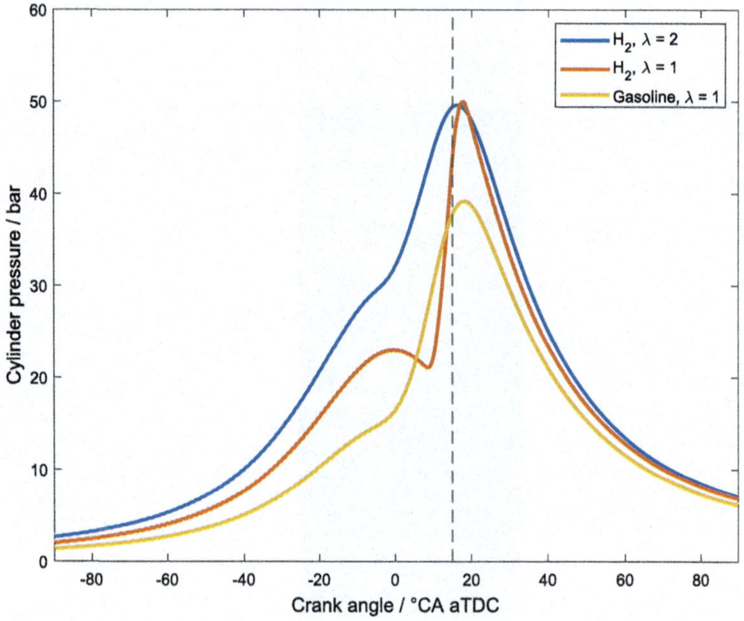

Fig. 4. Averaged pressure curves (100 cycles) for the three operations with MFB50 shown (8 bar IMEP and 1500 rpm)

From these pressure traces, it can be seen that the hydrogen mixtures have a shorter combustion delay than gasoline. It can also be seen that H_2 stoich has the highest pressure rise rate among the cases owing to its faster laminar flame speed. The peak pressure($P_{\max}$) for H_2 stoich and H_2 lean is higher in

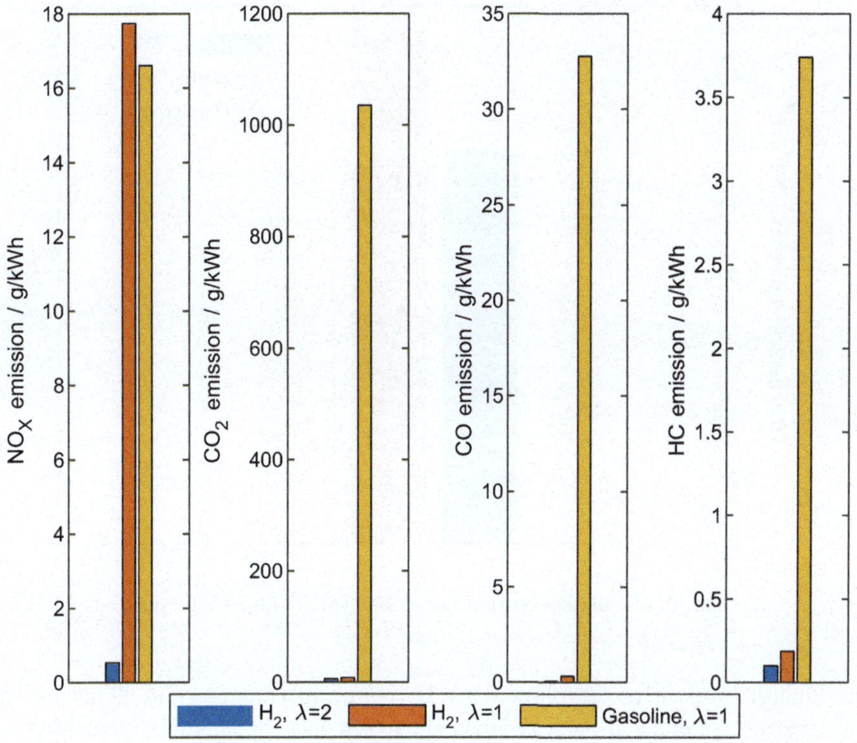

Fig. 5. Exhaust emissions (8 bar IMEP and 1500 rpm)

comparison to gasoline mainly due to the larger amount of air that hydrogen requires. However, $P_{\max}$ for both the hydrogen cases are similar only because the ignition time is retarded in H_2 stoich to maintain MFB50 similar to the other cases due to the short combustion delay at $\lambda = 1$.

The exhaust emission measurements for the corresponding operations are shown in Fig. 5. It can be seen that hydrogen operations have lower CO, CO_2, and HC emissions than gasoline. This is because hydrogen is a carbon-free fuel. This means that this small amount of these emissions has to come from the lubricating oil and not from the fuel itself. This confirms the requirement of this study. However, observations of the NO_x show another story. H_2 stoich has the highest NO_x emission of all three. This is because the hydrogen mixture at $\lambda = 1$ has the highest flame temperature (adiabatic flame temperature of 2100 °C compared to 1900 °C for gasoline mixtures at $\lambda = 1$) and, as a result, produces the highest amount of NO_x. However, at $\lambda = 2$ as in H_2 lean, the combustion is colder again and the NO_x emission is significantly lower.

Particulate emission, on the other hand, is higher than in gasoline for both H_2 stoich and H_2 lean with hydrogen operation as shown in Fig. 6. Since the particles must come from oil combustion, the graph shows that H_2 stoich burns

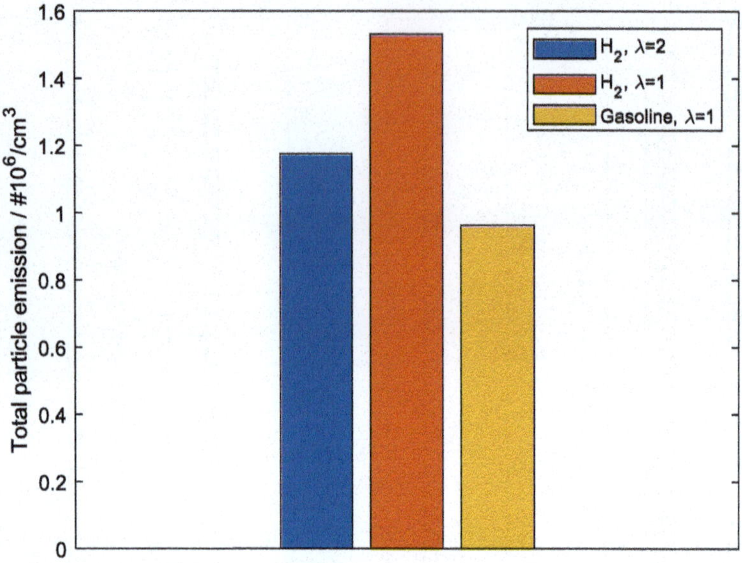

Fig. 6. Particulate emissions (8 bar IMEP and 1500 rpm)

the highest amount of oil followed by H_2 lean and then gasoline. This satisfies the theory explained in Sect. 2. Since H_2 stoich has the smallest quenching distance, the flame propagates closer to the cylinder walls and also deeper into the ring lands, leading to the ignition of oil droplets that are present on the cylinder walls and also in the piston ring lands. During exhaust measurements with hydrogen operation, sudden peaks in particulate emissions were also observed. This phenomenon increased in frequency with higher loads and also as the lambda reduced toward 1. The optical observation of the flame propagation phenomenon described in the following section provides further insights into this topic.

5.2 Flame Propagation Along Piston Ring Lands

Figure 7 shows a high-speed capture of a typical top land flame propagation (TLFP) by the visible camera during gasoline operation. Marker 1 on the image series shows the homogeneous flame propagates from the combustion chamber through the piston top land towards the gap in the first ring. This flow is generated due to the pressure difference between the combustion chamber and the 1st ring land.

It was seen that this flame propagation follows the peak pressure in the combustion chamber in a cycle. It was also observed that sometimes the homogeneous flames ignite the oil droplets (marker 2) in the top land region, which then travel with them, as in the cycle shown in Fig. 7. Once it reaches the 1st ring gap, it gets extinguished (marker 3). No flame propagation was seen in the second ring land during gasoline operation even at higher loads.

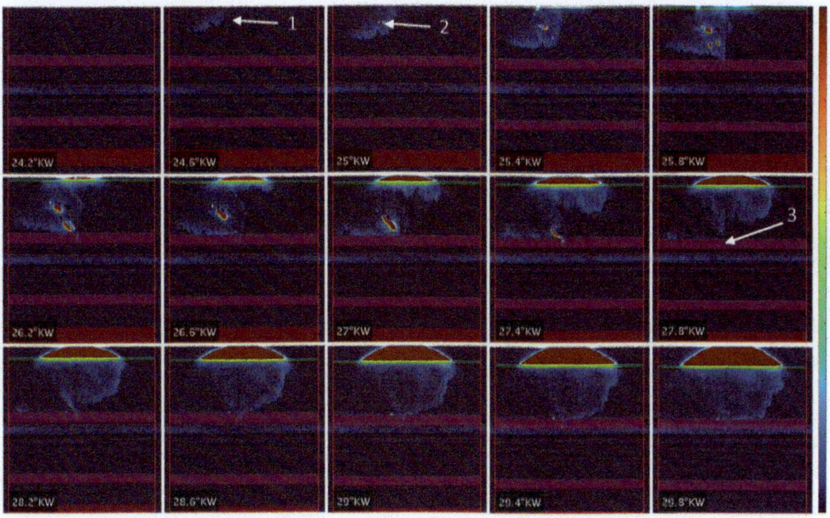

Fig. 7. A cylce depicting typical top land flame propagation for gasoline stoichiometric operation. Operational parameters: 15 bar IMEP and 1500 rpm, MFB50 = 23 °C aTDC

Here, since the images were captured with a black and white camera, the diffusive flame and the homogeneous premixed flame can be differentiated by their emission intensity. Homogeneous premixed flames are generally of lower intensity when compared with the diffusive flames produced due to oil combustion. As mentioned in the visualization section, the raw images are processed so that the lower-intensity emission of the premixed flame can be detected easily (see the emission intensity on the right). So, the light blue signals in the top land section are the homogenous premixed flames and the high-intensity emissions come from the diffusive flame which are visualized as bright red. This can be confirmed with the experimental observations performed by [11] with a high-speed color camera, where the homogeneous flame is detected as a dim blue flame while the diffusive flames are bright yellowish-orange.

Figures 8, 9 and 10 show a series of images captured at different time-steps for three cycles with stoichiometric hydrogen, lean hydrogen and gasoline operations. All these cycles are observed for the same operational points mentioned in Sect. 3.4. Each of these image series represents the typical top and ring land flame propagation seen for different mixtures.

Figure 8 depicts the top and ring land flame propagation phenomena with hydrogen $\lambda = 1$ operation. Marker 1 shows a faint appearance of the flame front in the view-field. It then reaches and propagates through the top land extremely fast compared to the other cases. In contrast to the typical phenomena in gasoline operation ($\lambda = 1$) shown earlier in Fig. 7, the flame burns through the 1st ring gap into the 1st ring land. The ignitable mixture in the 1st ring land is ignited by the flame arriving at the 1st ring gap. Marker 2 shows this phenomenon and it can be seen that this process emits a strong signal and

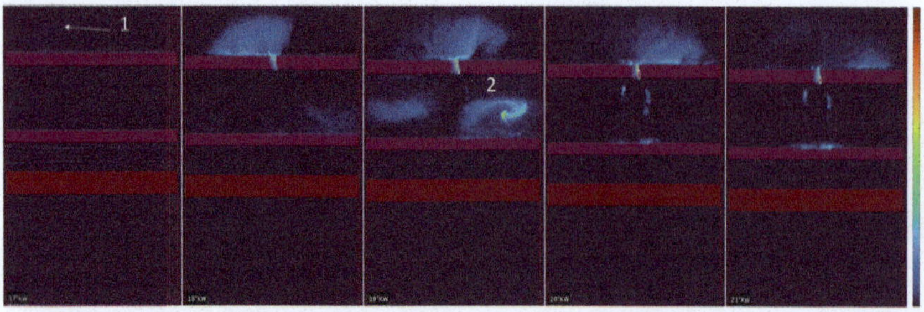

Fig. 8. Piston land flame propagation for hydrogen stoichometric operation. Operational parameters: 8 bar IMEP and 1500 rpm

generates detectable vortex-like structures on both sides below the ring gap in the 1st piston ring. These vortexes are formed due to the flow generated due to the pressure differences mentioned previously. From the images, motion blurs were often seen at stoichiometric hydrogen operation typically generated by bright diffusion flame. These emissions must be from oil droplets/particles ignited by the homogeneous flame which is then carried by the flow in the 1st ring land.

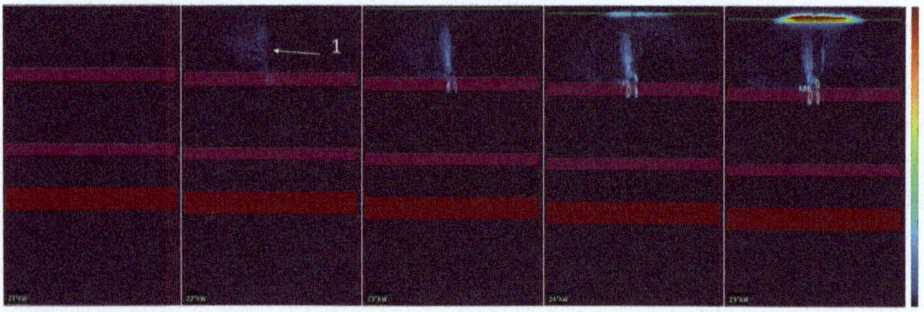

Fig. 9. Piston land flame propagation for hydrogen $\lambda = 2$ operation. Operational parameters: 8 bar IMEP and 1500 rpm

Similar to the previous images series, Fig. 9 shows the flame propagation phenomenon with lean hydrogen ($\lambda = 2$) operation. In contrast to Fig. 8, it can be noticed that the flame starts propagating through the first piston top land shortly before the combustion chamber is visible, as shown by marker 1 in the figure. It was also observed that the intensity of emission in the 1st ring land is significantly less in comparison to stoichiometric H_2 mixtures and also difficult for the visible camera to detect. In addition, there were typically no diffusive flames observed in the 1st ring land which means that there is less (if at all) oil ignition/combustion. This could be explained by the fact that the flame temperature is lower in this case than in H_2 stoich and as a result is unable to ignite the oil droplets present here.

It must be noted that flame propagation in the first ring lands in lean hydrogen operation was not as consistent as in the case of stoichiometric mixtures. The frequency of these propagations typically reduced as the load reduced and became more frequent as the load increased. This is also in good agreement with the flame quenching theory explained in Sect. 2.1.

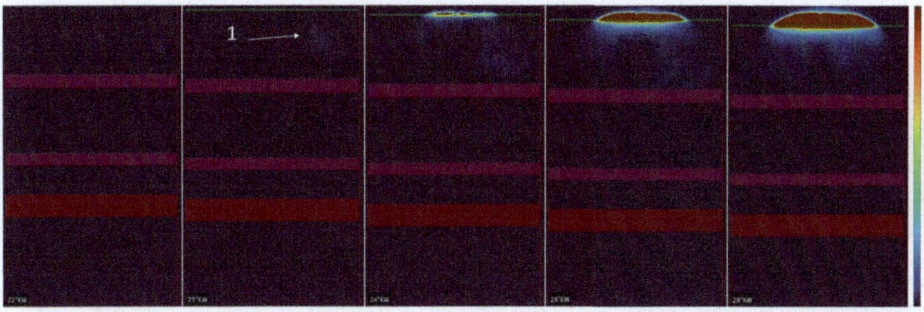

Fig. 10. Piston land flame propagation for gasoline stoichiometric operation. Operational parameters: 8 bar IMEP and 1500 rpm

To present the whole picture and for a fair comparison, Fig. 10 shows TLFP at gasoline operation at the same operation point. As indicated by marker 1, the mixture is the slowest to arrive (1 °C later than lean hydrogen case) even though the MFB50 is kept the same in all the cases. It can be seen that the flame propagation signal is extremely weak and signifies the requirement for the post-processing. As explained in the case of Fig. 7, both the homogeneous premixed flame and the diffusive flame are quenched at the ring gap of the 1st piston ring, and no flame is detected in the 1st ring land. In addition, it has to be kept in mind that out of the three mixtures, stoichiometric gasoline mixture creates the lowest peak pressure as is evident from the pressure curves. Lower pressure means larger quenching distance.

The observations presented above show the higher tendency of the hydrogen mixtures to burn through the top and ring lands of the piston and this follows the literature presented in Sect. 2. From these optical observations, the flame propagation phenomenon for each of these mixtures were characterized and compared with the emission measurements. From this, a correlation between flame propagation in the piston top and ring lands and the particle emission was obtained. It was seen that the higher the intensity of flame propagation in the ring lands, the higher the particulate emission was.

5.2.1 Abnormal Ignition Phenomenon During the experimental investigations with hydrogen operation, several abnormal phenomena were detected in the piston ring lands. These phenomena were interesting since they lend validity to the possibility of oil as a source for preignition or knocking as described in studies [10].

Flow reversal in Piston top land. Figure 11 depicts such an abnormal phenomenon in which the flame propagation through the ring gap in the first ring is disrupted and reversed for a short moment due to a sudden change in pressure. The images shown are taken at a time resolution of 0.2°crank angle. Marker 1 shows the bright emission from the diffusive flame of the oil droplet ignited by the premixed flame propagation through the top land.

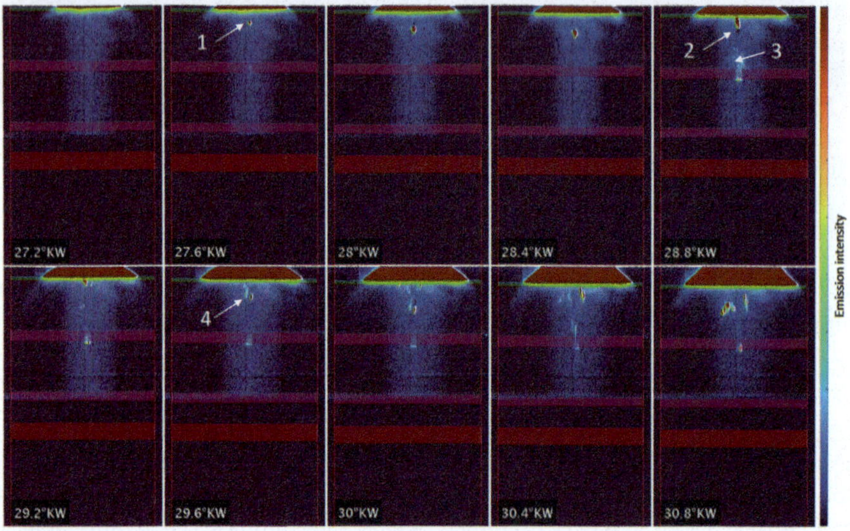

Fig. 11. Flow reversal in the piston top land. Operational parameters: 6 bar IMEP, 1500 rpm, MFB50 = 17 °C aTDC and $H_2, \lambda = 1$

Marker 2 shows the flow disruption as the oil droplet moves back towards the combustion chamber. Since the flame propagation through the ring gap is driven by the pressure difference between the combustion chamber and the first ring land, a sudden and momentary reversal in this flow should result from a sudden reversal in this pressure difference. In this case, this is accompanied by a bright emission in the ring gap as shown by marker 3. This sudden bright emission could be an explosion in the ring gap or behind the 1st ring that introduces a sudden pressure change. After this, as shown by marker 4, the flow returns to the normal condition. Such a phenomenon could, in extreme cases, lead burned particles to be introduced into the combustion chamber and this could be a source for knocking and preignition.

Flame propagation in second ring land. Another interesting phenomenon that was observed was the flame propagation in the 2nd ring land. Figure 12 shows an example of such a phenomenon. Marker 1 shows images with flame in the second ring land, which appears to travel from the right of the window. In the subsequent images, it moves towards the left and then disappears. It is important to note that there is no flame detected in the top land or in the first ring land. Four frames after the flame in the 2nd ring land moves out of the viewfield to the left, there is a sudden appearance of brightness (homogeneous flame) in the first ring land that appears to originate from the left of the viewfield (marker 2). This is atypical to the normal phenomenon. Since there is no flame in the top land, it can be certain that this is not ignited by the flame propagating from the combustion chamber.

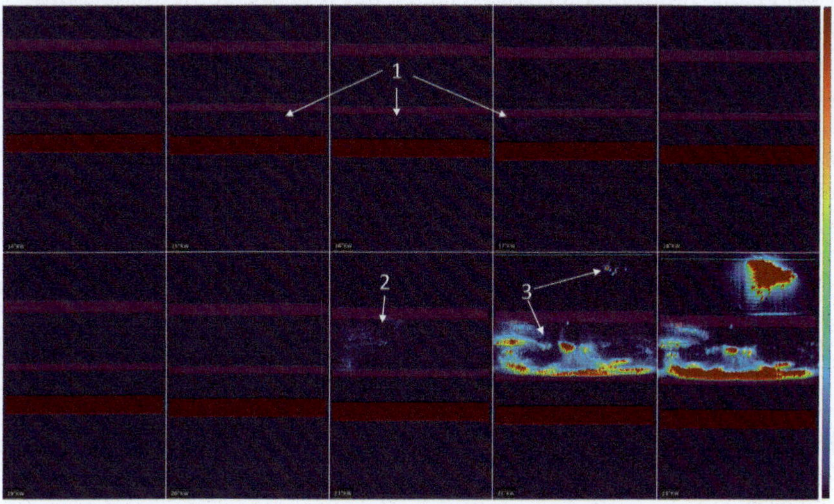

Fig. 12. Flame propagation in the second ring land. Operational parameters: 11 bar IMEP, 1500 rpm, MFB50 = 15 °C aTDC, and $H_2, \lambda = 1.6$

In the subsequent frames, as indicated by marker 3, bright emissions are visible in the top and first ring land. This is the diffusion flame from the ignition of the oil droplets. This is another phenomenon that could cause knocking or pre-ignition.

It was also seen from the experiments that the higher the load, i.e. the higher pressure inside the combustion chamber, the higher the frequency and intensity of occurrence of these phenomena. These could also lead to a higher probability of occurrence of abnormal combustion phenomena like knocking or pre-ignition. Although a direct correlation between these statistic ignition phenomena in the top and ring lands and irregular combustion in the main chamber has not yet been found.

6 Conclusion

This paper tries to explain the emission results obtained at different operational points and correlate the optical investigations. Through experiments in an optically accessible engine, the gasoline and hydrogen operations were studied, focusing on flame propagation in the piston lands. Through comparison, it was observed that hydrogen displays extremely different phenomena in the ring lands because of its properties that are vastly different from those of gasoline. Since it has a small quenching distance, it burns easily through the top land and even the first and second ring land of the piston.

From the comparison of the observed phenomenon to the emission measurement, a correlation was found regarding the top land flame propagation (TLFP), ring land flame propagation (RLFP), and particle emission. In simple words, the higher the intensity of flame propagation in the piston ring lands or the stronger the combustion here, the higher the particle emission. However, this only forms a small part of the puzzle and requires further investigations to understand the phenomenon that causes the emission so that they can be reduced, mitigated, or eliminated. These include time-correlated measurements of exhaust and optical emissions, more detailed investigations under different engine conditions and the use of measurement tools that provide a deeper insight into particulate emissions from H_2-ICEs.

Along with these, several abnormal phenomena were observed, which could lead to knocking or preignition. A better understanding of these phenomena is the key to understanding abnormal ignition in hydrogen engines, as well as to reducing emissions towards the development of zero-impact-emission engines.

Acknowledgments. The authors gratefully acknowledge the support of colleagues at Esslingen University of Applied Sciences, Mr. Christoph Spang and Mr Tim Schwellinger.

References

1. Eichlseder, H., Wallner, T., Freymann, R., Ringler, J.: The potential of hydrogen internal combustion engines in a future mobility scenario. SAE Tech Pap (2003). https://doi.org/10.4271/2003-01-2267
2. Stepień, Z.: A comprehensive overview of hydrogen-fueled internal combustion engines: achievements and future challenges. Energies **14**(20) (2021). https://doi.org/10.3390/en14206504
3. Klell M, Eichlseder H, Trattner A (2022) Hydrogen in automotive engineering, 1st edn. Springer, Springer Fachmedien Wiesbaden GmbH (2022). https://doi.org/10.1007/978-3-658-35061-1
4. Falfari, S., Cazzoli, G., Mariani, V., Bianchi, G.M.: Hydrogen application as a fuel in internal combustion engines. Energies **16**(6) (2023). https://doi.org/10.3390/en16062545
5. Medda, M., Calia, V., Di Sacco, M., Gullino, F., Rossi, V., Silvestri, N., Tonelli, R., Zaffino, F.: Challenges and Opportunities in developing a Hydrogen High Specific Power SCE in the roadmap towards zero net GHG (2023)
6. Catapano, F., Di Iorio, S., Magno, A., Sementa, P., Vaglieco, B.M.: A complete assessment of the emission performance of an SI engine fueled with methanol, methane and hydrogen. Energies **17**(5) (2024). https://doi.org/10.3390/en17051026

7. Thawko, A., Tartakovsky, L.: The mechanism of particle formation in nonpremixed hydrogen combustion in a direct-injection internal combustion engine. Fuel **327**, 125187 (2022). https://doi.org/10.1016/j.fuel.2022.125187
8. Thawko, A., Eyal, A., Tartakovsky, L.: Experimental comparison of performance and emissions of a direct-injection engine fed with alternative gaseous fuels. Energy Convers. Manage. **251**, 114988 (2022). https://doi.org/10.1016/j.enconman.2021.114988
9. Apicella, B., Catapano, F., Di Iorio, S., Magno, A., Russo, C., Sementa, P., Stanzione, F., Tregrossi, A., Vaglieco, B.M.: Comprehensive analysis on the effect of lube oil on particle emissions through gas exhaust measurement and chemical characterization of condensed exhaust from a DI SI engine fueled with hydrogen. Int. J. Hydrogen Energy **48**(58), 22277–22287 (2023). https://doi.org/10.1016/j.ijhydene.2023.03.112
10. Lejsek, D., Seboldt, D., Geiler, J., Leick, P., Frank, M., Stapf, K.: Mixture formation and combustion in a passenger car engine with low pressure H2 direct injection. In: 19th Symposium "Sustainable Mobility, Transport and Power Generation" (2023)
11. Schöler, J.: Optical investigation of the flame propagation into the piston lands under high load and knocking conditions. In: ENCOM (2019)
12. Friedman, R., Johnston, W.C.: The wall? Quenching of laminar propane flames as a function of pressure, temperature, and air? Fuel ratio. J. Appl. Phys. **21**(8), 791–795 (1950). https://doi.org/10.1063/1.1699760. https://pubs.aip.org/aip/jap/article-pdf/21/8/791/18309461/791_1_online.pdf
13. Ishizawa, S.: An experimental study on quenching crevice widths in the combustion chamber of a spark-ignition engine. In: Symposium (International) on Combustion vol. 26(2), pp 2605–2611 (1996). https://doi.org/10.1016/S0082-0784(96)80094-4
14. Schöler, J.: Visualisierung von Mehrphasen-Phänomenen in der Kolbengruppe eines direkteinspritzenden Ottomotors. Ph.D. thesis (Mar 2024). https://doi.org/10.17185/duepublico/81497 . https://duepublico2.uni-due.de/receive/duepublico_mods_00081497
15. Jung, Y., Lee, M.J., Kim, N.I.: Direct prediction of laminar burning velocity and quenching distance of hydrogen-air flames using an annular stepwise diverging tube (ASDT). Combust. Flame **164**, 397–399 (2016)
16. Fukuda, M., Korematsu, K., Sakamoto, M.: On quenching distance of mixtures of methane and hydrogen with air. Bull. JSME **24**(193), 1192–1197 (1981)
17. Schefer, R.W., Kulatilaka, W.D., Patterson, B.D., Settersten, T.B.: Visible emission of hydrogen flames. Combust. Flame **156**(6), 1234–1241 (2009). https://doi.org/10.1016/j.combustflame.2009.01.011
18. Eisen, S.-M.: Visualisierung der dieselmotorischen Verbrennung in einer schnellen Kompressionsmaschine. Ph.D. thesis, Technische Universität München (2003). https://mediatum.ub.tum.de/601871
19. Müller, T.: Simultane Visualisierung von Öl- und Kraftstoffschichten in der Kolbengruppe eines direkteinspritzenden Ottomotors durch laserinduzierte Fluoreszenz. Ph.D. thesis (December 2018)

A Method and a Set of Tools for In-Service Operation Emulation

Tim Martini[1](✉), André Kempf[2], Michael Auerbach[1], and André Casal Kulzer[3]

[1] LFA Esslingen University, Kanalstraße 33, 73728 Esslingen a.N, Germany
{tim.martini,michael.auerbach}@hs-esslingen.de
[2] Mercedes-Benz AG, Mercedesstraße 120, 70372 Stuttgart, Germany
andre.a.kempf@mercedes-benz.com
[3] IFS University of Stuttgart, Pfaffenwaldring 12, 70569 Stuttgart, Germany
andre.kulzer@ifs.uni-stuttgart.de

Abstract. The implementation of new methods and tools is essential to achieve the desired shift-left concept of vehicle development. The off-car development method outlines a data-driven approach for the continuous adaptation of vehicle software in accordance with the evolving overall system. With the intention of utilising the exploratory nature of road testing in early development stages, off-car robustness testing and automated off-car calibration are integrated into the product development process. These tools serve to bridge the gap between conceptual design and road testing. It is necessary to provide suitable stimuli for simulations and test benches in order to imitate road testing. To enhance the focus on customer applications, this paper uses test drives to emulate in-service operations. Using a big data infrastructure, an algorithm is presented that enables the efficient determination of probabilistic stimuli.

Keywords: Off-car development · Customer-oriented · Shift-left

1 State of the Art in Stimuli Computation and Research Motivation

In order to conduct an off-car imitation of road testing, it is necessary to provide at least one time-speed and one time-slope advance profile as stimuli for simulations and test benches. In the majority of well-known studies on the calculation of stimuli for simulations and test benches, speed measurement reports from test drives are aggregated into compressed, equivalent advance profiles. Depending on the conceptual framework, the measurements are either directly utilised or probability-based pseudo-database calculations are performed in a preceding step [1]. The probability-based, synthetic expansion of the database enables a reduction in the number of necessary test drives. Furthermore, the existing paradigm differentiates between the utilisation of micro-trips (MTs), sequences, clusters or modal values for the combination of the basic data. MTs are defined

A. Öchsner (ed.), *Collaborative Research Advancing Engineering Solutions for Real-World Challenges 2*, Proceedings in Technology Transfer,
https://doi.org/10.1007/978-3-031-79093-5_4

as speed intervals between two periods of standstill. When using sequences, clusters, or modal values, the sections are provided with additional information and detached from the framing of standstill phases based on MTs. For instance, modal sections are divided into phases with identical signs of acceleration, as well as cruising and standstill times. The sections are then selected for calculation using one of four methods: random selection [2], selection based on incremental properties [3], probability-based selection [4], or a combination of these methods. The designated stimuli are then subjected to a post-processing evaluation, during which the optimal variant is selected. Further investigations implement the section selection within the context of threshold acceptance [5] or genetic algorithms [1, 6], with the objective of enhancing the quality [7, 8].

In order to reinforce the customer-oriented development approach, the off-car development method (OCM, [9]) employs an in-service operation emulation. As the series-production control unit records only aggregated data, a further data source is required to provide the time series for calculating the stimuli. This database is provided by test drives conducted by development vehicles. As the established paradigm compresses the database to equivalent stimuli, which represent the characteristics itself, it is essential to guarantee that there is an equivalence between the database and the in-service operation in order for this work to be applicable.

Figure 1 illustrates the probability functions of recorded speeds (v), accelerations (a) and slopes (s) of the used series of test drives, as well as the in-service operation of a performance hybrid vehicle. The speed and acceleration functions are derived from the speed-acceleration frequency distribution. The boundaries of the bins are defined by the parameters of the current series-production software. The bin numbering system runs from the smallest negative value to the largest positive value. In the context of the acceleration data, bin number eight represents the standstill and cruise phases, while the slope bin numbers 10 and 11 represent slight negative and slight positive inclines respectively. The speed bin number 1 represents standstill times. The in-service operation is represented by connected points, with the database provided through the test drives marked by squares. In general, the lower speed bin numbers are overrepresented in the database, while the upper speed bin numbers are underrepresented. Furthermore, the test drives show a greater proportion of the higher slope bin numbers and the acceleration bin numbers 6–10 also deviate from the in-service operation.

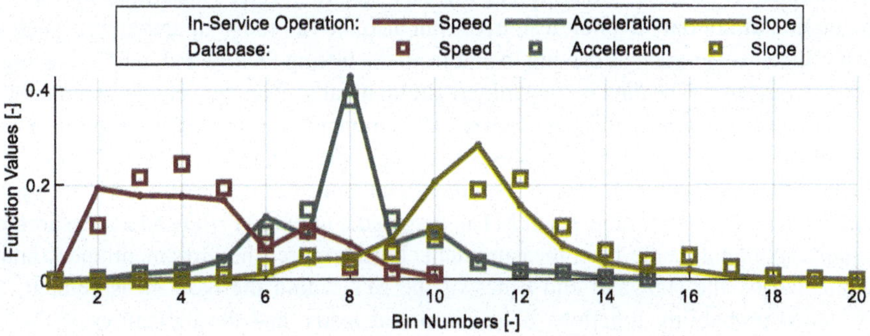

Fig. 1 Comparison of the characteristics of the database and the in-service operation

As the imposed loads (speed, acceleration, slope) of the database do not correspond to the in-service operation, it is not possible to achieve equivalent compression in accordance with the established paradigm for application within the OCM framework. In order to align off-car testing and off-car calibration with the eventual in-service operation, this work presents a new method for calculating stimuli utilising a database with divergent characteristics. The probability functions of the imposed loads are employed as a benchmark for comparison. The objective is to identify and filter relevant MTs in the database in order to achieve, in relation to the known equivalent compression, a shift in the stimuli towards a characterisation that is more representative of the in-service operation.

The following section outlines the recently developed method, including the employed tools and the approach to explicit filtering. Explicit filtering is utilised to define stimuli that are equivalent to an in-service operation, based on aggregated data and test drive measurements.

2 In-Service Operation Emulation: Calculation Method and Tools

In-service operation emulation is a method for shifting the characteristics of test-drive-based stimuli in order to align them with a freely selectable reference. The characteristics that are specified are limited to variants that are sufficiently represented in the database. Within the framework of OCM, documented measurement runs from on-road tests are segmented into MTs and characteristic parameters are computed. The data are then employed for the emulation of the in-service operation, utilising the principle of explicit filtering. In order to consider local characteristics, a probabilistic implementation of the chronology and the distribution of standstill phases follows the compilation of MTs. The following section describes the data capture and processing.

2.1 Database

In order to emulate the in-service operation, development test drives are used as the basis for data collection. To measure and archive the journeys, an existing big data infrastructure is employed. For this purpose, vehicles are equipped with data loggers. A measurement is initiated and terminated at each ignition cycle. This allows for the documentation of not only explicit test drives but also the full range of activities associated with vehicle usage, such as driving to and from the testing facility and regular driving in everyday settings. The data is transmitted automatically from the vehicle to a database.

In order to ensure the efficiency of the calculation process, upload routines have been implemented. The data is then processed in order to prepare it for analysis. This involves dividing the measurements into MTs, calculating specific parameters, and storing the data in a memory data frame. Each MT is assigned a unique MT ID and a measurement ID corresponding to the base measurement. Furthermore, the aforementioned data is augmented by the start and end times of the MT within the base measurement, the associated probability functions of the imposed loads, and the indication of the bin number at which the functions reach their maximum. Additionally, further maximum, minimum, and average values are stored.

For the presented work, a database of 1,300,000 MTs is used. The data was recorded over a period of 46,000 h and a distance of 2,900,000 km by 150 vehicles in 518,000 measurements. The calculation of stimuli, utilising the principle of explicit filtering, is subsequently based on the memory data frame. Consequently, the calculation is made independent of time-based measurements, and it can be carried out in any suitable environment (cloud, remote computer, local computer) in an efficient manner.

2.2 Principle of Explicit Filtering

The calculation of stimuli is based on the developed principle of explicit filtering, utilising the memory data frame. Figure 2 displays an exemplary database consisting out of three speed profiles in the left-hand column and the corresponding probability functions in the right-hand column. The black-coloured probability functions represent the entire speed profiles. Each of the measurements is comprised of three MTs. In order to illustrate the principle, in addition to the probability functions of the overall measurements, the probability functions of the individual MTs are plotted in grey and in blue. The red-coloured function serves as an exemplary target, representing the in-service operation. Given that the entire measurements at the bin numbers 1, 2, 4, 5 exhibit pronounced characteristics, they are unsuitable for fulfilling the specified requirements.

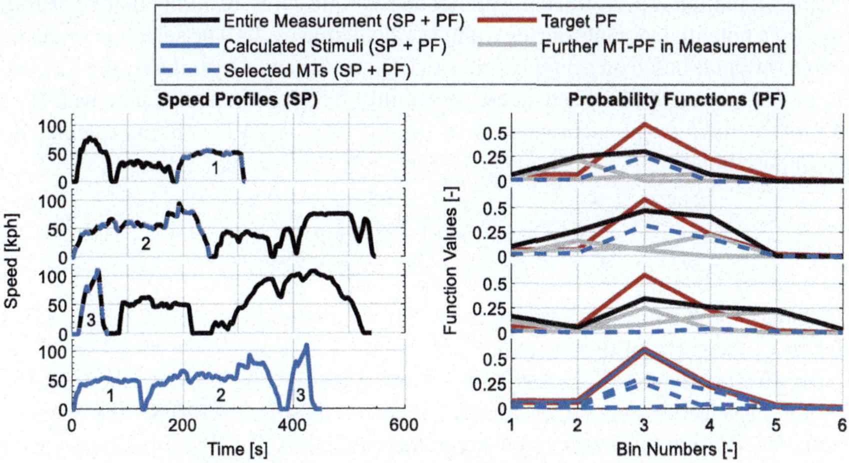

Fig. 2 Visualisation of the principle of explicit filtering

In order to emulate the target, it is necessary to select and filter the appropriate MTs (blue) from the dataset. Selecting a MT from each of the measurements (1, 2, 3) allows the red-coloured target to be achieved. The further MTs in the database are not utilised (grey-coloured PF). The principle is derived from the characteristics of the database used. It is common practice for test vehicles to undergo functional tests, the frequency of which is not representative of the use that would be expected during in-service operations. However, the vehicles are also driven on public roads in an in-service-representative manner. The algorithm developed searches the database for the relevant MTs and utilises them for the calculation of the stimuli.

The principle of explicit filtering is illustrated in pseudo-code in Algorithm 1. It describes how relevant MTs are filtered based on the probability functions of the imposed loads and combined into stimuli equivalent to the in-service operation. The filtering process continues until a specified distance is reached. The selection of individual MTs applied here corresponds to a combined approach of incremental filtering and randomized selection.

Algorithm 1 : Calculation of stimuli equivalent to an in-service operation
Input: Memory data frame (mdf) with measurements of the database, target probability functions and target distance
Return: List of selected MTs that are equivalent to the target in total and the quality achieved

```
1: Randomised selection of an initial MT
2: Repeat
3:        Calculate current stimuli probability functions (PF) of the imposed loads
4:        Calculate current distance of the stimuli
5:        Calculate differences between in-service PF and stimuli PF of the imposed loads
6:        Calculate difference between target distance and current distance
7:        Filtering mdf on MTs with max. support points equal to max. difference support points (line 5)
8:        Filtering mdf on MTs with distance < distance difference (line 6)
9:        Randomised selection of further MT from filtered mdf
10: While current distance < target distance
11: Calculate quality criterion of the created stimuli
12: Return list with selected MTs and achieved quality
```

The calculation is performed in parallel on multiple cores, resulting in the generation of multiple potential variants during a single calculation cycle. The selection of the most suitable variant is based on a quality criterion. This is achieved by utilising the Kullback–Leibler divergence (KLD) for discrete probability functions in accordance with Eq. (1) and a linear function in accordance with Eq. (2) [10]. A KLD value is calculated for each imposed load.

$$\mathrm{KLD(p, q)} = \sum_{x \in X} p(x) \log \frac{p(x)}{q(x)} \tag{1}$$

with: $p(x)$ = probability function stimulus, $q(x)$ = probability function target, X = bin numbers of the probability functions.

The distance achieved is a result of the mdf filtering (line 7 and 8), the subsequent random MT selection and the termination criterion. In order to consider the distance in the quality criterion it is evaluated using a linear function. This is calibrated to increase the value of the ratings above and below the target distance, in a scale consistent with the KLD values. The individual terms for speed, acceleration, slope and distance are aggregated to form an arithmetic mean and then treated as a scalar quality (line 11).

$$y(x) = m \cdot (x - x_1) + b \tag{2}$$

with: x = current distance, y = quality criterion of distance, m = gradient, x_1 = target distance, b = y-intercept

Results of the Explicit Filtering

The presentation of the results is also based on the probability functions of the imposed loads. Figure 3 presents a sample calculation of explicit filtering for a target distance of

2000 km. The in-service operation is illustrated like in Fig. 1, with the emulation marked with squares. The quality value of the variant is 0.01242, representing a successful shift in the stimuli character towards an in-service operation. The individual probability functions and distance attain the following values:

• KLD_v_100% : 0.00050 • KLD_s_100%

•KLD_a_100% : 0.03360 • y_100% : 0.00903

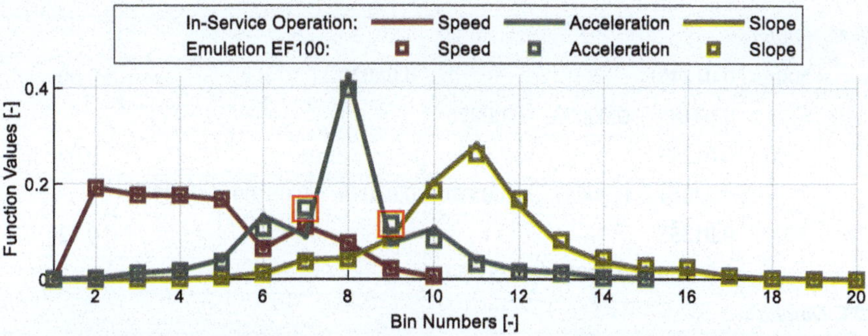

Fig. 3 Result of the explicit filtering, visualisation of the probability functions of the imposed loads, target distance: 2000 km, 100% database

While the values for speed and slope greater than two orders of magnitude are reduced (see Table 1, row C), the improvement in acceleration is less significant. In Fig. 3, the values of the bin numbers 7 and 9 are identified as the main source of concern (red-coloured squares). The characteristics of the database are preserved in these areas. The cause of the issue has been identified as an insufficient representation of in-service operation characteristics within the database in this area.

In order to address the issue, the objective is to reduce the incidental use of MTs with high probabilities within the affected areas. Since the algorithm considers peaks only (line 7) and the field characteristic is insufficiently represented, the database is subjected to a filtering process. Consequently, all MTs are excluded whose acceleration bin values at numbers 7 and 9 are the second- and third-highest in value. This leads to a reduction of the database by 70%. The results for the reduced database are presented in Fig. 4. The quality of the variant is 0.01030 with part values as follows:

- KLD_v_30% : 0.00230 • KLD_s_30% : 0.00558
- KLD_a_30% : 0.02503 • y_30% : 0.00829

The filtering of the database results in an improvement in the replication of the acceleration bin values of numbers 7 and 9. With regard to the individual values (bin numbers 7 and 9), a relative improvement of 87 and 98% is achieved. However, the significant reduction of the database causes quality degradation in the replication of the remaining bin values. Consequently, the quality of the acceleration reconstruction is enhanced by 25% in comparison to the 100% database. The reconstruction of the slope

Table 1 Summarised overview of the achieved quality values of calculated stimuli

Target distance [km]		Quality values [−]						Multi-MTs [%]
		Calculated stimuli versus in-service			Database versus in-service			
		Total			Total			
		a	v	s	a	v	s	
100% database								
A	500	0.01345			**0.08030**			0.00
		0.03185	0.00073	0.00776				
B	1000	0.01243						0.00
		0.03064	0.00087	0.00579	0.04078	0.06109	0.13902	
C	2000	**0.01355**						0.32
		0.03360	**0.00050**	**0.00654**				
30% Database								
D	500	0.00958			**0.14252**			4.72
		0.02353	0.00198	0.00323				
E	1000	0.01005						6.90
		0.02129	0.00372	0.00513	0.15276	0.08584	0.18895	
F	2000	**0.01097**						13.02
		0.02503	**0.00230**	**0.00558**				

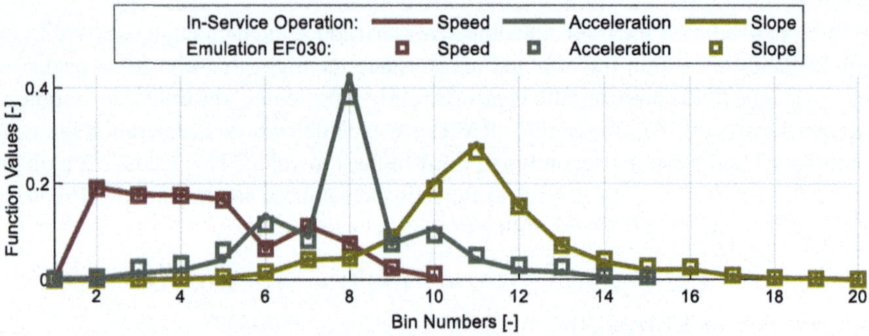

Fig. 4 Result of explicit filtering, visualisation of the probability functions of the imposed loads, target distance: 2000 km, 30% database

is not affected, but the reconstruction quality of the speed is reduced by a factor of 3.6. Nevertheless, this maintains a high standard of quality. Overall quality is improved by the filtering process.

To provide a summarised overview of the power of explicit filtering, Table 1 presents the quality values for each of three target distances, based on the 100% and the filtered 30% database. The left-hand column of the table displays the values of the in-service operation in comparison to the calculated stimuli while the right-hand column displays the values in comparison to the used database. The total value is derived from the three achieved KLD values for speed, acceleration and slope. As the target distances are well reproduced, the quality of the distance assessment is not considered in order to facilitate a more accurate comparison.

The explicit filtering principle enables a successful shift of the stimuli towards a characterisation of the in-service operation. With regard to the 100% database, an improvement in quality of 83% can be achieved through the utilisation of the complete basis at a target distance of 2000 km. In the case of the filtered basis, the improvement is in excess of 85%. The improved quality of the filtered database results in increased multiple use of individual MTs (see Table 1, multi-MTs) due to the reduced amount of data. The overall quality is also satisfactory given the shorter target distances, which also reduce the need for multiple use of individual MTs.

The utilisation of the test vehicles on a daily basis results in the continual expansion of the database. As the database expands, it is expected that the shift towards the in-service characteristic will become even more pronounced. Additionally, the reuse of data through filtering should become less common. Moreover, an additional work could be conducted to determine whether adapting the algorithm to the collecting conditions of the database would enhance the quality. For instance, the implementation of additional bin values into the MT filtering process could be a potential avenue for improvement.

The previously presented method enables the generation of a set of MTs that is globally equivalent to an in-service operation. However, the selection does not contain any information about the transitions between the MTs. In the light of the aforementioned selection, it is necessary to define local conditions, such as the chronological sequence and the standstill periods between MTs. As the in-service operation data lacks both chronological ordering and information regarding the distribution of periods of standstill phases, these must be generated synthetically in a subsequent phase.

2.3 Chronology and Standstill Distribution

The algorithmic approach, designated as Algorithm 1, is based on the explicit filtering principle and combines relevant MTs from a database to match a specified characteristic. The outcomes describe the utilized MTs for the stimuli, yet lack any indication of the temporal sequence. To address the chronology within the stimuli, which is not documented in the current series-production control unit, the MTs are arranged probabilistically through Monte Carlo simulations [11]. In order to determine the transition probability of a MT to the subsequent one, the MTs are divided into distance or dynamic clusters. The following section illustrates the method using distance clusters. The cluster definition and assignment of the utilized MTs are conducted using a K-means algorithm [12]. Approximately 80% of the MTs are located within the first cluster, situated at a distance of 0.4 km. An additional 10% are distributed across the second cluster, centred at 3.4 km. The third and fourth clusters are situated at 8.2 and 15.6 km respectively.

Lastly two and three MTs are also allocated to the fifth (36.4 km) and sixth (75.8 km) clusters respectively.

A Monte Carlo simulation is employed to shuffle the sequence of MTs 10^4 times. The transition probabilities between the clusters are then determined on the basis of the resulting sequences generated by the random experiments. Figure 5 illustrates the typical outcome of this process. The plot is read as follows: each colour represents a current cluster. With the given probability of the colour, the subsequent MT is part of the following cluster number on the abscissa.

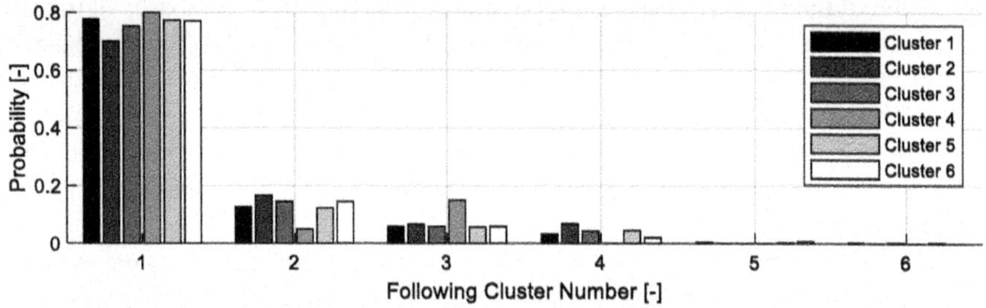

Fig. 5 Results of the Monte Carlo simulation: transition probabilities between MT distance clusters

The final configuration of the MTs for the stimuli is achieved by sequentially combining the MTs, with the probability function of the current cluster determining the subsequent MT in each calculation step. Given the high proportion of MTs in cluster 1, each class exhibits the highest likelihood that a current MT will be followed by an MT from cluster 1.

Similarly to the imposed loads, the periods of standstill are determined in relation to the frequency of the in-service operation. In order to select the MTs using Algorithm 1, the standstill periods of the target are set to zero. As test drives frequently entail wide periods of standstill, these are not incorporated into the MTs but rather synthesised. The absolute standstill time for the calculated stimuli is determined by the computation frequency in the series-production control unit and the counted entries within the speed bin number one ($v < 0.1$ kph) and the acceleration bin number eight ($-0.2 < a < 0.2$ m/s^2) in the speed-acceleration frequency distribution. Since no documentation is provided for the duration of individual standstills in the in-service operation, the total standstill duration is randomly distributed between the MTs.

3 Conclusion

A method for the emulation of in-service operations is presented. The employed tools allow the calculation of stimuli for simulations and test benches based on aggregated data and test drive measurements. The developed algorithm uses the principle of explicit filtering to determine stimuli characteristics that differ from those derived from the database used. To ensure efficient calculation, upload routines have been implemented using a big data infrastructure and specific parameters for each measurement journey have been documented for each micro-trip. The information lost during the documentation of the in-service operation, namely the chronological sequence and the distribution of standstill phases, are probabilistically implemented in the stimuli. Within the OCM framework, the calculated stimuli serve to bridge the gap between conceptual design and road testing using off-car testing and automated off-car calibration. This approach allows for early stages of development to benefit from the exploratory nature of road testing, utilizing simulations and test bench measurements. The results presented demonstrate that with a sufficient frequency of the required instances, a shift of the stimuli towards predefined characteristics can be achieved. The discrepancy between the recorded loads of the in-service operation and the stimuli for the off-car tests is reduced by over 80% compared to an equivalent compression of the used database. In order to achieve an additional enhancement in quality, it is possible to further adapt the filter logic to align with the characteristics of the data collection of the employed database. Through the design of the method, a significant element is provided for the implementation of a successful shift-left concept and a customer-oriented data-based design in automotive software development.

References

1. Förster, D., Inderka, R.B., Gauterin, F.: Data-driven identification of characteristic real-driving cycles based on k-means clustering and mixed-integer optimization. IEEE T. Veh. Technol. **69**, 2398–2410 (2020). https://doi.org/10.1109/TVT.2019.2963272
2. Pfriem, M., Gauterin, F.: Development of real-world driving cycles for battery electric vehicles. WEVJ **8**, 14–24 (2016). https://doi.org/10.3390/wevj8010014
3. André, M.: The ARTEMIS European driving cycles for measuring car pollutant emissions. Sci. Total. Environ. Total. Environ. **334–335**, 73–84 (2004). https://doi.org/10.1016/j.scitotenv.2004.04.070
4. Wang, Z., Zhang, J., Liu, P., Qu, C., Li, X.: Driving cycle construction for electric vehicles based on markov chain and monte carlo method: a case study in Beijing. Enrgy. Proced. **158**, 2494–2499 (2019). https://doi.org/10.1016/j.egypro.2019.01.389
5. Liessner, R., Dietermann, A., Bäker, B., Lüpkes, K.: Generation of replacement vehicle speed cycles based on extensive customer data by means of markov models and threshold accepting. SAE Int. J. Alt. Power. **6**, 165–173 (2017). https://doi.org/10.4271/2017-26-0256
6. Perhinschi, M.G., Marlowe, C., Tamayo, S., Tu, J., Wayne, W.S.: Evolutionary algorithm for vehicle driving cycle generation. J. Air. Waste. Manage. **61**, 923–931 (2011). https://doi.org/10.1080/10473289.2011.596742
7. Dai, Z., Niemeier, D., Eisinger, D.: Driving cycles: a new cycle-building method that better represents real-world emissions. Department of Civil and Environmental Engineering, University of California, Davis, vol. 570 (2008)
8. Giakoumis, E.G.: Driving and Engine Cycles. Springer International Publishing, Cham (2017)
9. Martini, T., Winke, F., Arnold, M., Schmitz, J., Auerbach, M., Bargende, M., Kulzer, A.: Methodik und Tools zur Betriebsstrategieentwicklung für Performance-Hybride. 9. In: AutoTest Technical Conference 2022
10. Suh, C.: Convex optimization for machine learning. Now Publishers, Boston-Delft (2022)
11. Imkamp, T., Proß, S.: Einstieg in stochastische Prozesse. Springer, Berlin (2023)
12. scikit-learn: KMeans (2024). https://scikit-learn.org/stable/modules/generated/sklearn.cluster.KMeans

Statistical Size Influence and Support Effect of Electrical Steel Strips

Patrick Schwarz[1](✉), Konstantin Naumenko[2], and Peter Häfele[1]

[1] Faculty of Mobility and Technology, Esslingen University of Applied Sciences, Kanalstraße 33, 73728 Esslingen, Germany
{patrick.schwarz,peter.haefele}@hs-esslingen.de
[2] Faculty of Mechanics, Institute of Mechanics, Otto-von-Guericke-University, Universitätsplatz 2, 39106 Magdeburg, Germany
konstantin.naumenko@ovgu.de

Abstract. Electrical steel strips are functional materials employed to enhance the efficiency of motors. The rotor design incorporates small bridges that are critical areas susceptible to failure. For the fatigue assessment of those bridges both statistical size effect and dynamic support effect has to be taken into account. Stress-controlled tests are conducted on NO30-15 using unnotched specimens of varying sizes as well as notched specimens. The material exhibits significant statistical size and support effects. Various methods for determining the Weibull exponent are evaluated based on these test results. By applying concepts of material-mechanical support effects, it is shown that determining the Weibull exponent in the limit lifetime regime results in strong agreement between theoretical predictions and experimental data.

Keywords: Fatigue test · Electric steel strips · Statistical size effect · Support effect

1 Introduction

In recent years, the demands of electric motors have progressively increased with a significant trend towards more efficient drive concepts. These challenges encompass not only motor design but also the selection of appropriate materials. Non-grain-oriented electrical steel has become the material of choice for the rotors of synchronous machines due to its ability to minimize remagnetization losses in alternating magnetic fields. This specific category of materials is characterized by a microstructure consisting of coarse grains, ranging in size from 20 to 200 μm, and a silicon content between 1 and 4.4% [1]. To further reduce eddy current losses, the rotor of an electric machine is constructed from individual sheets with thicknesses less than 1 mm [1].

A. Öchsner (ed.), *Collaborative Research Advancing Engineering Solutions for Real-World Challenges 2*, Proceedings in Technology Transfer,
https://doi.org/10.1007/978-3-031-79093-5_5

In addition to the reduced thickness of the rotor lamellae, pockets are incorporated into the component to enhance motor efficiency via permanent magnets. The so called saturation bridge between two pockets is the critical fatigue failure point due to magnetic and centrifugal forces [2]. Very small transition radii associated with high stress gradients and a thin cross-section of the bridge characterize this area. Therefore two problems for the fatigue assessment have to be addressed: A reduction in the highest loaded material volume increases the fatigue strength. Studies from [3, 4] show that a decreasing specimen size correlates with an increasing fatigue strength, which is referred to as statistical size influence. This effect is not only observed in unnotched specimen but also arises from an inhomogeneous stress distribution in the notched cross-section. The stress distribution reduces the range of failure-critical stress, leading to higher fatigue strength than theoretically predicted by the notch concentration factor [5]. In this resulting support effect of notched specimens, a high proportion is attributed to the statistical support effect in advanced engineering concepts [6, 7]. The statistical support effect is based on the mechanism of statistical size influence.

The focus of these investigations is to determine the statistical size effects of electrical steel and their application to the material—mechanical support effect. In this context, various methods for determining material parameters that characterize size effects are evaluated.

2 Size Influence and Statistical Support Effect

As mentioned, the fatigue strength decreases with increasing component volume which is referred to as the size effect. According to [8], this effect includes four different size influences:

- Technological size influence
- Stress gradient size influence
- Statistical size influence
- Surface size influence.

The technological size effect encompasses the impact of machining processes on components of varying sizes. For instance, heat treatment imparts different characteristics to a large metallic component compared to a small one. In smaller components, cooling is uniform across the volume, resulting in a more homogeneous microstructure within the cross-section. The stress gradient size effect pertains to the inhomogeneous stress distribution within a component, induced by geometric notches or bending stress. This inhomogeneous distribution limits the range within which failure-critical stress operates. The statistical size effect indicates that the probability of a failure-causing defect increases with the component's size. The surface size effect implies that the likelihood of machining-related defects rises as the component's surface area expands. Given that fatigue-induced cracks typically initiate at a components' surface, the distribution of surface defects is crucial for the fatigue strength. In practice, these four effects are interrelated and influence each other [8, 9].

Notches are less damaging under cyclic load than under monotonous load. This can be elucidated more precisely by analysing local stresses, where it is observed that the SN-curves of notched specimens lie above those of unnotched specimens [5]. This phenomenon is known as the support effect and is accounted mathematically using various approaches. In the material—mechanical approach, as utilized in [6, 7], the support number η_σ is the product of the statistical support number η_{st}, the deformation support number η_{dm}, and the fracture-mechanical support number η_{fm}, see Eq. (1) [6, 7].

$$\eta_\sigma = \eta_{st} \cdot \eta_{dm} \cdot \eta_{fm} \tag{1}$$

The statistical support number η_{st} is based on the concept of size influence and the Weibull weakest—link model [10]. The support effect is described by the ratio between the reference surface of an unnotched specimen $A_{ref,st}$ and the highest loaded surface of a notched structure $A_{\sigma,st}$, see Eq. (2).

$$\eta_{st} = \left(\frac{A_{ref,st}}{A_{\sigma,st}}\right)^{\frac{1}{\kappa_{st}}} \tag{2}$$

The Weibull exponent κ_{st} is a material parameter that describes the scatter of fatigue strength and represents the shape parameter of the Weibull distribution, see Eq. (3). $P(\sigma_{ai})$ denotes the failure probability of a stress amplitude σ_{ai} relative to the mean value of the fatigue strength σ_E. In addition to the statistical support effect, κ_{st} determines the highly stressed surface of the notched structure through the integral of the normalized stress distribution $g(x, y, z)$, see Eq. (4). Similar considerations can also be applied to the highly stressed volume of a structure.

$$P(\sigma_{ai}) = 1 - e^{-\left(\frac{\sigma_{ai}}{\sigma_E}\right)^{\kappa_{st}}} \tag{3}$$

$$A_{\sigma,st} = \int_A g(x, y, z)^{\kappa_{st}} dA \ with \ g(x, y, z) = \frac{\sigma(x, y, z)}{\sigma_{max}} \tag{4}$$

The deformation support effect considers the macro-support effect according to Neuber [11]. This effect requires an increase in stress at the notch ground, leading to the plastic flow of the crystal elements. As the load further increases, the grains deform, and the load is redistributed to neighboring grains. This model employs the parameters of the cyclic stress–strain curve according to Liu and Zenner and depends on the statistical support number, see Eq. (5) [6, 12]. σ_W represents the alternating fatigue strength of the material with the plastic strain amplitude $\varepsilon_{pl,w}$.

$$\eta_{dm} = \sqrt{1 + \frac{E \cdot \varepsilon_{pl,w}}{\sigma_W} \cdot (\eta_{st})^{\frac{1}{n'}-1}} \tag{5}$$

The fracture mechanics support number depends on η_{st} and η_{dm}. If the model is used in the nominal stress concept, both the notch concentration factor K_t and the related stress gradient χ' or G_σ are included, see Eq. (6). $R_{m,bm}$ describes the reference tensile strength of steels with 680 MPa. When used in the local stress concept, only the related stress gradient is required [6], see Eq. (7).

$$\eta_{fm} = \frac{1}{\frac{\eta_{st}\cdot\eta_{dm}}{K_t} + \sqrt{\frac{2-3\cdot K_t^{-1}+K_t^{-3}}{4\cdot G_\sigma\cdot 0,1\cdot\left(\frac{R_{m,bm}}{R_m}\right)^2}}} \tag{6}$$

$$\eta_{fm} = \frac{5+\sqrt{G_\sigma}}{5\cdot\eta_{st}\cdot\eta_{dm} + \frac{R_m}{R_{m,b,m}}\cdot\sqrt{\frac{7.5+\sqrt{G_\sigma}}{1+0.2\cdot\sqrt{G_\sigma}}}} \tag{7}$$

3 Methods for Determining the Weibull Exponent

The Weibull exponent is considered a measure of the scatter in the fatigue strength of a material. According to [13], the accuracy of this estimate depends on the number of specimens tested. Therefore, a reliable estimation of fatigue strength scatter requires as large number of specimens and extended test durations (several million cycles). Similarly, results from [14, 15] indicate that tests on electrical steel typically yield a single mixed horizon (load amplitudes showing both runouts and fractures), resulting in an uncertain estimation of the Weibull exponent. In this study, four different methods for determining the Weibull exponent will be presented and compared.

Method 1 aims to estimate the Weibull exponent using results from stress—controlled tests (hereinafter referred to as Method HCF). Comparable to the bead string method from [13], the test results are transformed to a number of cycles. This transformation yields a distribution of load amplitudes $\sigma_{ai,trans}$ of varying magnitudes, which are then sorted in ascending order and assigned a probability of occurrence $P_F(\sigma_{a,i,trans})$ using Eq. (8). The parameter i represents the rank of the sorted trial, and n denotes the total number of trials. By using the logarithmized pairs of values from $\sigma_{ai,trans}$ and $P_F(\sigma_{a,i,trans})$, κ_{st} is estimated using Eq. (9) and linear regression.

$$P_F\left(\sigma_{a,i,trans}\right) = \frac{3\cdot i - 1}{3\cdot n + 1} \tag{8}$$

$$P_F\left(\sigma_{a,i,trans}\right) = 1 - e^{-\left(\frac{\sigma_{a,i,trans}}{\overline{\sigma}_{a,trans}}\right)^{\kappa_{st}}} \tag{9}$$

Method 2, also named as Method LCF, is based on [16]. It utilizes results from strain—controlled tests to construct a corresponding damage parameter curve. The damage value P_i is derived from the stress and strain data obtained during the test. In the next step, the equivalent damage value $P_{50\%}(N_i)$ is determined using the number of cycles from the test and the damage parameter curve. The ratio of these two damage values, $P_{rel,i}$, is calculated using (10). Similar to Method HCF, $P_{rel,i}$ is sorted, and the probability of its occurrence is estimated using Eq. (8). The Weibull exponent κ_{st} is then estimated through regression analysis of the logarithmized pairs of values.

$$P_{\text{rel,i}} = \frac{P_i}{P_{50\%}(N_i)} \tag{10}$$

The method T_S (Method TS) is based on the relationship between the scattering in the direction of cycles $T_N = N_{90\%}/N_{10\%}$ and in the direction of stress $T_S = \sigma_{E90\%}/\sigma_{E10\%}$. The correlation is made via the inclination of the SN-curves in the limit lifetime regime k (inverse Basquin coefficient) see Eq. (11). The Weibull exponent is estimated using Eq. (12) [17].

$$T_S = T_N^{\frac{1}{k}} \tag{11}$$

$$\kappa_{st} = \frac{1.3151}{\log(T_s)} \tag{12}$$

In the Micro method (Method Mic.), the ratios of the fatigue strength of different large test specimens are determined and equated with η_{st}. When using unnotched specimens, it can be assumed that the highly loaded surface corresponds to the test surface [16]. κ_{st} can be determined with Eq. (2).

4 Material, Specimens and Test Scope

The analysed material is the non-grain-oriented electrical steel NO30-15. The nominal thickness of the sheets is $t = 0.3$ mm and the material has an average grain size of $d_k = 0.085$ mm [14]. The tensile strength of the material is $UTS = 570$ MPa with a yield strength of $YS = 439$ MPa. The elongation at break ε_{AF} of NO30-15 is 21%. All results were determined longitudinally to the rolling direction.

Four different specimen geometries are tested to investigate the statistical size effect under quasi- monotones loading. The tensile specimens are illustrated in Fig. 1a. Micro-specimens, depicted in Fig. 1d, feature three different widths B of the test area: $B = 2$ mm, $B = 1$ mm, and $B = 0.5$ mm. Five tensile specimens (Fig. 1a) were used for the tests and ten tests per geometry were carried out on the micro specimens.

Stress-controlled fatigue tests are conducted at room temperature with loading oriented longitudinally to the rolling direction. Three specimen geometries without geometric notches (Fig. 1c) and with $B = 2$ mm and $B = 0.5$ mm are tested. Additionally, tests are performed on two notched geometries (Fig. 1b) with notch radii $R_A = 4$ mm and $R_A = 0.35$ mm. The load ratio is $R_\sigma = 0.1$. A resonance testing machine with a shut-off criterion $\pm$ 1 Hz and a servo-electric testing machine with a shut-off criterion of $\pm$ 0.5 mm increase in displacement during the test are utilized. Tests are terminated and evaluated as runouts when the limit of $N_G = 10^7$ cycles is reached. The statistical evaluation is according to DIN 50100:2016 [13]. All specimens are laser-cut, the edge conditions are similar to those described in [15].

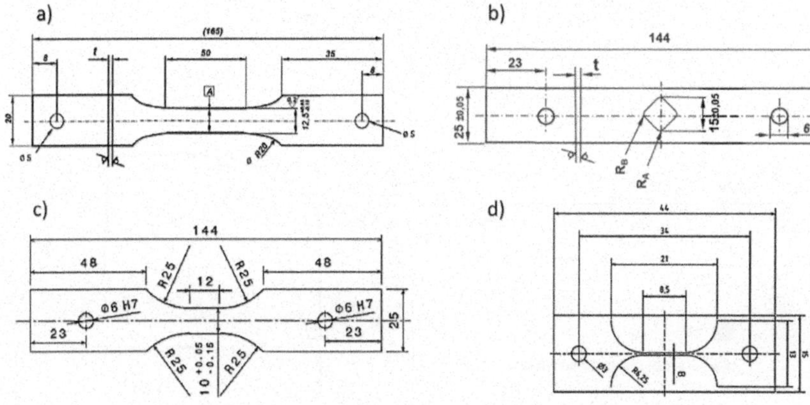

Fig. 1 **a** Tensile specimen B = 12.5 mm, **b** notched fatigue specimens, **c** unnotched fatigue specimen B = 10 mm, **d** micro specimens with width B = 2, 1 and 0.5 mm

The highly stressed surface is determined using the SPIEL method [18]. Ansys Workbench 2024 R2 is used to calculate the related stress gradient, notch factor and the maximum loaded surface.

5 Results of the Tensile and Fatigue Tests

The results of the tensile tests, depicting tensile strength plotted against specimen width, are illustrated in Fig. 2a, with exemplary measurement results shown in Fig. 2b. There is no significant difference in tensile strength observed between tests conducted with $B =$ 12.5 mm and $B = 2$ mm. However, specimens with $B = 1$ mm and $B = 0.5$ mm exhibit approximately 5% higher tensile strength compared to the other two specimen variants. As specimen width decreases, there is an increase in scatter. Specifically, specimens with $B =$ 12.5 mm show a scatter of approximately 1 MPa, whereas specimens with B = 0.5 mm exhibit a scatter of approximately 26 MPa.

In fatigue tests on micro specimens, there is no significant difference between the results obtained with $B = 2$ mm and $B = 0.5$ mm specimens. Micro specimens exhibit approximately 20% higher fatigue strength ($\sigma_{\text{E,micro,50\%}} \approx 221$ MPa) compared to specimens with B = 10 mm ($\sigma_{\text{E,B=10 mm,50\%}} = 185$ MPa). Similar to tensile tests, the scatter increases with decreasing material volume i. e. decreasing width B ($T_{\text{N,B=10 mm}} = 3.3$, $T_{\text{N,B=2 mm}} = 4.4$, $T_{\text{N,B=0.5 mm}} = 14.1$). The SN-curve is shallower with micro specimens ($k_{\text{mikro}} \approx 27$), see Fig. 3a.

Comparing notched and unnotched fatigue specimens, there is no significant influence of geometric notches on the slope of the SN-curve ($k \approx 7$). Specimens with a notch radius of $R_{\text{A}} = 4$ mm experience a decrease in fatigue strength of approximately 40%, while specimens with $R_{\text{A}} = 0.35$ mm exhibit a decrease of around 70%.

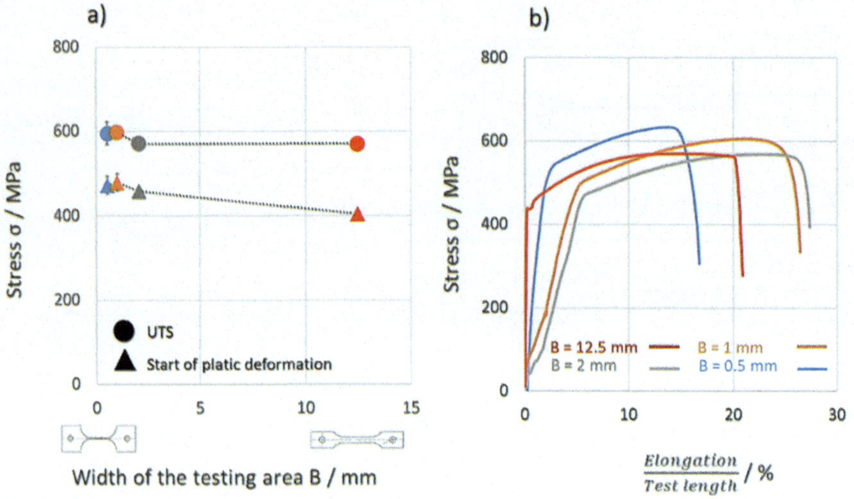

Fig. 2 Results of the tensile tests. **a** Tensile strength across specimen width B, **b** exemplary test results with specimens of different widths

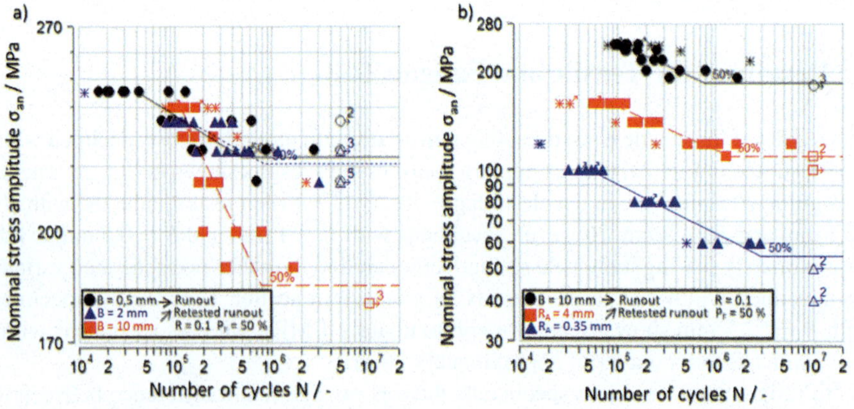

Fig. 3 Results of stress-controlled tests, along the rolling direction R = 0.1 PF = 50%. **a** Unnotched specimens with B = 0.5 mm, B = 2 mm and B = 10 mm, **b** results of unnotched and notched fatigue specimens

6 Size Influence with NO30-15 Under Monotonous and Cyclic Load

Direct comparison between the results of tensile and cyclic tests reveals that the influence of specimen size is more pronounced with NO30-15 under cyclic loading. Studies in [3] draw a similar conclusion for stainless steel. Despite a comparable cross-sectional area ratio (approximately 8 in [3] and approximately 5 in this study), NO30-15 (around 20%) exhibits a higher statistical size effect compared to stainless steel (approximately

9%) [3]. These findings indicate that both the magnitude and sensitivity of the size effect depend on the material properties [3, 19, 20].

The sensitivity towards the specimen size depends on whether the load is cyclic or monotonous. In tensile tests, significant differences only emerge for widths B < 2 mm, whereas under cyclic loading, specimens with B = 2 mm would exhibit a noticeable size effect. It is important to note that the size effect observed under monotonous loading influences the size effect observed under cyclic loading. The differences become clear in the representation of the logarithmic distribution function of $B = 0.5$ mm and $B = 2$ mm, see Fig. 4.

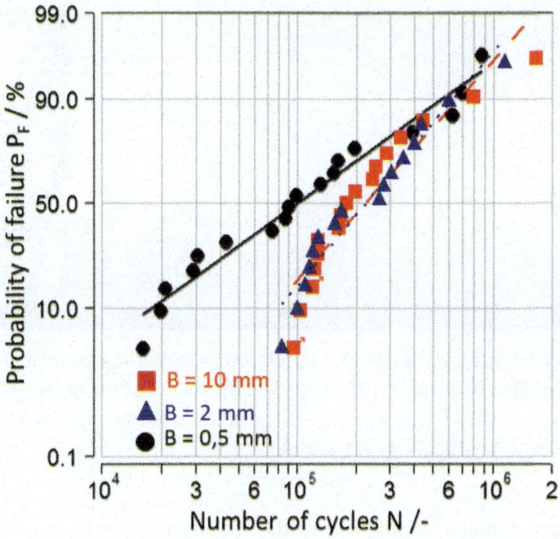

Fig. 4 Representation of the distribution functions of NO30-15 specimens with different test widths

While the SN curves for $B = 0.5$ mm and $B = 2$ mm are almost parallel, see Fig. 3, their distribution functions exhibit differences. The influence of specimen size on tensile strength has a more pronounced effect on the distribution function compared to its influence on fatigue strength. Observations of size influence on fatigue strength indicate diminishing differences with increasing load amplitude. Beyond a load amplitude of approximately 230 MPa, there is no significant disparity among the SN-curves of the three specimen sizes. This can be attributed to increased plastic deformation under higher loads, which mitigates the detrimental effects of material defects, akin to the reduction of edge effects discussed in [15].

7 Estimation of the Weibull Exponent and Statistical Support Effect at NO30-15

To assess methods for estimating the Weibull exponent, calculated support effects will be compared with test results from specimens with notched cross-sections. To determine the stress integral, it must be clarified whether surfaces or the volume must be used. For this purpose, the location of crack initiation is analysed for specimens with $B = 10$ mm and for notched specimens, seen Fig. 5.

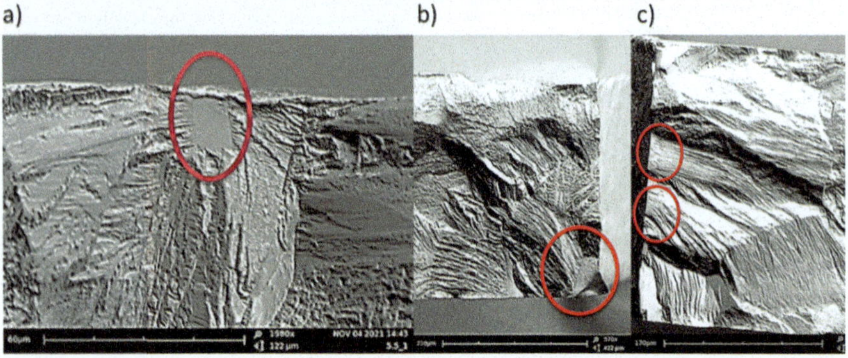

Fig. 5 Fatigue fracture surfaces NO30-15 **a** Crack across the rolling surface of a specimen with B = 10 mm, $\sigma_a = 230$ MPa, N = 4.5 × 105, **b** crack at the corner of the notch ground of a specimen with RA = 4 mm, $\sigma_a = 150$ MPa, N = 1.7 × 105, **c** crack at the notch ground of a specimen with RA = 0.35 mm, $\sigma_a = 100$ MPa, N = 4.0 × 104

No cracks were detected inside any of the specimens, justifying the selection of the surface approach. When choosing the surface for notched specimens, options include selecting either only the notch ground or both the notch ground and the adjacent rolled surface. Figure 5b illustrates that several specimens feature cracks in the corner of the notch base, leading to ambiguity in surface determination. In Fig. 6, the deviation of the statistical support effect is plotted against the surface considered. $\eta_{st,surface}$ denotes consideration of the entire specimen surface, while η_{st_notch} considers only the surface of the notch base. Weibull exponents were determined using the four different methods presented in Sect. 3, with results from [15] applied to the Method LCF. Differences in the statistical support effect across different surfaces depend on stress distribution and the Weibull exponent. While there are negligible differences for specimens with $R_A = 0.35$ mm (< 0.3%), deviations are observed for mild notches (between 2 and 3.5%). This variation is attributed to the stress distribution depth effect; mild notches result in a smoother stress distribution towards the center of the specimen, where the rolled surface contributes to the highly stressed area. The differences in the support number also decrease with higher Weibull exponents, as the differences in the surfaces under consideration become smaller.

Figure 7 compares the results of the stress-controlled tests with the calculated support effects for the nominal and local stress concept. The maximum highly stressed surfaces

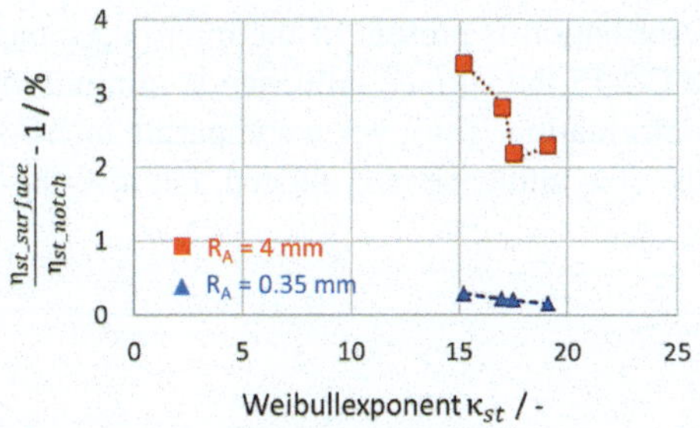

Fig. 6 Difference in the statistical support number depending on the surfaces considered

were determined considering all surfaces. The Weibull exponents were analysed according to the methods described in Sect. 3. The calculation of the support effect is based on Eq. (1–7). The parameters for η_{dm}, see Eq. (5) were taken from [15].

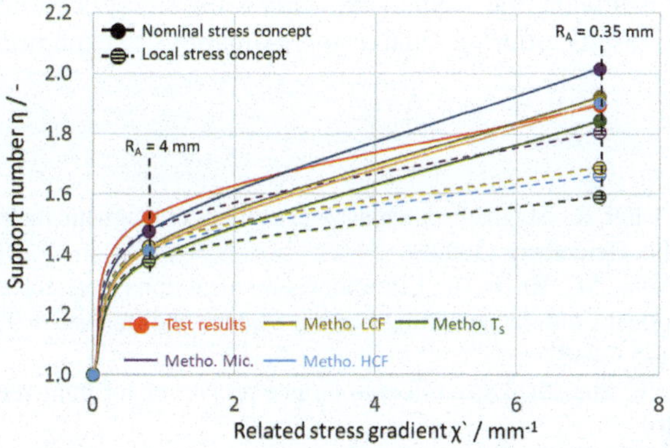

Fig. 7 Comparison of calculated support effect and support effect from fatigue tests

When calculating the support effect, differences arise between the nominal and local stress concepts for specimens with sharp notches (R_A = 0.35 mm). The local stress concept predicts smaller support effects regardless of the Weibull exponent estimation. This is due to an underestimation of the fracture mechanical support effect, which does not contribute for the small notch radius (0.35 mm) [18]. For the mild notch (R_A = 4 mm), η_{fm} does not contribute in either of the two concepts to η_σ.

In the nominal stress concept, the accuracy of the support effect model is higher for $R_A = 0.35$ mm notches. The deviation between Method LCF. and Method HCF. is less than 1% and both methods overestimate the test results (< 0.5%). Both methods produce comparable estimations of κ_{st} and are both suitable for the material NO30-15.

For specimens with the mild notch ($R_A = 4$ mm), the Method Mic. shows the highest accuracy. However, it should be noted that the method in the nominal stress concept overestimates the support effect of the material by approx. 6%. The overestimation is due to an inaccuracy in the comparison between micro specimens and fatigue specimens ($B = 10$ mm). While the fatigue strength of the fatigue specimens is in the range of linear elastic load, plastic deformation is present in the micro specimens. Studies from [13, 14] show that with NO30-15 the surface influence is minimised with increasing load amplitude via plastic deformation. Thus, when comparing micro-specimens and fatigue specimens, not only the size influence is evaluated, but also the reduction of the edge influence.

8 Summary

The material NO30-15 reacts sensitively to variations in specimen size and shows statistical size effects under both monotonous and cyclic loading. The statistical size effect is particularly pronounced under cyclic loading.

Statistical size effects in conjunction with material-mechanical support concepts achieve good agreement between the theoretical and experimental results. The accuracy depends on the selected calculation concept, as fracture mechanics approaches are weighted differently.

Methods for estimating the Weibull exponent using stress or strain-controlled tests yield comparable results, offering reliable approximations of experimental data.

References

1. Hubert, A., Schäfer, R.: Magnetic domains: The analysis of magnetic microstructures, vol. 3. Springer, Berlin, Heidelberg (2009)
2. Häfele, P., Thum, M., Knerr, T.: Life-time-assessment for rotors for electric drives. In: Third International Conference on Engineering Science and Innovative Technology (ESIT), Thailand, pp 1–5 (2018)
3. Tomaszewski, T.: Statistical size effect in fatigue properties for mini-specimens. Materials **13**, 1–14 (2020)
4. Heckel, K., Köhler, J.: Experimentelle Untersuchungen des statistischen Größenflusses im Dauerschwingversuch an ungekerbten Stahlproben. Z. Werkstoffech. **6**, 52–54 (1975)
5. Rösler, J., Harders, H., Bäcker, M.: Mechanical Behaviour of Engineering Materials: Metals, Ceramics, Polymers, and Composites. Springer Fachmedien Wiesbaden GmbH, Braunschweig (2007)
6. Forschungskuratorium Maschinenbau (FKM): Analytical Strength Assessment of Components. (7) VDMA Verlag GmbH, Frankfurt am Main (2020)
7. Fiedler, M., Wächter, M., et al.: Richtlinie Nichtlinear. Rechnerischer Festigkeitsnachweis unter expliziter Erfassung nichtlinearen Werkstoffverformungsverhaltens. VDMA Verlag GmbH, Frankfurt am Main (2019)
8. Kloos, K.: Einfluss des Oberflächenzustands und der Prüfgröße auf die Schwingfestigkeitseigenschaften. VDI-Verlag, Düsseldorf (1976)
9. Götz, S.: Weiterentwicklung eines bruchmechanischen Konzepts zur formzahlfreien Abschätzung der Dauerfestigkeit gekerbter Strukturen am Beispiel verschiedener Sinterstähle. Dissertation, Tu Dresden, Dresden (2012)
10. Weibull, W.: A Statistical Theory of Strength of Materials. Generalstabens Litografiska Anstalts Färlag, Stockholm (1939)
11. Neuber, H.: Über die Berücksichtigung der Spannnungskonzentration bei Festigkeitsberchnungen. Konstruktion **20**, 245–251 (1968)

12. Liu, J., Zenner, H.: Berechnung von Bauteilwöhlerlinien unter Berücksichtigung der statistischen und spannungsmechanischen Stützziffer. Mat.-wiss. u. Werkstofftech. **26**, 14–21 (1995)
13. DIN Deutsches Institut für Normung e. V.: DIN 50100:2016-12, Load Controlled Fatigue Testing-Execution and Evaluation of Cyclic Tests At Constant Load Amplitudes on Metallic Specimens and Components. Beuth Verlag GmbH, Berlin (2016)
14. Schwarz, P., Häfele, P.: Rissinitiierungsort von Elektroband in Abhängig keit der plastischen Verformung. In: Werkstoffprüfung 2022, Werkstoffe und Bauteile auf dem Prüfstand, pp. 307–311, Dresden (2022)
15. Schwarz, P., Häfele, P.: Influence of the machining conditions on the lifetime under cyclic loading of non-grain-oriented electrical steel strips. In: 2023 13th International Electric Drives Production Conference (EDPC), Regensburg, 1–6 (2023)
16. Dittmann, F.: Lebensdauervorhersage von Strukturen aus Feinblech mit dem Örtlichen Konzept unter Berücksichtigung des Kanteneinflusses. Dissertation, TU Darmstadt, Darmstadt (2021)
17. Haibach, H.: Betriebsfestigkeit: Verfahren und Daten zur Bauteilberechnung, vol. 3. Springer, Berlin (2006)
18. Müller, C., Wächter, M., Thumers, R., et al.: Bestimmung der statistischen Stützzahl mithilfe von FE-Simulation für die Praxis. In: Neue Entwicklung für die Bauteilfestigkeitsnachweise. DVM-Tagung, Berlin, pp. 49–58 (2018)
19. Hirose, T., Sakasegawa, H., Kohyama, A., et al.: Effect of specimen size on fatigue properties of reduced activation ferritic/martensitic steels. J. Nucl. Mater. **283**, 1018–1022 (2000)
20. Zhu, S. P., Ai, Y., Liao, D., Correia, J. A., et al.: Recent advances on size effect in metal fatigue under defects: a review. Int. J. Fract. Mech. 1–23 (2021)

Enhancing Lifetime of High-Voltage Traction Inverters Through Damage Avoidance Strategies Using Advanced Control and Modulation Techniques

Christoph Sachs[1,2](✉), Felix Gliese[1], and Martin Neuburger[1]

[1] Faculty of Mobility and Technology, University of Applied Sciences Esslingen, Robert-Bosch-Straße 1, 73037 Göppingen, Germany
{christoph.sachs,felix.gliese,martin.neuburger}@hs-esslingen.de

[2] Institute for System Dynamics, University of Stuttgart, Waldburgstraße 17/19, 70563 Stuttgart, Germany

Abstract. Enhancing the longevity of high-voltage traction inverters is critical for the reliability of future electric vehicles. This paper presents innovative damage mitigation strategies targeting converter components with the highest failure rates. By evaluating two distinct inverter topologies, the research outlines the damage mitigation potential for both semiconductor switches and DC-link capacitors. Advanced control and modulation techniques, such as the selective application of discontinuous pulse-width modulation methods, are used to underscore their efficacy in mitigating damage. Utilizing an extensive dataset from real-world driving conditions, the study comprehensively assesses component degradation and operational longevity. Both explicit and implicit damage avoidance strategies are introduced, with a focus on preemptive and real-time protective measures. An eight-step calculation framework is employed to analyze power losses, thermal behavior, and damage fractions, yielding valuable insights for enhancing the reliability and lifespan of modern power electronic systems.

Keywords: Traction inverter · Damage mitigation · Inverter reliability

1 Introduction

The relentless pursuit of efficiency and reliability in power electronic systems, particularly inverters, has spurred extensive research and innovation. Inverters are critical components in a wide range of applications, from renewable energy systems to electric vehicles, where they convert DC power to AC power, enabling the functioning of various electronic devices. This paper examines the intricacies of inverter technologies, focusing on component reliability, advanced topologies,

A. Öchsner (ed.), *Collaborative Research Advancing Engineering Solutions for Real-World Challenges 2*, Proceedings in Technology Transfer,
https://doi.org/10.1007/978-3-031-79093-5_6

control and modulation techniques, and damage avoidance strategies. A significant aspect of inverter reliability is the understanding of the failure rates of electronic components. Power devices such as IGBTs and MOSFETs, as well as DC-link capacitors, often experience high failure rates due to the severe electrical and thermal stresses they endure in high-power applications. These components are pivotal to the performance and longevity of inverters, necessitating thorough investigation and mitigation strategies to enhance their reliability. This study examines two distinct inverter topologies: the B6 inverter and the B6^2 inverter. The B6 inverter is well-known for being the most cost-effective inverter design. In contrast, the B6^2 inverter offers enhanced performance and control flexibility through the use of a more complex configuration that necessitates additional switches and enables the utilization of full DC-link voltage.

The application of advanced control and modulation techniques, particularly various pulse width modulation (PWM) strategies, is essential for the optimisation of inverter performance. This study evaluates the continuous PWM (CPWM) and discontinuous PWM (DPWM) techniques for the B6 inverter, highlighting their respective advantages regarding potential damage mitigation. Furthermore, it investigates the impact of distinct modulation strategies for the B6^2 inverter on component stress and system efficiency. Implicit and explicit damage avoidance algorithms are crucial for maintaining the operational integrity of power electronic systems. Implicit strategies proactively prevent the system from entering damaging states by adhering to safe operational limits, while explicit strategies involve concrete calculations to optimize control parameters and mitigate component degradation. This comprehensive analysis utilizes real driving data to simulate operational conditions, providing a robust framework for assessing component damage and reliability. The methodology includes detailed power loss calculations for semiconductor switches and DC-link capacitors, alongside advanced thermal modeling to predict component temperatures and their impact on system performance. All relevant simulation assumptions and parameters are presented in Table 1 for ease of reference, allowing the reader to gain a comprehensive overview throughout this document.

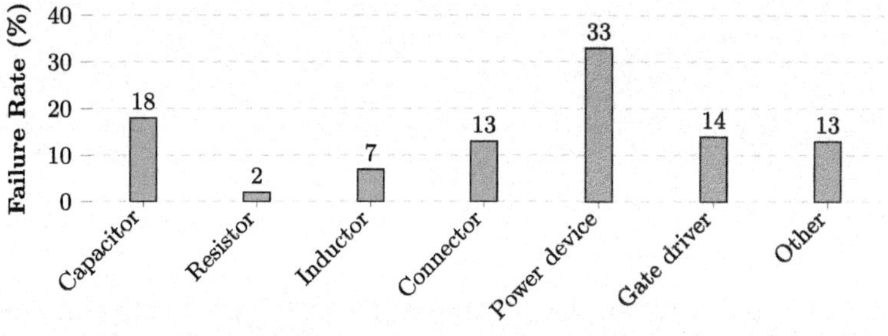

Fig. 1. Failure rates of electronic components according to [1]

Through this investigation, the paper aims to contribute to the development of more reliable and efficient inverter technologies, addressing key challenges in component reliability, control strategies, and damage avoidance in power electronic systems.

1.1 Failure Rates

In power conversion systems, component reliability significantly impacts overall performance. Figure 1 illustrates failure rates across various electronic components based on [1], with power devices (33%) and capacitors (18%) emerging as the primary contributors to inverter failures. Power devices, including IGBTs and MOSFETs, show the highest failure rate due to intense electrical and thermal stresses in high-power applications. DC-link capacitors, crucial for high-frequent energy storage and supply, are prone to degradation under significant thermal and electrical strain. Other components, like connectors, gate drivers, and inductors, contribute to failures to a lesser extent.

1.2 Topologies

In this study, the inverter topologies B6 and B6^2 are examined. The B6 inverter, illustrated in subfigure (a) of Fig. 2 is the most used standard three-phase inverter topology used for controlling variable speed drives due to its low costs and control-simplicity. It is connected to a star-point motor configuration. Shown in subfigure (b), the B6^2 inverter on the other hand is an advanced configuration designed to offer enhanced performance and control flexibility. It requires two times the amount of switches and is connected to an open-winding motor configuration. It is capable of using the full DC-link voltage instead of $V_{\text{DC}}/\sqrt{3}$, assuming common-mode injection for the B6 inverter topology and leads to lower fed currents through the switches to reach equal output power.

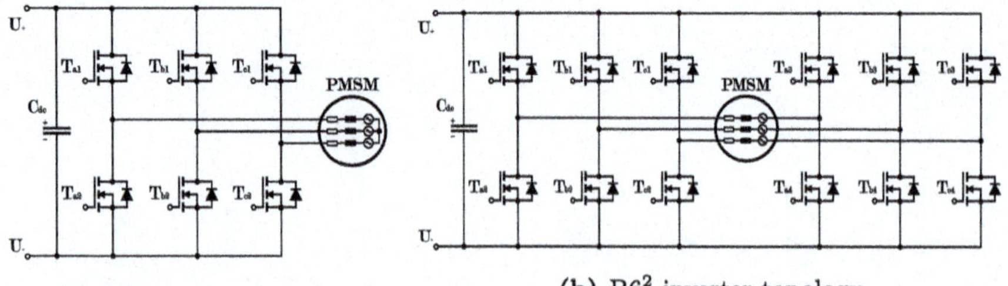

(a) B6 inverter topology

(b) B6^2 inverter topology

Fig. 2. Inverter topologies with MOSFETs: **a** B6 inverter with star-point motor configuration, **b** B6^2 inverter with open-winding motor configuration

1.3 Advanced Control and Modulation Techniques

The investigation includes various pulse width modulation (PWM) techniques [2] for the B6 inverter and different modulation strategies for the $B6^2$ topology. Further strategies like changing the space vector pointer scheme or including switching frequency variations are not investigated herein.

Continuous pulse width modulation is a modulation technique where the switching signals are continuously modulated to produce the desired AC output. The modulating signal continuously varies, ensuring that the inverter switches are operating throughout the entire period of the output waveforms illustrated by Fig. 3 for SPWM (a) and SVPWM (b). SVPWM is created by overmodulating the signal with a zero sequence V_0 which leads to increasing the leg voltage to $V_{DC}/\sqrt{3}$. The main advantages of CPWM include high linearity between the control signal and the output voltage, lower harmonic distortion in the output waveform, and stable operation over the entire modulation index range M.

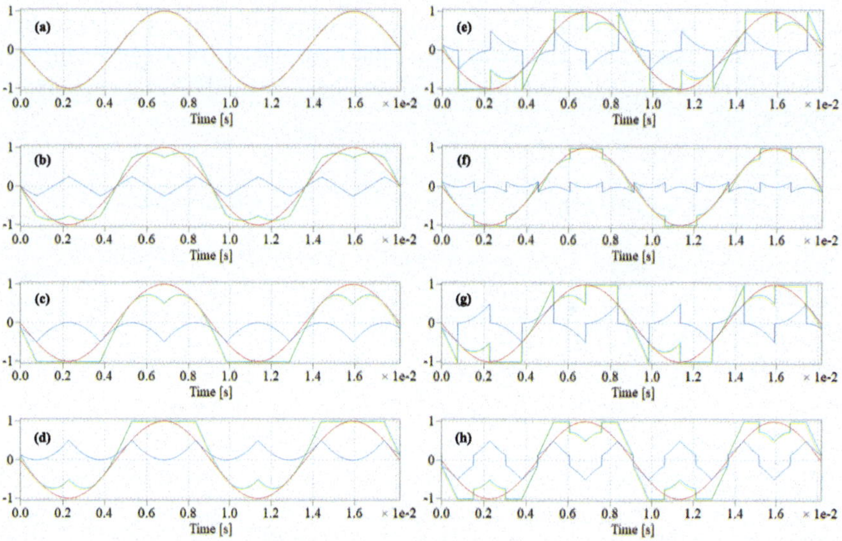

Fig. 3. Modulation types used for the B6 inverter represented for one of the half-briges: **a** SPWM, **b** SVPWM, **c** DPWMMIN, **d** DPWMMAX, **e** DPWM 0, **f** DPWM1, **g** DPWM2, **h** DPWM3. The red line indicates the reference sine wave, whereas the blue line indicates the zero-sequence V_0. The green line is the sum of the two lines and represents the resulting actuation signal for the switches. A constant signal value of 1 or −1 indicates that the half-bridge is clamped for a specific time frame to the voltage levels $U+$ or $U-$

DPWM, on the other hand, introduces periods during which the inverter switches are not operating. This is achieved by changing the zero sequence signal V_0, as illustrated in subfigure (c)–(h) in Fig. 3 for different DPWM types. There are various types of DPWM techniques, each differing in how and when they introduce these clamping periods. For instance, during DPWM1 in subfigure (f), the half bridge clamps the output during the positive and negative peak phases of the modulating signal. The primary advantages of DPWM include reduced switching losses due to fewer switching events and lower thermal stress on the semiconductors. However, DPWM can introduce higher harmonic distortion in the output waveform compared to CPWM, leading to increased harmonic motor losses, and is more complex to implement due to the need for precise timing and control to achieve the clamping periods.

For the B6[2] inverter topology, the damage mitigation algorithm focuses on the DC-link capacitor. Therefore, two distinct modulation types are compared. As the standard, the unipolar modulation (a) illustrated in Fig. 4 controls both inverter halves by inverting the input signal of the modulator for the right inverter half as done in [3]. The unfolder modulation, on the other hand, actuates the left inverter half in a six-step mode and changes the input signal by subtracting the input sinus with six-step signal, leading to equal voltages across the windings of the open-winding motor configuration. The control of the input signals for the right and left inverter halves is switched after a period to ensure equal thermal impact on the switches. This modulation strategy changes the DC-link currents $I_{C_{dc},RMS}$, leading to reduced heat in the DC-link capacitor. As derived in [4],

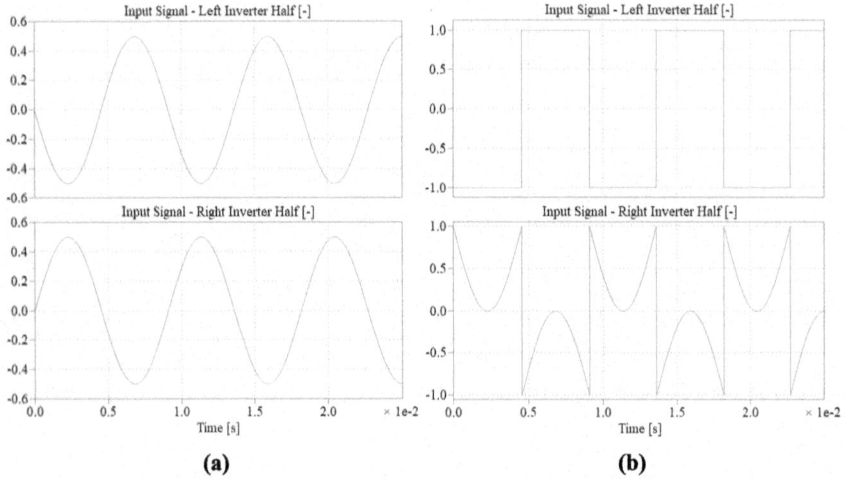

Fig. 4. Modulation types used for the B6^2 inverter represented for one of the full-briges: **a** unipolar modulation and **b** unfolder modulation

$$I_{C_{dc},RMS} = I\sqrt{M\left[\frac{\sqrt{3}}{4\pi} + \cos^2(\varphi)\cdot\left(\frac{\sqrt{3}}{\pi} - \frac{9}{16}M\right)\right] + K_{\mathrm{I}}} \tag{1}$$

for the unipolar mode with $K_{\mathrm{I}} = 0$ and for the unfolder mode (b) with

$$\begin{aligned} K_I = & \frac{\sqrt{3}}{8\pi}M\sin\left(2\varphi + 3\,\mathrm{acos}\left(\frac{2}{\sqrt{3}M}\right)\right) \\ & + \frac{1}{M^2}\left[\sin(2\varphi)\cdot\frac{9M^2-16}{12\pi} + \cos(2\varphi)\cdot\sqrt{3M^2-4}\frac{9M^2-16}{8\pi}\right] \\ & + \frac{3}{2\pi}\sqrt{3M^2-4} - \frac{3}{\pi}\,\mathrm{acos}\left(\frac{2}{\sqrt{3}M}\right) \end{aligned} \tag{2}$$

The capacitor currents are illustrated in Fig. 5. It can be seen that the capacitor current maximum $I_{C_{\mathrm{dc,a}},\mathrm{RMS,max}} = 0.83I$ at a phase angle of $\varphi = 0$ and a modulation index of $M = (10\sqrt{3}+14)/9\pi$ using the unipolar modulation strategy (a) might lead to enormous heat impact on the DC-link capacitor, whereas the equal operation point does not stand under severe current stress using the unfolder mode $I_{C_{\mathrm{dc,b}},\mathrm{RMS}} = 0.46I$ in (b).

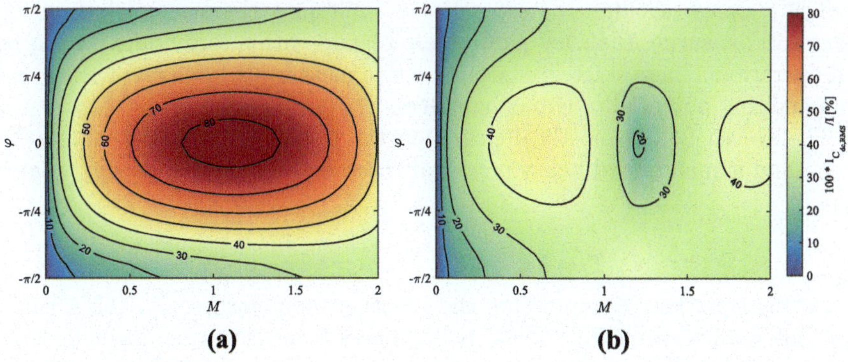

Fig. 5. $I_{C_{\mathrm{dc}},\mathrm{RMS}}$ as a function of phase shift φ and modulation index M regarding two modulation types for the B6^2 inverter: **a** unipolar mode, and **b** unfolder mode

1.4 Reference Drive Cycle

The worldwide harmonized light vehicles test procedure (WLTP) is not used in this study due to its limitations in accurately predicting the lifetime outcomes of system components. Instead, real driving data is utilized, offering more reliable simulation outputs. To ensure reliable predictions, a vehicle model is used along with over 500,000 data points (approximately 140 h of stacked driving cycles).

This extensive dataset allows for the accurate modeling of various operational scenarios and the assessment of their impact on component damage. The operation conditions are calculated based on an inverse motor model, which has been derived from measurement data. Therefore, the electric target values can be fed back to the inverter to ensure the desired operation point is achieved, as indicated by Eq. (3).

$$\begin{pmatrix} \vec{u}_{dq} \\ \vec{i}_{dq} \\ f_f \\ cos(\varphi) \end{pmatrix} = \vec{f}(M, n) \tag{3}$$

Further information about the vehicle model and the 300kW motor is provided in Table 1.

2 Explicit and Implicit Damage Avoidance Algorithms

This section explores the critical aspect of damage avoidance in power electronic systems, particularly inverters, by categorizing the protective measures into explicit and implicit calculation steps.

2.1 Implicit Damage Avoidance

Implicit damage avoidance strategies are built into the system's operational framework, ensuring that key parameters remain within safe limits during normal operation. These strategies are proactive measures that prevent the system from entering potentially damaging states.

The switch failure mostly depends on the maximum junction temperature $T_{j,\max}$ and is maintained below a critical threshold using the following relationship:

$$T_{\mathrm{j}} = T_{\mathrm{hs}} + R_{\mathrm{th}} \cdot P_{\mathrm{switch}} \leq T_{j,\max} \tag{4}$$

In this equation, T_{j} represents the junction temperature, T_{hs} (55°C) is the heat sink temperature, R_{th} (0.0867 K/W) denotes the chip area-dependent thermal resistance, P_{switch} is the power dissipated, and $T_{j,\max}$ (175°C) is the maximum allowable junction temperature. Controlling the DC-link voltage ripple is essential for ensuring all the other components connected to the battery are not affected regarding their operation. The voltage ripple is kept below a maximum threshold:

$$\Delta U = \left| \frac{Q_{\mathrm{ac}}}{C_{\mathrm{DC}}} \right| = \left| \frac{\int i_{\mathrm{c}} \cdot dt}{C_{\mathrm{DC}}} \right| \leq \Delta U_{\max} \tag{5}$$

Here, ΔU represents the DC-link voltage ripple, Q_{ac} is the capacitor charge, C_{dc} denotes the DC-link capacitance, i_{c} is the capacitor current, and $\Delta U_{\max}$ (± 20 V) is the maximum allowable voltage ripple. Considering the bearing-voltage ratio (BVR) is crucial for reducing the potential of motor bearing discharges, which are one of the most common failure types in electrical traction motors.

The BVR for the B6 inverter is calculated by

$$\Delta U_{CM} = \frac{U_a + U_b + U_c}{3} \leq BVR \tag{6}$$

whereas the BVR of the B6^2 inverter sums up by

$$\Delta U_{CM} = \frac{U_a + U_b + U_c + U_d + U_e + U_f}{6} \leq BVR \tag{7}$$

In these equations, ΔU_{CM} represents the common-mode voltage, and U_a, U_b, U_c, U_d, U_e, U_f are the line voltages between the switches of the inverter topologies. The BVR utilized here is limited by $V_{dc} \cdot 0.05$ as commonly defined, e.g. in [5] based on a capacitance network depending on the motor insulation design.

2.2 Explicit Damage Avoidance

Explicit damage avoidance strategies mean the concrete calculation of damage influences caused by the operating point and a variation of the control strategy. This leads to damage parameters D_{switch} and D_{cap}, which are used for the SOD calculation.

Figure 6 illustrates an 8-step calculation and optimization framework used to determine the explicit damage fractions in traction inverters, depending on various modulation strategies and optimal switching frequencies. The process begins with step 0, where system properties are defined, including topology, thermal properties, battery electric vehicle (BEV) type, motor type, and the choice of active components. These foundational parameters set the stage for the calculations.

In step 1, the mission profile is established, providing critical input data such as operation conditions, velocity v and the angle α of the road for the longitudinal vehicle model. This data is essential for determining the mechanical demands on the system. Following this, step 2 involves using the mission profile to calculate the necessary mechanical outputs through the inverse vehicle model, yielding torque M and rotational speed n. Next, in step 3, the inverse motor model translates the mechanical outputs into electrical parameters, such as d-q axis voltages u_{dq}, d-q axis currents i_{dq}, fundamental frequency f_f, and power factor $\cos(\phi)$. These parameters are crucial for subsequent power and thermal calculations. Step 4 sees the controller determining optimal operation settings for damage mitigation and efficiency, including selecting the switching mode, modulation scheme, and switching frequency f_{sw}. The controller is capable of considering the state of damage (SOD) of the DC-link capacitor and the switches, as well as the state of charge (SOC) of the battery. In step 5, system power losses are calculated, encompassing losses in the motor P_{motor}, switches P_{switch}, and the capacitor P_{capa}. These calculations are fundamental for assessing the thermal load and efficiency of the system. Details regarding efficiency calculation are not further detailed in this study, as several other papers have focused on these points, like [3,6]. A quantitative categorisation of the efficiency

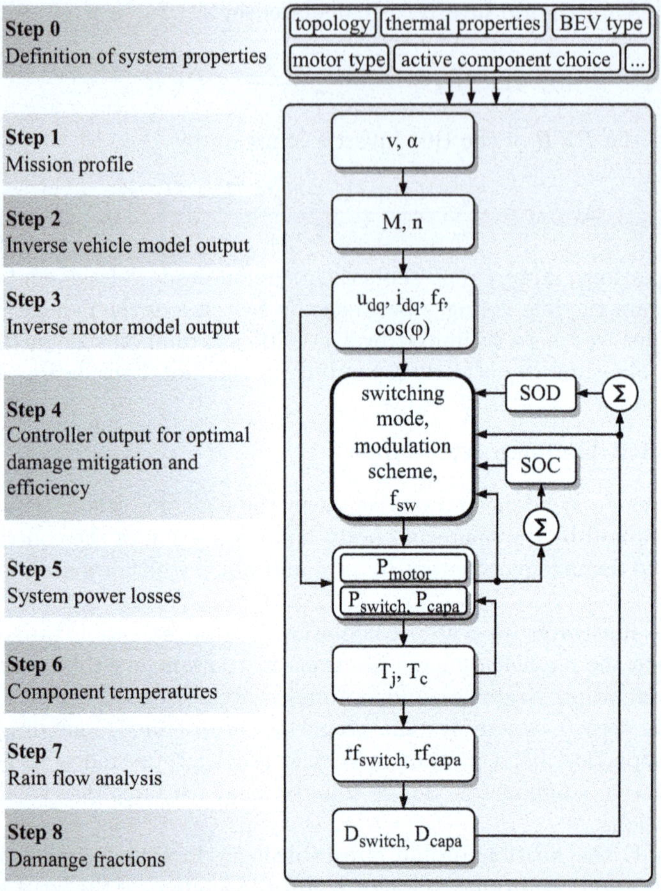

Fig. 6. Processing and calculation framework to determine the explicit damage fractions depending on the modulation strategies

impact of the new strategies will be provided in the conclusion. Step 6 involves evaluating the component temperatures based on the calculated power losses, including the junction temperature T_j of the switches and the temperature T_c of the capacitor. Step 7 performs a rainflow analysis to assess the thermal cycling experienced by the components, helping to count the temperature changes with low computing effort. Finally, step 8 calculates the damage fractions for switches D_{switch} and capacitors D_{capa}, quantifying the degradation and wear of components over time and providing a measure of their expected lifetime under given operating conditions.

2.3 Power Loss Calculation

The accurate calculation of power losses in inverter components, particularly DC-link capacitors and semiconductor switches, is crucial for assessing the overall efficiency and reliability of power electronic systems. The power losses in semiconductor switches, specifically IGBTs, are composed of two main components: conduction losses and switching losses. Conduction losses occur when the IGBT is in its on-state and are a function of the collector current and the collector-emitter voltage. These losses can be approximated using the following equation:

$$P_{con} = v_{ce0} \cdot i_c + r_{on} \cdot i_c^2 \tag{8}$$

where P_{con} represents the conduction losses, v_{ce0} is the zero-current collector emitter voltage, i_c is the collector current, and r_{on} is the collector emitter on-state resistance. Switching losses, on the other hand, occur during the transition between on and off states. These losses depend on various factors, including the switching frequency, DC-link voltage, and the current being switched. The switching losses can be expressed as:

$$P_{sw} = f_{sw} \cdot (E_{on} + E_{off}) \tag{9}$$

Here, P_{sw} represents the switching losses, f_{sw} is the switching frequency, and E_{on} and E_{off} are the energy losses during turn-on and turn-off transitions, respectively. In our study, a constant switching frequency of 10 kHz is considered for the switches during operation. The total power loss in a semiconductor switch is the sum of these two components:

$$P_{switch} = P_{sw} + P_{con} \tag{10}$$

The power losses in DC-link capacitors are primarily due to the high-frequency currents that these components must buffer. These losses are calculated by considering the AC ripple current components at various frequencies and the frequency-dependent equivalent series resistance ESR of the capacitor. The power loss in a DC-link capacitor can be expressed as:

$$P_{cap} = \sum_{f_{\min}}^{f_{\max}} \left[I_{ac}(f_h)^2 \cdot ESR(f_h)\right] \tag{11}$$

In this equation, P_{cap} represents the total power loss in the capacitor, and f_{min} and $f_{\max}$ define the frequency range of interest, set to capture all relevant loss components. $I_{ac}(f_h)$ is the RMS value of the AC ripple current at harmonic frequency f_h, and $ESR(f_h)$ is the equivalent series resistance of the capacitor at frequency f_h.

Table 1. Overview of parameters, variables and assumptions for the simulation

Category	Parameter	Value	Unit (Description)
Thermal data			
	T_{hs}	50	°C (Heat sink temperature)
	R_{th}	0.0867	K/W (Heatsink resistance)
Operation constraints			
	$T_{j,max}$	175	°C (Max. junction temperature)
	BVR	$0.05 \cdot V_{dc}$	V (Bearing voltage ratio)
	ΔU_{max}	±20	V (Max voltage ripple)
DC-link			
	C_{dc}	500	μF (Capacitance)
	V_{dc}	800	V (DC-link voltage)
Modulation			
	M ($B6^2$)	0 – 2	- (Modulation index)
	M (B6)	0 – 1.15	- (Modulation index)
	φ	$-\pi/2$ to $\pi/2$	rad (Phase angle)
	f_{sw}	10	kHz (Switching frequency)
Motor data			
	P_{max}	300	kW (Output power)
	f_f	0 – 660	Hz (Motor frequency)
	M_{max}	±580	Nm (Max. torque)
	n_{max}	18000	min^{-1} (Max. speed)
Vehicle model			
	A	2.2	m^2 (Front area)
	Cd	0.24	- (Aerodynamic parameter)
	ρ_{air}	1.25	kg/m^3 (Air density)
	R_r	0.01	- (Friction resistance)
	g	9.81	m/s^2 (Gravitational acceleration)
	m	1930	kg (Mass)
	G	11.4	- (Gear ratio)
	η_{gear}	1	- (Gear numbers)
	J_{axle}	0	$kg{\cdot}m^2$ (Axle inertia)
	J_{EM}	0	$kg{\cdot}m^2$ (Motor inertia)
	N_{EM}	1	- (Motor number)
Switch model			
	v_{ce0}	1.67	V (C-E saturation voltage)
	r_{on}	0.0037	Ω (On resistance)
	E_{on}	78	mJ (Turn-on energy loss)
	E_{off}	198	mJ (Turn-off energy loss)
	$R_{th1,sd}$	0.013	K/W (Thermal resistances)
	$R_{th2,sd}$	0.082	K/W
	$R_{th3,sd}$	0.09	K/W
	$R_{th4,sd}$	0.05	K/W
	τ_{sd1}	0.001	s (Thermal time constants)
	τ_{sd2}	0.032	s
	τ_{sd3}	0.26	s
	τ_{sd4}	1.6	s
	$C_{th,sd}$	$\tau_{sd}/Rth,sd$	J/K (Thermal capacitance)
	A	$6 \cdot 10^{11}$	- (Lifetime material constant)
	β_1	-3.8	- (Coffin manson exponent)

(continued)

Table 1. (*continued*)

Category	Parameter	Value	Unit (Description)
Capacitor model			
	$ESR(f_h)$	3..8	$m\Omega$ (Capacitor resistance)
	$R_{\text{th1,cap}}$	0.48	K/W (Thermal resistances)
	$R_{\text{th2,cap}}$	-12.37	K/W
	$R_{\text{th3,cap}}$	6.01	K/W
	$R_{\text{th4,cap}}$	8.14	K/W
	$R_{\text{th5,cap}}$	0.87	K/W
	τ_{cap1}	51.21	s (Thermal time constants)
	τ_{cap2}	393.21	s
	τ_{cap3}	286.41	s
	τ_{cap4}	572.86	s
	τ_{cap5}	7.41	s
	$C_{\text{th,cap}}$	$\tau_{\text{cap}}/R_{\text{th,cap}}$	J/K (Thermal capacitance)
	L_O	$8 \cdot 10^8$	– (Nominal lifetime parameter)
	V_O	900	V (Nominal lfetime voltage)
	T_O	110	°C (Nominal lifetime temperature)
	E_a	1.2	eV (Activation energy)
	n_1	−9.5	– (Voltage stress exponent)
	K_b	1.380649×10^{-23}	J/K (Boltzmann constant)

2.4 Temperature Calculation

We employ a Foster network model to represent the thermal impedance of the components under investigation. A 4th-order Foster network is utilized for the semiconductor switches, while a 5th-order network is applied to the DC-link capacitors. The thermal impedance $Z(t)$ is expressed as:

$$Z(t) = \sum_{i=1}^{K} R_{th}^{i}(1 - e^{\frac{-t}{R_{th}^{i} C_{th}^{i}}}) \tag{12}$$

In this equation, K represents the order of the Foster network (4 for switches, 5 for capacitors, as shown in Table 1), R_{th}^{i} and C_{th}^{i} denote the thermal resistance and capacitance of the i-th RC pair of the network, respectively. To efficiently compute the temperature change in DC-link capacitors and semiconductor switches, we implement the thermal model as a one-dimensional digital filter. This approach allows for real-time temperature estimation with reduced computational complexity, according to [9]. The filter is described by the following equation:

$$y(n) = \sum_{i=0}^{M} b_i \cdot x(n-i) - \sum_{i=1}^{N} a_i \cdot y(n-i) \tag{13}$$

Here, $y(n)$ is the output at time step n, $x(n)$ represents the input at time step n, while b_i and a_i are the filter coefficients. The parameters M and N define the filter order. The filter coefficients are derived using a Butterworth filter design approach. To bridge the continuous-time thermal model with the discrete-time implementation, we employ the bilinear transformation $z = (e^{sT/2})/(e^{-sT/2})$. This allows us to express the transfer function $H(j\omega)$ in terms of the z-transform variable:

$$H(j\omega) = H\left(\frac{2}{\Delta t} \cdot \frac{z-1}{z+1}\right) = \sum_{i=1}^{K} \frac{R_{th}^{i}}{1 + j\omega \cdot R_{th}^{i} C_{th}^{i}} \tag{14}$$

In this equation, Δt represents the sampling period, and z is the complex variable in the z-domain. Combining the Foster network model, digital filter implementation, and bilinear transformation, we arrive at the following discrete-time equation in accordance with [9] for our temperature calculation in DC-link capacitors and semiconductor switches:

$$T(n) = \sum_{i=1}^{K} \left[\frac{2\tau_i - \Delta t}{2\tau_i + \Delta t} \cdot T_i(n-1) + \frac{R_{th}^{i} \Delta t}{2\tau_i + \Delta t} \cdot (P(n) + P(n-1)) \right] \tag{15}$$

In this formulation, $T(n)$ is the calculated temperature at time step n, $T_i(n-1)$ represents the temperature contribution of the i-th RC pair at the previous time step, and $P(n)$ and $P(n-1)$ are the power losses at the current and previous time steps, respectively. The time constant of the i-th RC pair is given by $\tau_i = R_{th}^{i} C_{th}^{i}$. This approach enables efficient and accurate temperature estimation in DC-link capacitors and semiconductor switches, taking into account both the thermal characteristics of the materials and the dynamic nature of power losses during inverter operation. The method's ability to capture the thermal behavior of these critical components contributes to more accurate predictions of their degradation over time, ultimately leading to improved reliability and performance in power electronic systems.

2.5 Damage Calculation

Using thermal and electrical data, we estimate the cumulative damage over time to identify the most life-conserving operational modes. For the switch damage calculation, the Coffin-Manson model [8] is used, which describes the cycle strength of a material in relation to the number of temperature cycles. While alternative approaches such as Bayerer or Package Failure models allow for more precise damage predictions, they also lead to significantly higher computational efforts. The switch damage is calculated using the following equations:

$$N_f = A \cdot (\Delta T)^{\beta} \tag{16}$$

N_f represents the number of cycles to failure, A is a material-dependent constant, ΔT is the temperature change, and β is the Coffin-Manson exponent.

$$D_{switch} = \sum_{i=1}^{N_{cyc}} \frac{n_i}{N_f^i} \tag{17}$$

By applying a rainflow algorithm as a cycle-counting variant, as done in e.g., [8], the number of cycles to failure is determined. Here, D_{switch} is the cumulative switch damage, N_{cyc} is the total number of cycles, n_i is the number of cycles at a specific stress level, and N_f^i is the number of cycles to failure at that stress level.

For the capacitor damage calculation, we use a Prokopowicz-Vaskas model [10], which combines an inverse power-law for external field and an Arrhenius relation for absolute temperature impact:

$$D_{cap} = \sum_{i=1}^{N_{cyc}} \frac{1}{L} \tag{18}$$

D_{cap} represents the cumulative capacitor damage and L is the lifetime under specific conditions. The lifetime L is calculated as:

$$L = \left[\frac{V}{V_0}\right]^{n_1} \cdot \exp\left[\frac{E_a}{K_b} \cdot \left(\frac{1}{T} - \frac{1}{T_0}\right)\right] \cdot L_0 \tag{19}$$

V is the applied voltage, V_0 is the rated voltage, n_1 is the voltage stress exponent, E_a is the activation energy, K_b is the Boltzmann constant, T is the operating temperature, T_0 is the rated temperature, and L_0 is the lifetime under rated conditions.

The state of damage calculation is based on the actual damage estimation and predicted damage fractions, depending on the control algorithm:

$$SOD_i(t + \Delta t) = SOD_i(t) + D_i \tag{20}$$

SOD_i is the state of damage for component i, and D_i is the damage increment. Post-processing details regarding the consideration of parameter uncertainties, the calculation of total uncertainty through Monte Carlo simulations, the integration of probability density functions into the cumulative density function, the calculation of cumulative system failure, and the determination of B_x values are not further explained in this paper, as these are common statistical methods in the field of cumulative lifetime estimation. For an in-depth understanding of these methods, readers are referred to [1,7].

3 Impact of Advanced Control Strategies for Damage Avoidance

This section investigates how different control strategies influence the damage outcomes of the system. Various possibilities for further reducing damage and enhancing system efficiency are explored.

Figure 7 shows a comparative analysis of the semiconductor damage impact over a driving cycle using **a** SVPWM and **b** DPWM0. The plots, arranged from top to bottom, depict the semiconductor switch power loss P_{sd}, the semiconductor temperature T_{sd}, the semiconductor damage impact due to temperature

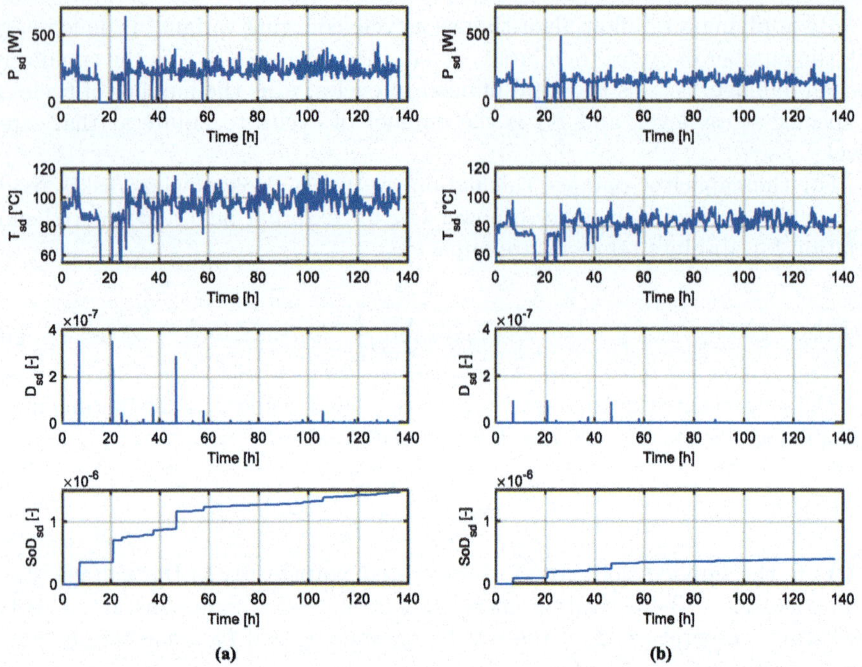

Fig. 7. Plots from top to bottom: Semiconductor switch power loss, semiconductor temperature, semiconductor damage impact due to temperature changes and accumulated damage of the switch, all based on the driving cycle with **a** SVPWM and **b** DPWM0

changes D_{sd}, and the accumulated state of damage SoD_{sd} of the switch over time. For both SVPWM and DPWM0 modulation strategies, the power loss and temperature fluctuate significantly over the driving cycle. The damage impact D_{sd} exhibits spikes, indicating moments of high stress on the semiconductor. However, the accumulated damage SoD_{sd} for DPWM0 shows a slower growth rate compared to SVPWM, highlighting the effectiveness of DPWM0 in mitigating overall semiconductor damage over the given driving cycle. Investigations of alternative DPWM methods indicate that the damage spikes occur at different operation points, which allows further optimization of the control strategy.

Figure 8 illustrates the impact of different modulation strategies on DC-link capacitor stress over the mission profile. Subplots show power loss P_{cap}, temperature T_{cap}, and damage D_{cap} for unipolar (a) and unfolder (b) modulations. Unipolar modulation exhibits higher peak values and more frequent fluctuations in all parameters. Notably, P_{cap} reaches up to 50 W for unipolar vs. 25 W for unfolder, while T_{cap} peaks at 110°C vs. 75°C, respectively. D_{cap} shows an order of magnitude difference, with unipolar modulation causing significantly more damage accumulation. These results suggest unfolder modulation may extend capacitor lifetime significantly.

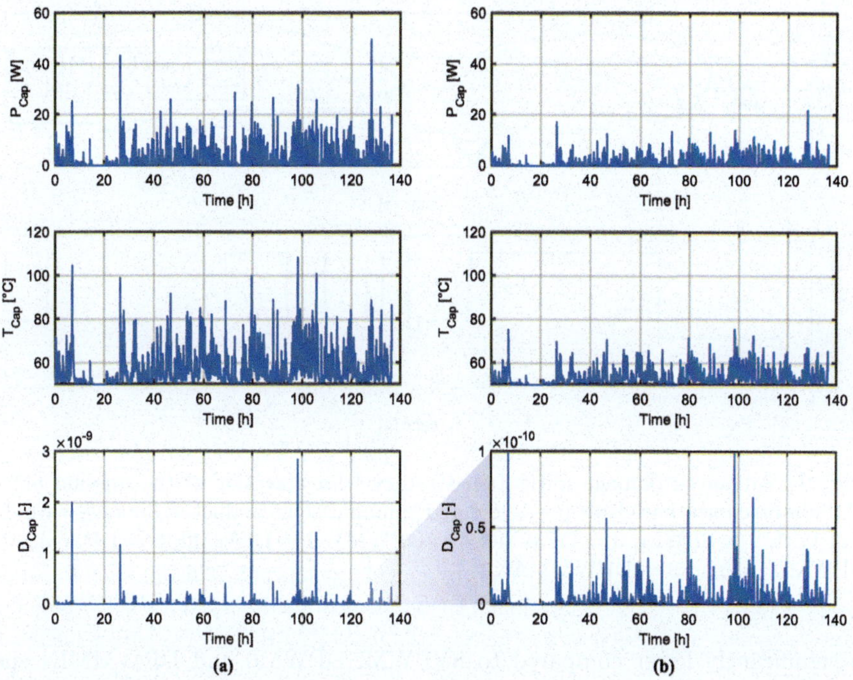

Fig. 8. Capacitor damage impact based on the driving cycle with **a** unipolar modulation andb unfolder modulation using the B6[2] inverter

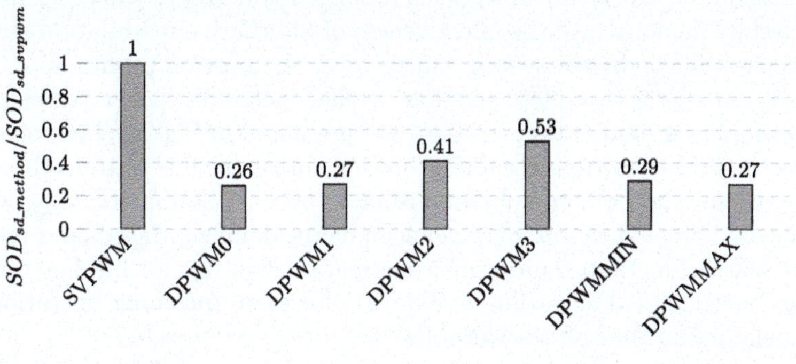

Fig. 9. Damage mitigation potential of DPWM strategies based on the state of damage over the complete mission profile in relation to the conventional SVPWM strategy for B6 inverters

The semiconductor damage output of various PWM modulation strategies compared to SVPWM is depicted in Fig. 9 using the B6 topology. The plot displays the ratio $SoD_{\text{sd, method}}/SoD_{\text{sd, svpwm}}$ for each modulation strategy over the complete mission profile. SVPWM serves as the reference with a normalized damage output of 1. The damage output for other modulation strategies

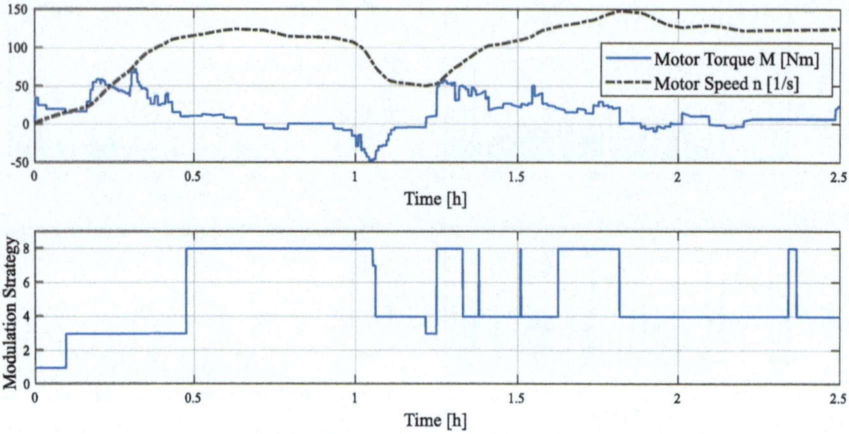

Fig. 10. Advanced damage mitigation strategy I for a section of the mission profile: Modulator choice based on pre-calculated minimal semiconductor damage with 1 = SVPWM, 2 = SPWM, 3 = DPWM0, 4 = DPWM1, 5 = DPWM2, 6 = DPWM3, 7 = DPWMMIN, and 8 = DPWMMAX

is significantly lower compared to SVPWM. DPWM0 and DPWMMIN show the lowest damage ratios, with values of 0.26 and 0.29, respectively. DPWM1 and DPWMMAX both have a damage ratio of 0.27. DPWM3 has a slightly higher damage ratio of 0.53, due to the division of the clamping phase into $\pi/12$ steps, which does not result in as much damage reduction as the other methods. These results demonstrate the effectiveness of alternative modulation strategies in reducing semiconductor switch damage over the mission profile.

A section of the complete mission profile, including motor torque, speed, and the employed modulation strategy, are presented in Fig. 10. The top subplot presents the torque and speed values, showing noticeable fluctuations throughout the operation. The bottom subplot displays the modulation strategy, which is dynamically adjusted to minimize semiconductor damage. Based on a minimum damage search function, the optimal modulation strategy for lifetime optimization can be calculated according to Eq. (21) for each upcoming operation point and depending on the control variables.

$$\text{argmin}\{D_i(n, M, mod, ..)\} \tag{21}$$

The modulation strategy transitions between various types in the figure. In this section of the mission profile, DPWM1, DPWM1, and DPWMMAX are the most frequent modulation types, which fits to the simulation outcome presented in Fig. 9.

The damage mitigation strategy II is illustrated in Fig. 11. This strategy checks if a pre-defined damage limit $D_{i,limit}$ is reached, and if yes, an alternative modulation strategy equal to the procedure of strategy I is chosen.

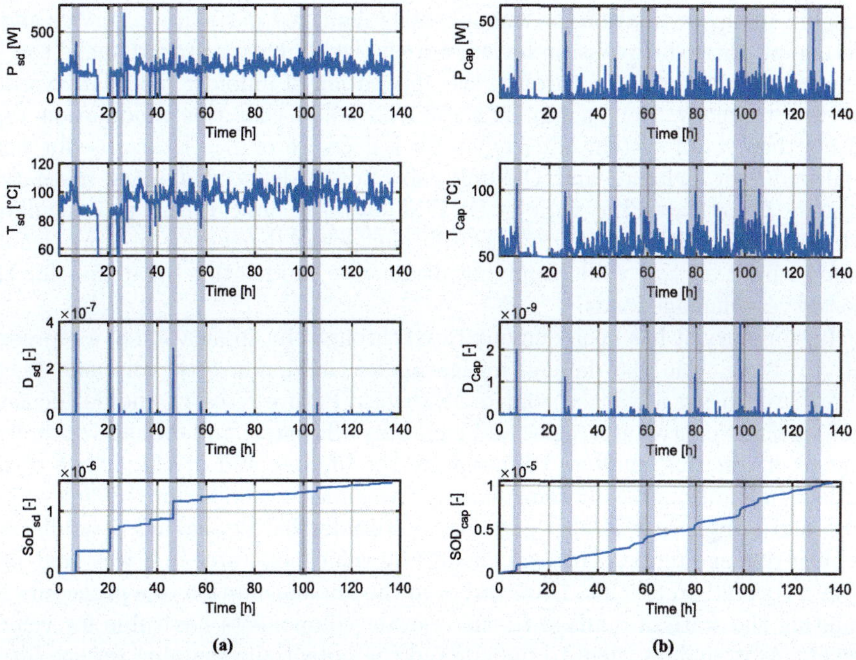

Fig. 11. Advanced damage mitigation strategy II: Modulator change indicated by colored bars depending on potential pre-calculated damage impact D_{sd} on semiconductor switches in **a** and pre-calculated damage impact D_{cap} on the DC-link capacitor in **b** using a B6 topology

$$D_i(n, M, \text{SVPWM}) > D_{i,\max}(n, M) \tag{22}$$

This approach reduces the number of modulator changes and avoids high efficiency decreases of the system. Moreover, comparing subfigures (a) and (b), it can be seen that some of the damage impacts on the switches and the DC-link capacitors occur during different operation points. This enables further optimization of the control strategy by either introducing weighting attributes based on the SOD of the relevant components or altering the pre-defined damage limit based on the individual component damage state.

4 Conclusion

When considering damage mitigation for semiconductor switches, damage can be significantly reduced through the complete utilization of DPWM0, as shown in Fig. 10. However, this approach decreases the overall system efficiency by 3% due to increased harmonic motor losses resulting from higher voltage distortion across the motor windings. This result is consistent across all DPWM strategies. It may be of interest in situations where short distances are traveled and recharging is possible at the destination, or where heavily damaged parts are in service and a lifetime-prioritized drive is preferred. For the DC-link capacitor, damage can be reduced by a factor of 3.5 through the application of unfolder modulation using the $B6^2$ topology. Here again, it should be noted that this modulation type affects the overall system efficiency with a decrease of 2%. Utilizable for both semiconductor switches and DC-link capacitors, advanced damage mitigation strategy II activates the various DPWM patterns only during pre-calculated extreme damage loads. This reduces efficiency losses depending on a pre-defined damage limit $D_{i,\max}$ while applying alternative modulation strategies for the most severe damage loads.

In summary, it has been demonstrated through the application of a representative driving cycle that damage mitigation for traction inverter components can be achieved using advanced control strategies. Future research should focus on refining these control strategies and exploring efficiency- and damage-optimized control of traction inverters to maximize the lifetime and driving range of the vehicle. In addition to combining damage avoidance strategies for DC-link capacitors and switches, it is also possible to consider battery-saving or additional implicit damage terms. This can ensure the vehicle operates in a way that protects its overall service life. However, even the implementation of a procedure for counting the state of damage to the relevant components has value for vehicle owners, as it enables them to understand the potential remaining service life of the inner power electronics components used for propulsion.

References

1. Lee, K.-B., Lee, J.-S.: Reliability Improvement Technology for Power Converters. Springer, Singapore (2017)
2. Hava, A.M., Kerkman, R.J., Lipo, T.A.: Simple analytical and graphical methods for carrier-based PWM-VSI drives. IEEE T. Power. Electr. **14**(1), 49–61 (1999)
3. Sachs, C., Stamer, F., Allgeier, J., Neuburger, M., Sawodny, O., Thrimawithana, D.: Optimization-based comparative system evaluation of single and dual traction inverters with focus on partial load efficiency and chip area. IEEE Ener. Conv., in press, (2024)
4. Cittanti, D., Vico, E., Armando, E., Bojoi, R.: Analysis and conceptualization of a 400V 100 kVA full-GaN double bridge inverter for next-generation electric vehicle drives. In: IEEE Transportation Electrification Conference & Expo (ITEC), pp. 740–747, Anaheim (2022)
5. Muetze, A., Oh, H.W.: Application of static charge dissipation to mitigate electric discharge bearing currents. IEEE T. Ind. Appl. **44**(1), 135–143 (2008)
6. Friedli, T., Kolar, J.W.: A semiconductor area based assessment of AC motor drive converter topologies. Appl. Power. Elect. Co. 336–342 (2009)
7. Blaabjerg, F., Wang, H., Vernica, I., Liu, B., Davari, P.: Reliability of power electronic systems for EV/HEV applications. P. IEEE **109**(6), 1060–1076 (2020)
8. GopiReddy, L.R., Tolbert, L.M., Ozpineci, B., Pinto, J.O.: Rainflow algorithm-based lifetime estimation of power semiconductors in utility applications. IEEE T. Ind. Appl. **51**(4), 3368–3375 (2015)

9. Touzelbaev, M.N., Miler, J., Yang, Y., Refai-Ahmed, G., Goodson, K.E.: High-efficiency transient temperature calculations for applications in dynamic thermal management of electronic devices. Int. J. Electron. **135**(3), 031001 (2013)
10. Liu, D.: Highly accelerated life stress testing (HALST) of base-metal electrode multilayer ceramic capacitors. https://nepp.nasa.gov/files/24685/Liu_2013_n219_NEPPETW_Pres_HALST.pdf. Accessed 17 July 2024

Advanced Control of Electro-pneumatic Quick-Exhaust Systems through Magnetic Force-Stroke Optimization of Solenoids

Christoph Sachs[1,2(✉)], Phillip Döbler[1,2], Sebastian Körner[1,2], and Martin Neuburger[1]

[1] Faculty of Mobility and Technology, University of Applied Sciences Esslingen, Robert-Bosch-Straße 1, 73037 Göppingen, Germany
{christoph.sachs,phillip.doebler,sebastian.koerner, martin.neuburger}@hs-esslingen.de

[2] Institute for System Dynamics, University of Stuttgart, Waldburgstraße 17/19, 70563 Stuttgart, Germany

Abstract. The improved magnetic force-stroke characteristics of electromagnetic solenoid valves provide benefits for quick-exhaust systems regarding their design, functionality, control strategies and costs. The quick-exhaust membrane, as the key component, is analyzed in terms of design, materials, and performance. A novel control strategy for solenoid valves is introduced, reducing system complexity by replacing an additional valve with a bypass mechanism. System modeling is performed using differential equations for motion and pressure state variables. The simulation model design is verified through the utilization of empirical data, whereby discrepancies are elucidated. This study offers insights into optimizing electro-pneumatic systems and highlights opportunities for more robust, cost-efficient and controllable quick-exhaust concepts.

Keywords: Quick-exhaust membrane · Quick-exhaust systems · Solenoid control

1 Introduction

Electro-pneumatic systems are crucial in commercial vehicle (CV) applications, particularly for high air flow systems such as truck braking systems [1]. These systems require efficient management of large volumes of compressed air, with quick-exhaust valves (QEVs) playing a crucial role in rapid air release. The increasing demand for advanced electro-pneumatic systems calls for innovations that enhance control while reducing costs. Traditional control strategies often rely on complex and expensive solenoid valves, increasing system complexity and cost. This paper introduces a novel quasi-proportional control method for

A. Öchsner (ed.), *Collaborative Research Advancing Engineering Solutions for Real-World Challenges 2*, Proceedings in Technology Transfer,
https://doi.org/10.1007/978-3-031-79093-5_7

solenoid valves in QEVs, incorporating a bypass mechanism that simplifies system architecture and reduces component count. A comprehensive system model is provided, including force balance and air mass flow equations, accounting for non-linear system attributes. Empirical tests are conducted to measure friction forces, enhancing model accuracy. Through simulation and analysis, we examine QEV performance under various operating conditions. This research aims to optimize electro-pneumatic systems in CV applications, focusing on improved efficiency, reliability, and cost-effectiveness. The findings contribute to the development of more robust and economical electro-pneumatic solutions for industries dependent on large-volume air management systems.

In this section, the functionality and purpose of QEVs are explained. Section 1.1 details the standard pneumatic circuit setup for controlling QEVs in CV applications. Essential for activating and deactivating QEVs are the control strategies of solenoid valves, which are elucidated in Sect. 1.2.

1.1 Quick-Exhaust Valves

QEVs are designed by inserting an elastomer, known as quick-exhaust membrane (QEM), into a housing with three pneumatic ports: a supply pressure source, an actuator at the working pressure port, and a silencer at the pressure relief port. The QEM is a single-material component, combining piston and sealing functionalities, enabling high exhaust air flow rates up to 5500 Nl/min according to the international ISO 8778:2003-03 [2]. Figure 1 shows different QEM types. The main difference of the QEMs is their material composition. Typically, fluoroelastomers (FKM), perfluoroelastomers (FFKM), or hydrogenated nitrile butadiene rubber (HNBR) are used to withstand high temperatures up to 130 °C, ensure functionality at low temperatures down to – 40 °C, and increase durability. Rubber components made from ethylene propylene diene monomer (EPDM) are not used due to their lack of resistance against most CV oils.

(a) HNBR (left) and FFKM (right) **(b)** FKM

Fig. 1. Pictures of QEMs made out of different rubber materials

The QEMs have eight guide ridges to increase mechanical stability, for instance, by preventing tipping during up-and-down movements. The injection points are located in the center of the component's upper side and between the guide ridges. In picture (b) of Fig. 1, wear of the gasket seat geometry on the lower side of the QEM can be clearly recognized. The circular press-in sign emerges due to high forces when the QEM hits the valve seat and during the sealing state.

By controlling the surrounding pressure of the QEM, the switching states *a)*, *b)*, and *c)* can be generated [3], which lead to a connection of pressure chambers ①, ② and ③, so that a mass flow $\dot{m}$ between these chambers occurs, as illustrated by Fig. 2 for the following states.

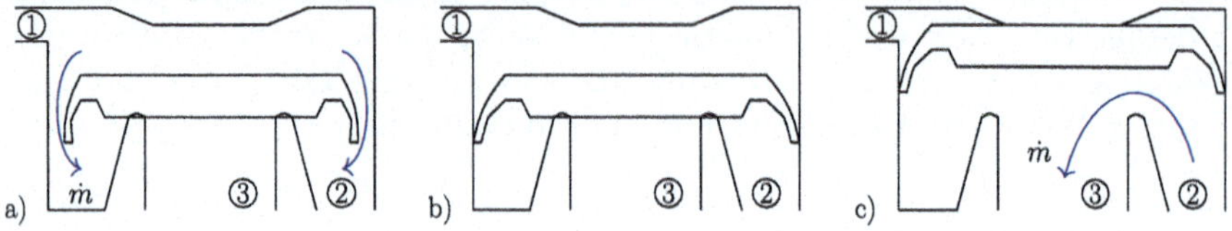

Fig. 2. QEM: **a** filling function, **b** sealing function, **c** quick-exhaust function

(a) *Filling Function*

The elastic QEM lips deform due to occurring pressure differences, leading to a mass flow from ① to ②, where ① is the supply pressure port and ② is the working pressure port. Actuators such as cylinders can be connected to the working port ②.

(b) *Sealing Function*

The system is in a stationary rest position, so that neither filling nor venting is possible. The pressure ports ② and ③ are sealed from each other. This state is desired after the set-point pressure level has been achieved in port ②. It is necessary to have the same pressure levels above and below the QEM to achieve the stable state (b).

(c) *Quick-Exhaust Function*

The QEM is located at the upper stop, which is caused by exhausting the air above the QEM. Due to the pressure from the lower side, the QEM accelerates to the upper stop. This causes rapid venting from the working port ② to ③, allowing the working air pressure to be relieved quickly. The flow diameter depends mostly on the inner diameter of the mechanical interface to port ③.

To control the QEM position, it is necessary to vent and exhaust the pressure in port ①. The pressure in port ② can be calculated as a function of the pressure in port ①, while the pressure differences over time are influenced by the QEM states (a) and (c).

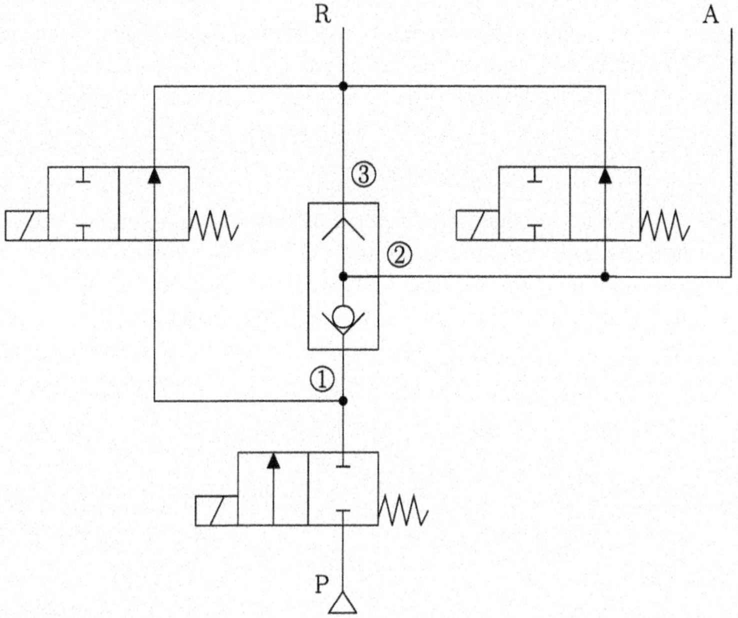

Fig. 3. Pneumatic circuit of a standard application with implemented QEM

The relevant pressure differences to enable movement and switching between states (a), (b), and (c) of the QEM are caused by a 2/2-NO (normally open) valve and a 2/2-NC (normally closed) valve, as depicted in Fig. 3. The QEV is shown in the middle of Fig. 3 with its ports ①, ②, and ③. The 2/2-NC valve deactivates the quick-exhaust function by pushing down the QEM, which is achieved by increasing the pressure above the QEM, while the 2/2-NO valve is activated to ensure no exhausting air flow. To activate the quick-exhaust function, the pressure above the QEM must be relieved by deactivating the 2/2-NO valve. Another 2/2-NO valve is required to enable small pressure set-point decreases, as activating the QEM for small set-point changes leads to significant pressure undershoots. For controlling the QEM proportional solenoid valves or 4/3-NC valves are not used due to higher costs, primarily driven by the complexity of such solenoid valves.

1.2 Solenoid Actuators and Control Strategies

The solenoid valve coils are electrically connected with field-effect transistors (FETs) and controlled using pulse-width modulation (PWM), which plays an essential role in providing the control input $u(t) = \dot{m}$ for the QEV as explained in Sect. 2.3. Sophisticated techniques such as hit-and-hold control are not detailed within this document, but are explained in, e.g., [4]. In the case of a direct current solenoid with a standard magnetic force curve (i.e., with a flat armature and armature counterpart), the magnetic energy change is essentially determined by the change in the magnetic resistance of the air gap between the armature and armature counterpart. The occurring magnetic force F_m is proportional to the applied current and reciprocally proportional to the square of the air gap length for standard magnetic circuits of solenoid switching valves [5].

$$F_m \sim \frac{1}{\delta^2} \tag{1}$$

Figure 4 depicts the magnetic force functions of a solenoid valve with a standard and an adapted magnetic circuit. As seen, the magnetic force increases in a certain stroke area, causing an additional stable armature position in the solenoid force equations, apart from the upper and lower stops. This armature position causes a mass flow that is smaller than the flow at the maximum stroke position and can be used to slightly increase or decrease the pressure in an actuator.

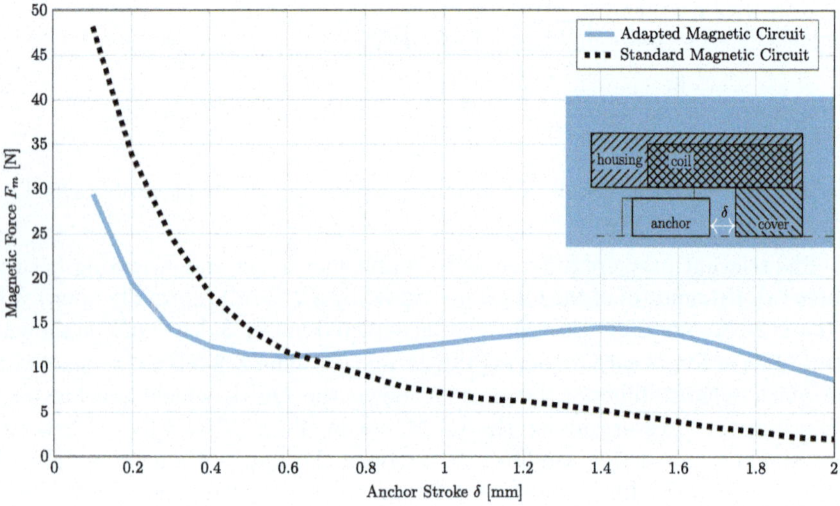

Fig. 4. Magnetic force functions of solenoids with standard and adapted magnetic circuit

The main force elements of a solenoid valve are the magnetic forces and a spring counterforce. The values of these forces decisively define the position of the armature in the valve. Proportional valves are designed in such a way that a defined opening cross-section. Is achieved for each defined current value. In a quasi-proportional switching valve, the adapted solenoid force curve results in a defined opening cross-section. range between the upper and lower switching positions. The background here is an interSect. in the force balance between springs and magnetic force. The limitation of solenoid curve adaptation is that the magnetic energy must remain constant [5].

$$W_m = \int_{0}^{\inf} F_m(\delta) \cdot d\delta = const. \tag{2}$$

One disadvantage is that at an end-stop armature position of $\delta_1 = 0.1\,\text{mm}$, considerably less magnetic force can be generated to push the armature away. When designing solenoid valves, the magnetic force diagram is limited by a working range that is smaller than the physically possible armature stroke to minimize the stroke and ensure an anti-remanence gap.

2 System Modelling

For the simulation of physical systems, it is useful to build up system models based on the system's differential equations. This is done for the motion and pressure state variables within Sect. 2.2. Section 2.3 summarizes the air mass flow equations and explains the mass flow dependencies inside the QEM itself. This chapter is completed by Sect. 2.4 to check if non-linear system attributes can be linearized. The simulation results are analyzed in Chap. 3. The pneumatic system functionality is based on the QEV and the quasi-proportional control of solenoid actuators, as explained in Sect. 2.1.

2.1 Control Strategy

According to Fig. 2, the QEM can be described by three states. These system states can be amended by a fourth state, utilizing quasi-proportional control of a solenoid valve, as depicted in Fig. 5. This allows for economisation by omiting an expensive valve by inserting a bypass. Summarized, two instead of three solenoid valves need to be implemented in the system. The following QEM states can be achieved if a pneumatic system equal to Fig. 6 is built up.

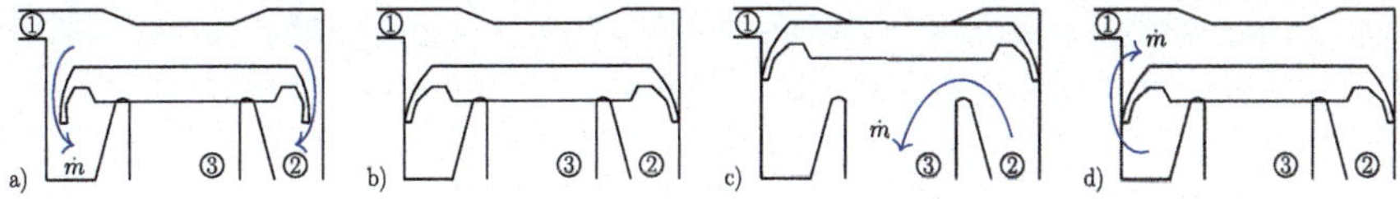

Fig. 5. QEM: **a** filling function, **b** sealing function, **c** quick-exhaust function, **d** exhaust-function (bypass-exhausting)

(a) Filling function according to Sect. 1.1.

The elastic QEM lips deform due to occurring pressure differences and lead to a mass flow from ① to ②, where ① is the supply pressure port and ② the working pressure port. Actuators like cylinders can be connected to the working port ②.

(b) Sealing function according to Sect. 1.1.

The system is in a stationary rest position, so there is neither ventilation nor venting possible. The pressure ports ② and ③ are sealed from each other. This state is crucial after the set point pressure level has been achieved. It is necessary to have the same pressure levels above and below the QEM to achieve the stable state (b).

(c) Quick-Exhaust Function according to Sect. 1.1.

The QEM is located at the upper stop. This is caused by exhausting the air above the QEM. Due to the pressure from the lower side, the QEM is accelerated to the upper stop. This causes a rapid venting from the working port ② to ③, so that the working air pressure can be relieved quickly. The flow diameter is defined by the inner diameter of the mechanical interface to the port ③.

(d) Slow-exhaust function

The QEM reaches the lower stop. Nevertheless, a venting of the working volume ② to ③ through a mechanical bypass hole in the housing connecting ② and ① occurs. This is only a slow air pressure reduction. Further information is provided in Sect. 2.3.

The QEM switching states can be realized using the pneumatic circuit illustrated in Fig. 6. Thereby, a 2/2-NO valve with adapted magnetic force curve in accordance Sect. 1.2, a standard 2/2-NC valve, a bypass throttle and the QEV itself are necessary. The bypass throttle is basically a defined orifice diameter that does not change due to external disturbances and connects the working port ② directly to the port ① of the QEV. This has minor effects on the filling function (a), but a major influence on the exhaust functions and enables exhaust function (d).

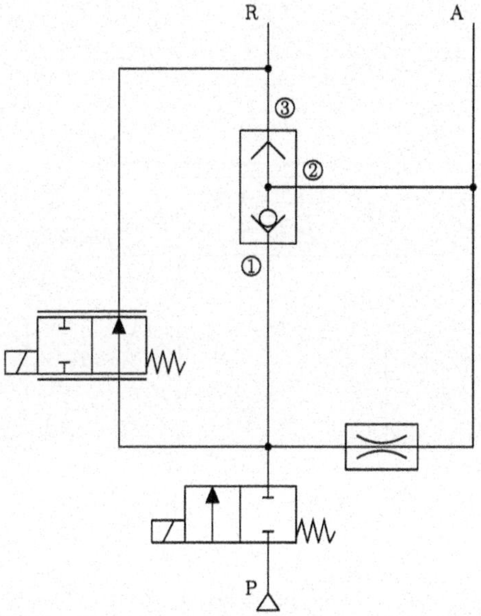

Fig. 6. Pneumatic circuit of a test application with implemented QEM

2.2 Force Balance

Based on the 1-DOF model of the QEM in Fig. 7, the system state variables of the

- pressure above the QEM $\xi_1 = p_1$
- pressure below the QEM $\xi_2 = p_2$
- QEM position $\xi_3 = x$
- QEM velocity $\xi_4 = \dot{x} = v = \dot{\xi}_3$.

are to be considered. The illustrated mass-spring-damper model represents the mechanical properties of the complete QEM. By applying Newton's principle of force balance $\sum \vec{F} = \vec{0}$, the time variation of velocity can be determined as a function of the following state variables.

$$\begin{aligned} l < x < 0 \Rightarrow m \cdot \ddot{x} &= F_{p_1} - F_{p_2} + F_g \pm F_\mu \\ \ddot{x} &= \frac{A_1}{m} \cdot p_1 - \frac{A_2^*}{m} \cdot p_2 + g \pm \frac{F_\mu}{m} \\ x \geq 0 \Rightarrow m \cdot \ddot{x} &= F_{p_1} - F_{p_2} - F_c - F_d + F_g \pm F_\mu \\ \ddot{x} &= \frac{A_1}{m} \cdot p_1 - \frac{A_2}{m} \cdot p_2 - \frac{c_k}{m} \cdot x - \frac{d_k}{m} \cdot \dot{x} + g \pm \frac{F_\mu}{m} \end{aligned} \tag{3}$$

- Where the force F_{p_1}, is caused by the pressure p_1 above the QEM in chamber (1) and depends on the upper cross-sectional area A_1.

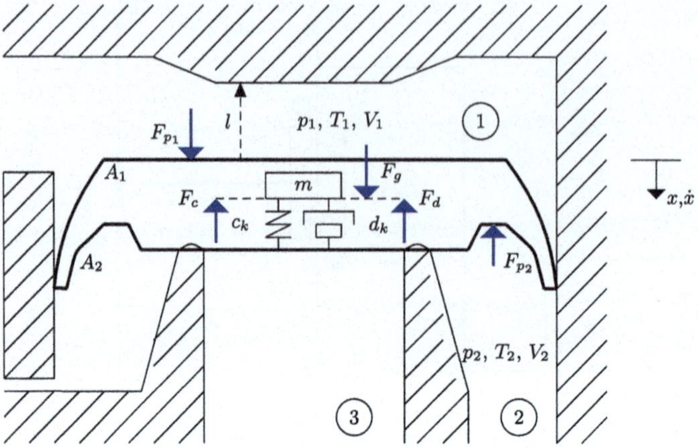

Fig. 7. 1-DOF model of the QEM with highlighted applied forces

- And the force F_{p_2}, due to the pressure p_2 below the QEM in chamber (2), depending on the ring area A_2 defined by the QEMs total cross-sectional area A_1 and the required area of the valve seat to enable the required quick-exhaust flow rate.

If the QEM has been lifted up from the valve seat at $x = 0$, the area A_2 increases to $A_2^* = A1$. At the same moment, the dynamic pressure below the QEM decreases, because chamber (3) and (2) are becoming connected, and a pressure equalisation starts to build up. The remaining pressure difference is neglected within the simulation model. This is justified by the assumption that the QEM moves without disturbances to its upper end stop and therefore has no major influence on the system state variables ξ_1 and ξ_2.

- The indentation forces $F_c(x)$ and $F_d(\dot{x})$ emerge when the QEM hits the seat geometry depending on approximated spring and damping constants c_k and d_k.

According to Fig. 7, spring and damping constants c_k and d_k can be considered to model the elastic repel behavior while the QEM hits the valve seat. Since only the transient range between the end positions is considered, the model is simplified by the indentation compliance and internal damping.

- The weight force F_g depends on the weight m of the QEM and the gravitational constant g.

Therefore, the orientation of the module needs to be considered. In following empirical investigations, the orientation of the illustrated 1-DOF model in Fig. 7 is be turned by 90°, which leads to disregarding the gravitational forces.

- The friction component F_μ is depending on temperature T, pressure difference Δp between chamber (1) and (2), as well as the velocity $\dot{x}$ of the QEM.

To receive an approximation of the static friction force values of the QEM, an empirical test is performed. The unsymmetrical lip geometry necessitates to differentiate between an upwards and a downwards movement. For measuring hold friction values, the QEM is assembled in its installation space while a stamp from above moves downwards until the stamp and QEM have contact with each other. The force increases slowly and is measured if a QEM movement is recognized. For upward movement, this simplified test can not be carried out, as the hold friction value mainly depends on the pressure in chambers (1) and (2) because the QEM lip is pressed radially to the inner housing diameter by force built up by the pressure difference of p_1 and p_2. This effect is neglected, whereby its compromising influences on the simulation results are considered and discussed in Sect. 3.2 (Fig. 8).

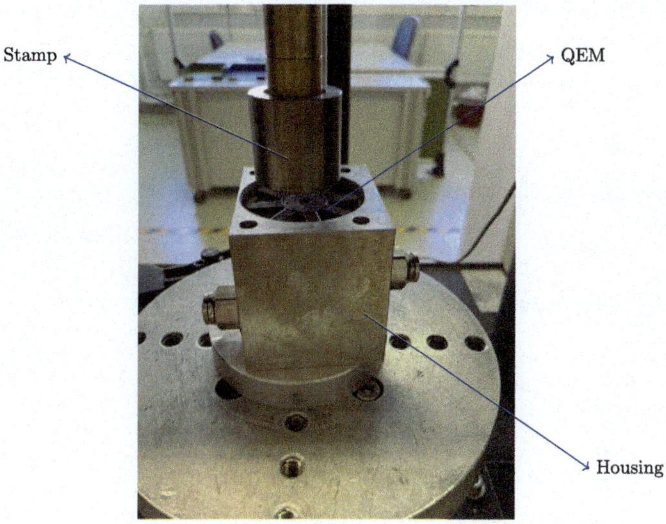

Fig. 8. Test setup for friction measurements

Table 1 summarizes the results of the friction force measurement. The mean friction force values are implemented in the dynamic system model to increase the model accuracy. The static friction forces are compared to occurring pressure forces so low that the influence can be neglected completely. The system's non-linear dynamic friction behavior is not considered further, due to the low total stroke and the high velocity of the QEM itself.

Considering that the bouncing forces F_c, F_d, the friction force F_μ and the weight force F_g are neglected, therefore the differential equations of $\dot{\xi}_3$ and $\dot{\xi}_4$ can be written as

$$\begin{aligned} \dot{\xi}_3 &= \xi_4 \\ \dot{\xi}_4 &= \frac{A_1 \cdot \xi_1 - A_2 \cdot \xi_2}{m} \end{aligned} \tag{4}$$

Table 1. Friction force measurement values shown in mean and standard deviation values

Measurement No.	mean(F_μ) in **N**	$\sigma(F_\mu)$ in **N**
1	0.66	± 0.03
2	0.64	± 0.03
3	0.64	± 0.02

To calculate the differential equations of the pressure changes $\dot{p_1}$ and $\dot{p_2}$, it is necessary to derive the thermodynamic relationships and match them with the geometric system properties. Airflow balances is analyzed within Sect. 2.3. In the following equations, $\dot{m}_1$ defines the total amount of air that is inserted or exhausted in chamber ① and $\dot{m}_2$ represents the total amount of air that passes the chamber ②. Starting with the volume V_1 in chamber ①, it is defined as a constant volume V_{10} and a changing volume $A_1 \cdot x$ based on the QEM movement.

$$V_1 = V_{10} + A_1 \cdot x \tag{5}$$

by derivating the changing volume is written as

$$\dot{V_1} = A_1 \cdot \dot{x} \tag{6}$$

From the general relation

$$\dot{m}(p, T) = V \cdot \rho(p, T) \tag{7}$$

and the ideal thermodynamic equation of state

$$p = R_s \cdot \rho \cdot T = \frac{n \cdot R \cdot T}{V} \tag{8}$$

whereby R_s is the general gas constant, n is the amount of substance, T is the temperature in the balance chamber, ρ is the density of the working fluid, R is the ideal gas constant, p is the pressure in the chamber, and V is the volume in the balance chamber, Eq. (9) can be formulated.

$$\begin{aligned} \dot{m} &= \dot{V} \cdot \rho + V \cdot (\frac{\partial \rho}{\partial p})_{|T} \cdot \dot{p} + V \cdot (\frac{\partial \rho}{\partial T})_{|p} \cdot \dot{T} \\ &= \dot{V} \cdot \rho + \frac{V}{R_s \cdot T} \cdot \dot{p} - \frac{p \cdot V}{R_s \cdot T^2} \cdot \dot{T} \end{aligned} \tag{9}$$

By setting the Eqs. (5), (6) and (9) equal, substituting (5) and (6) and rearranging according to the pressure derivation in the top chamber $\dot{p_1}$, the pressure differential equation solves to

$$\dot{p_1} = \frac{p \cdot \dot{T}}{T} + \frac{R_s \cdot T \cdot (\dot{m} - A_1 \cdot \dot{x} \cdot \rho)}{V_{10} + A_1 \cdot x} \tag{10}$$

Assuming that a isentropic process exists and air behaves like diatomic gases, applies

$$T \cdot p^{\frac{1-n}{n}} = K \tag{11}$$

By converting to temperature and differentiating over time, the results are as follows

$$\dot{T} = \frac{n-1}{n} \cdot K \cdot p^{-\frac{1}{n}} \cdot \dot{p} \tag{12}$$

As in Eq. (10) the ratio $T/\dot{T}$ occurs, the constant K is eliminated. Thus it results that

$$\frac{\dot{T} \cdot p}{T} = \dot{p} \cdot \frac{n-1}{n} \tag{13}$$

considering Eqs. (11), (12) and (13), it follows that

$$\dot{p_1} = \frac{n \cdot (R_s \cdot T \cdot \dot{m}_1 - A_1 \cdot \dot{x} \cdot p_1)}{V_{10} + A_1 \cdot x} \tag{14}$$

and accordingly

$$\dot{p_2} = \frac{n \cdot (R_s \cdot T \cdot \dot{m}_2 - A_1 \cdot \dot{x} \cdot p_2)}{V_{20} - A_1 \cdot x} \tag{15}$$

Due to the vast ratio of the constant volume V_{20} and stroke volume $A_1 \cdot x$ of chamber ②, as it is common in CV applications where QEVs are used,

$$\frac{V_{20}}{A_1 \cdot x} >> 100 \tag{16}$$

the influence of the changing volume $\dot{V}_2$ can be neglected and $\dot{p}_2$ solves to

$$\dot{p_2} = \frac{n \cdot R_s \cdot T \cdot \dot{m}_2}{V_{20}} \tag{17}$$

The system's differential equations can be calculated by

$$\dot{\vec{\xi}} = \begin{pmatrix} \dot{\xi}_1 \\ \dot{\xi}_2 \\ \dot{\xi}_3 \\ \dot{\xi}_4 \end{pmatrix} = \begin{pmatrix} \frac{n \cdot (R_s \cdot T \cdot \dot{m}_1 - A_1 \cdot \dot{\xi}_3 \cdot \xi_1)}{V_{10} + A_1 \cdot \xi_3} \\ \frac{n \cdot R_s \cdot T \cdot \dot{m}_2}{V_{20}} \\ \xi_4 \\ \frac{A_1 \cdot \xi_1 - A_2 \cdot \xi_2}{m} \end{pmatrix}, \vec{\xi}_0 = \begin{pmatrix} p_{10} \\ p_{20} \\ x_0 \\ v_0 \end{pmatrix} \tag{18}$$

The air mass flows $\dot{m}_1$ and $\dot{m}_2$ can be defined as functions depending on the pressure levels in the attached chambers. Section 2.3 shows how the mass flow equations are built up and on which parameters the mass flow function $\dot{m}_{i,j}$ for each mass flow path between chambers ⓘ and ② depends.

2.3 Air Flow Balance

As mentioned in Sect. 2.1, the control of the QEM can create different system states that lead to a connection of pressure chambers ①, ②, and ③, so that pressure equalization of these chambers can occur. The air mass flow $\dot{m}_1$ and $\dot{m}_2$ can be calculated considering Fig. 9. $\dot{m}_{1,1}$ and $\dot{m}_{2,1}$ represent the mass flow occurring due to activating the 2/2-NC valve, whereas $\dot{m}_{1,1}$ flows directly from port P to ① and $\dot{m}_{2,1}$ from the supply port P to $\dot{m}_{2,1}$. This effect during the filling process can be neglected based on empirical pressure increase measurements. For the slow exhaust function on the other side, it is essential to differentiate between those mass flows. $\dot{m}_{1,2}$ represents a leakage mass flow from chamber ① to the environment, whereby the leakage mass flow from chamber ② to the environment is neglected due to the large volume V_2 connected to the chamber. $\dot{m}_{2,3}$ is created if the QEM lifts, so that $x < 0$ and results in a quick exhaust airflow from chamber ② to ③. For $\dot{m}_{2,2}$, an FEM analysis of the elastic collapse behavior of the QEM lips can be approximated at different pressure differences between chambers ① and ②. The total mass flow $\dot{m}_2$ sums up from the mass flow $\dot{m}_{2,2}$ due to the collapse of the QEM lips and the mass flow $\dot{m}_{2,1}$ through the bypass according to Eq. (19). The collapse behavior of the QEM lips is described in Sect. 2.4. The mass air flow $\dot{m}$ can be calculated by the isentropic change of state of the compressed air.

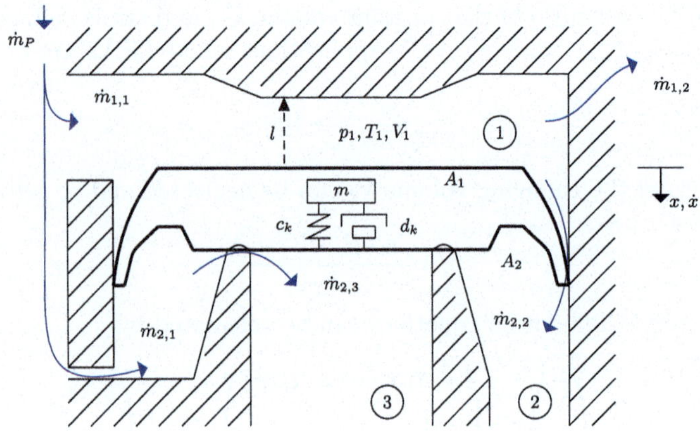

Fig. 9. 1-DOF Model of the QEM with highlighted mass flow paths

$$
\begin{aligned}
\dot{m}_P &= \dot{m}_{1,1} + \dot{m}_{2,1} \\
\dot{m}_1 &= \dot{m}_{1,1} - \dot{m}_{1,2} - \dot{m}_{2,2} \\
\dot{m}_2 &= \dot{m}_{2,1} + \dot{m}_{2,2} - \dot{m}_{2,3}
\end{aligned}
\tag{19}
$$

The slow-exhaust function is enabled by the mass flow reduction $\dot{m}_P$ of the solenoid valve due to quasi-proportional control. Due to the geometrical properties of the prototype, which are not exactly represented in Fig. 19, the airflow $\dot{m}_{1,1}$, which is based on a pressure difference between chamber (1) and the pressure level behind the exhaust port of the solenoid, is so small compared to the airflow $\dot{m}_{2,1}$ out of chamber (2) that it does not lead to dynamically reducing the pressure in chamber (2) considerable. This is based on the knot flow rule of $\dot{m}_P$, whereas $\dot{m}_{2,1}$ compensates the negative mass flow $\dot{m}_P$ and even $\dot{m}_{1,1}$ if the pressure in chamber (2) falls under a specific limit. This limit is in our case above the force limit, which leads to QEM movements to the upper stop. Further details are explained in Sect. 3.3 by comparing simulation results and empirical measurement results. Generally summarized, if the mass flow $\dot{m}_P$ rises above the limit of the bypass flow $\max(\dot{m}_{2,1})$, the mass flow $m_{1,1}$ becomes negative and the quick-exhaust function is activated. If the mass flow does not reach this limit, only the slow-exhaust function is activated. Further details are explained in Sect. 3.3 by comparing simulation results and empirical measurement results.

The mass flows are calculated based on [6] according to Eq. (20):

$$\dot{m} = \begin{cases} C_q \cdot A \cdot \Psi(b) \cdot \sqrt{\frac{1}{R_s \cdot T_1}} \cdot p_1 & 1 \geq b > 0.528 \\ C_q \cdot A \cdot \Psi_{max} \cdot \sqrt{\frac{1}{R_s \cdot T_1}} \cdot p_1 & 0 \leq b \leq 0.528 \end{cases} \tag{20}$$

where C_q is the correction factor for pneumatic flows and their losses at obstructions and other flow dependent resistances, $C_q = \epsilon \cdot \alpha$. The above flow equation is a function that has a maximum. However, this would also mean that before the extreme point (the ratio is smaller) the flux Ψ first increases slowly and then decreases. There are indications that, for example, with a fixed p_2 and an increasing p_1, the flux does not decrease at all. On the contrary, after reaching the maximum value, it remains at its maximum. This can be explained by physics, as the maximum air flow is limited, so it neither decreases nor increases. The flow equation describes how much can pass through the flow channel. In pneumatics, this is divided into the subcritical and supercritical ranges. The subcritical range describes the critical case when the flow rate becomes smaller, i.e. when the higher pressure tends towards the lower pressure. The medium does not flow faster than the speed of sound in standard pneumatic systems and therefore there is an upper limit which can be calculated by $\delta\psi \;/\; \delta\left(\frac{p_2}{p_1}\right)$. For ideal air there is a critical pressure $b = \frac{p_2}{p_1}$ =0.528 which gives a maximum air speed $c_{2,max} =$ 333 m/s considering [2]. The bypass cross-section $A_{2,1}$ that limits the bypass air flow $\dot{m}_{2,1}$ is calculated by $\frac{\pi \cdot d^2}{4}$ based on the circular flow profile, whereas this is not possible for the QEM, and its mass flow above the lips $\dot{m}_{2,2}$.

2.4 Elastic Lip Deformation

By a mathematical approximation of the QEM lip deformation due to a pressure difference Δp in chambers (1) and (2), a flow restricting area $A_{2,2}$ applied for $\dot{m}_{2,2}$ is determined

$$A_{2,2} = f(\Delta p) \tag{21}$$

whereas the pressure difference is written as

$$\Delta p = p_1 - p_2 \tag{22}$$

The simulation is based on the finite element method, utilizing the stated system properties inserted in Table 2. By simplifying the QEM as a hyperelastic solid model with C_{10} and C_{01} values according to a standard Mooney-Rivlin Model, which is state of the art for simulating rubber material component deformation [7]. For the simulation, the following points need to be taken into account:

- Deformation hysteresis: for a given deformation, the reaction forces always have the same value, regardless of the previous deformation.
- Force hysteresis: if the initial state is unloaded, it is resumed after any deformation when the loads are removed.
- The material behavior is cauchy-elastic. The speed at which deformation occurs does not affect the resistance the body offers to the deformation.
- The applied deformation work is completely stored in the body as distortion energy.

Table 2. Model properties of the FKM QEM

Attribute	Value
Shore hardness ShA	75
Temperature T	293.15 K
Max. applied pressure p_{max}	10 bar
Shear modulus GPa	2.182 N/mm
C_{10} value	0.872 N/mm^2
C_{01} value	0.219 N/mm^2

Figure 10 depicts the lip deformation due to forces depending on pressure differences between 0.001 and 1 bar. A mass flow restriction due to not moving QEM lips takes place if the pressure difference Δp is smaller than 100 mbar. There is no significant mass flow restriction above 100 mbar if filling valves with an overall orifice size of 3 mm or smaller are used, due to the QEMs bigger equivalent orifice size at this pressure difference. Considering how quickly 100 mbar are built up or reduced above the QEM, the overall mass flow restriction of the QEM lips is neglected to simplify the dynamic system model without a relevant loss of accuracy. This simplification of the QEMs mass flow equation is evaluated within Sect. 3.1.

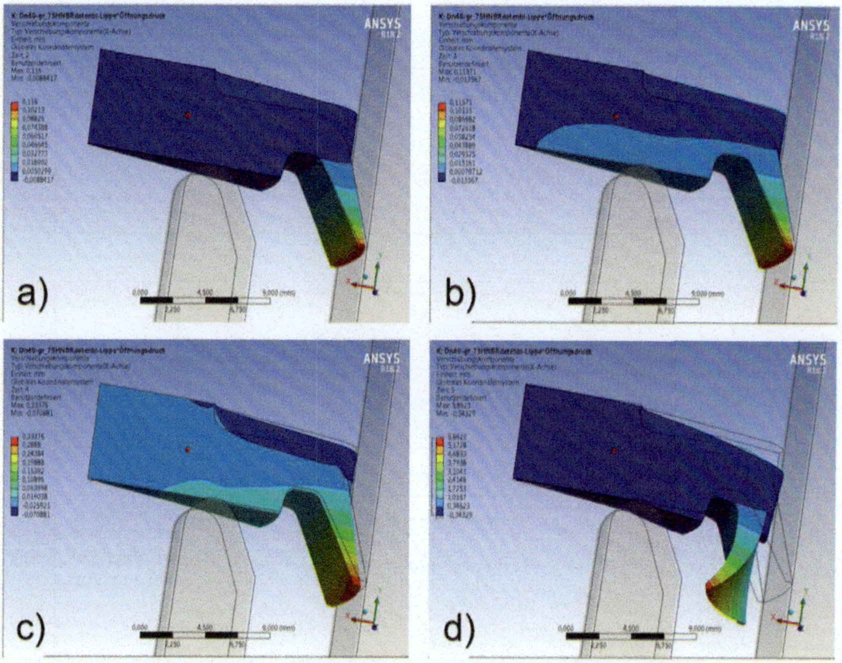

Fig. 10. False color display showing the elastic deformation of the QEM lips in a 16-th section due to applied pressure differences $\Delta p = p_1 - p_2$ of **a** 0.001 bar; **b** 0.01 bar; **c** 0.1 bar and **d** 1 bar

3 Simulation and Measurement Results

Figure 11 depicts the state variables of the QEM introduced in Sect. 2.2, occurring air mass flows and applied forces during pressure increase (QEM state a), sealing (QEM state b) and a quick exhaust (QEM state c) according to Sect. 2.1. The simulation results of the slow-exhaust function is explained in Sect. 3.3 by directly comparing them to measurement results. According to Fig. 11, the air mass flows $\dot{m}_1$ and $\dot{m}_2$ lead to an increasing pressure in chambers ① and ② by pushing down the QEM to the lower end stop and filling air into the actuator attached to chamber ② during 1 s and 2.5 s. The attached actuator is a 2.5 L volume in our simulation model and empirical laboratory setup. $\dot{m}_1$ and $\dot{m}_2$ are almost equal, due to the fact, that the $\dot{m}_2$ leads to decreasing the pressure in chamber ① and therefore, increasing the mass flow $\dot{m}_1$ until the pressure in chamber ① and chamber ② are equal. The QEM accelerates until the end stop is reached. Thereby, the bouncing effects during the QEM intrusion are not considered in our system model, because there are no effects on the pressure levels p_1 and p_2. The forces, which are mainly depending on the pressure above and below the QEM, are getting to equilibrium during the sealing phase between 2.5 s and 5.75 s. The negative airflow $\dot{m}_1$ decreases pressure p_1, which

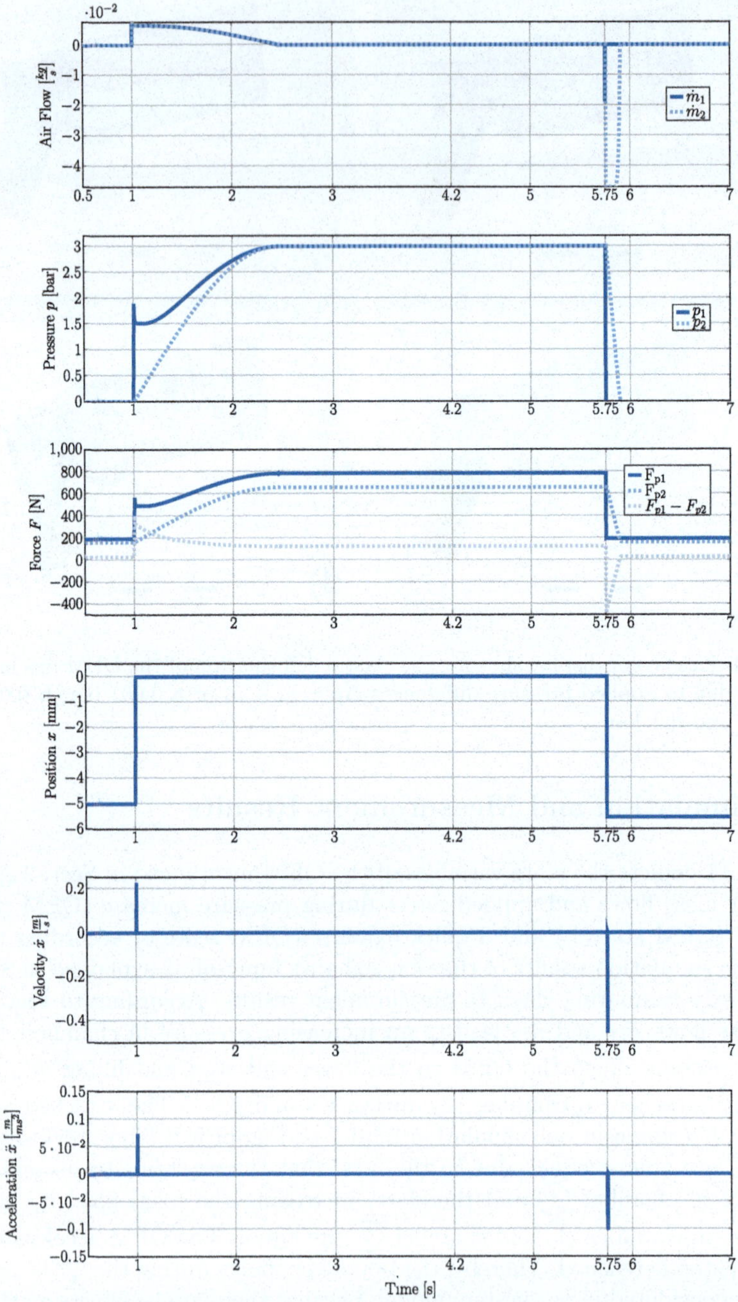

Fig. 11. Simulation results of state variables $\vec{\xi}$, mass flow $\dot{m}$ and forces F over time

leads to accelerating the other way around until it hits its upper stop. A resulting quick-exhaust mass flow $\dot{m}_2$ decreases the pressure in chamber (2) until it is equal to the pressure in the exhaust port. As described for the filling function, the QEM position change during the exhaust functions differs from simulation to empirical measurement as well. In summary, the simulation results show the expected behavior for the filling, sealing and quick-exhaust functions. To evaluate the accuracy of the system, empirical measurements are carried out. The differences between simulation and empirical measurement results are explained thoroughly in Sect. 3.2.

3.1 Empirical Investigations and System Setup

To validate the system model and verify the simulation results, it is necessary to conduct empirical research. The system setup, including data transfer and acquisition, is explained in Fig. 12. Sections 3.2 and 3.3 show measurement results while identifying the system responses by using the quick-exhaust function and the slow-exhaust function.

The overall system setup depicted in Fig. 12 is divided into an electric system setup, which is mainly placed on the lower stage of the Table, and a pneumatic system setup, which is placed on the upper stage (b). Thereby, the red volume represents the working pressure volume, and the beige-colored one represents the supply pressure volume. A pressure sensor is connected to each of the volumes to track set-point responses and control the supply pressure. The prototype contains the pneumatic components introduced in Fig. 6 of Sect. 2.1. A *MicroLabBox* from *dSpace* is used for data acquisition. The *MicroLabBox* controls power output stages that energize the solenoid valves installed in the prototype via FETs. Parameters like the supply pressure can be changed by the *dSpace ControlDesk 5.4* on the run. The four devices on the lower stage next to the *MicroLabBox* are pressure sensor signal amplifiers and power supply devices.

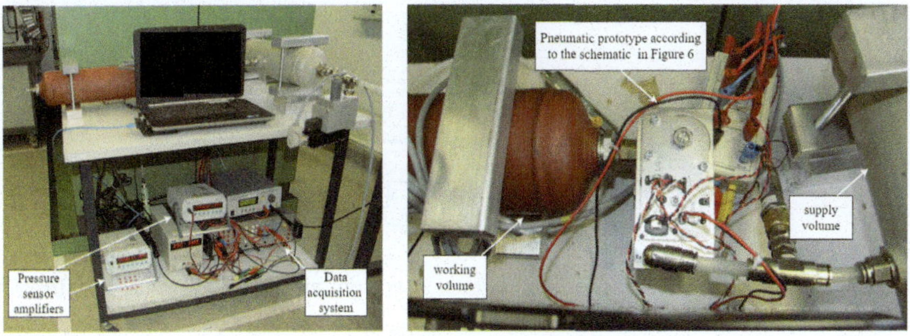

(a) Overview of the laboratory system setup **(b)** Pneumatic actuator connected to supply and work volumes

Fig. 12. Laboratory measurement setup

3.2 Quick-Exhaust Control

To verify the model accuracy for the filling function (a), the sealing function (b) and the quick exhaust function (c), it is necessary to compare the dynamic profile of the actual pressure curves for measurement p_2, simulation p_2^* and the pressure deviation $p = p_2 - p_2^*$. Figure 13 is separated into three subplots. The upper subplot shows empirical measurement results; the subplot in the middle shows simulation results; and the lower subplot illustrates the difference pressure function p. The profile starts by activating the vent and exhaust valves. Afterward, the vent valve is deactivated to seal in the pressure in the volume connected to ②. Finally, the complete deactivation of the vent valve occurs so that the QEM is pushed to the upper stop of chamber ① and rapid venting of the pressure in the working volume occurs.

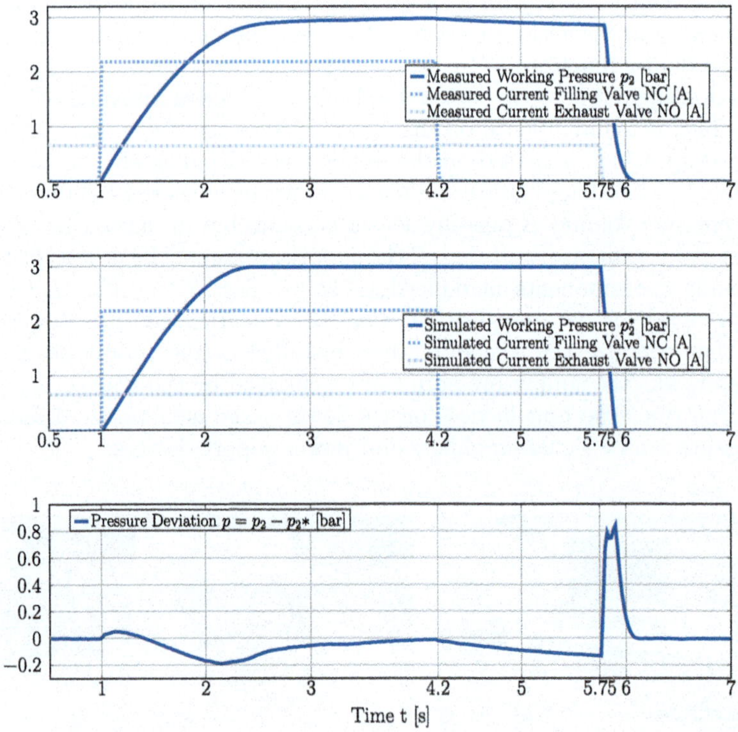

Fig. 13. Pressure increase and quick-exhaust function, simulation versus measurement

In the first section of venting between 1 s and 4.2 s, there is a filling of the volume with the supply pressure of 3 bar. The difference pressure curve indicates that overall pressure differences between simulation and measurement vary between ± 0.2 bar. This is caused by various reasons. Beginning with the measurement setup, the pressure sensor position and its attachment to the working volume have an influence on the measured pressure profile. This can be verified by repositioning the pressure sensor at the end of the working volume and decreasing the length of the tube as much as possible. On the other side, air leakage might be responsible for the time deviation until the supply pressure has been achieved. Furthermore, the pressure profiles indicate that neglecting the flow restricting QEM lip deformation according to Sect. 2.4 could lead to pressure deviations in an area where a pressure deviation of working pressure p_2 and supply pressure p_1 occurs. A flow analysis by reducing the pressure p_1 from 1 bar to 0 bar could provide information about the behavior of the QEM and verify the simulation results. The increasing pressure deviation during the sealing function (b) between 4.2 s and 5.75 s is caused by the thermodynamic relaxation of the pressure in the working volume according to Eq. (8). The increase in pressure in the working volume causes an increase in temperature, which is subsequently cooled at the inner wall of the volume. The cooling at constant volume in turn leads to a pressure reduction. Leakage in the working volume at ② could also play a role in sealing the pressure, as it presumable does during the pressure increase. Between 5.75 s and 6 s, the QEM moves from the lower stop to the upper end stop and enables a pressure decrease from chamber ② to ③ caused by the air mass flow $\dot{m}_{2,3}$. The simulated pressure decrease in chamber ② starts earlier than the empirical measurement depicts. This is partly based on the assumption that the QEM moves without interruption to the upper end stop. Further measurements of various quick-exhaust systems disproved this hypothesis. The air compression in chamber ② leads to an increasing pressure that is decreased in the simulation model because it is directly connected to a volume, which is not separated by an additional throttle. The actual pressure in chamber ② can not be decreased sufficiently to enable a continuous movement to the upper stop. Instead, the increasing pressure leads to accelerating the QEM back in the direction of the lower stop, which slows down the quick-exhaust process. To add this system behavior to the model, it is necessary to include a pneumatic throttle at the QEM mass flow $\dot{m}_{1,1}$, which is active during the quick-exhaust function. Moreover, when it comes to airflow effects, that could pull the QEM down in the direction of the lower stop. This is caused by a connection between pressure port ②, which is in this case pressurized, and port ③, which is not pressurized. Therefore, the push force from below is increased due to the extended pressure area A_2^*, but decreased due to potential suction effects of the pressure port ③. A computational fluid dynamics analysis is necessary to determine what flow effects occur and how the pressure balance between ports ② and ③ is established depending on the geometrical properties of the system. Apart from that, the position of the pressure sensor might also play an essential role, as explained for the filling function (a).

3.3 Slow-Exhaust Control

The slow-exhaust function (d) according to Fig. 5 leads to a decreasing pressure level in chamber ②, which is caused by a negative airflow $\dot{m}_{2,1}$ through the

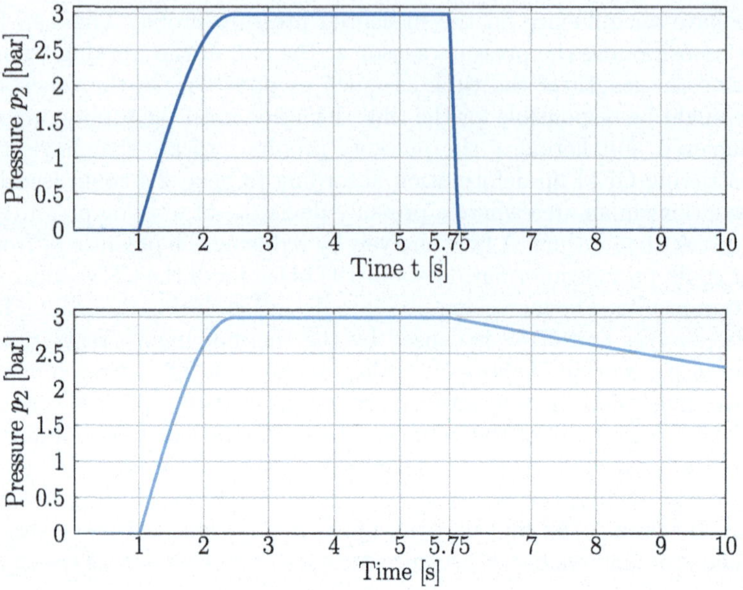

Fig. 14. Simulated pressure decrease with quick-exhaust (above) and slow-exhaust function (below)

bypass. The plots in Fig. 14 depict the slow-exhaust function compared to the quick-exhaust function by illustrating the pressure change in chamber ② after a 2/2-NC WV filling valve and a 2/2-NO WV venting valve are activated at 1 s to fill chamber ① and ② with supply pressure and deactivated at 5.75 s to exhaust chamber ② and reduce the pressure p_2 slowly (plot below) or quickly (plot above). During the slow pressure reduction, the 2/2-NO WV is not fully deactivated. The current is reduced slightly to enable quasi-proportional control, which causes an armature position between the solenoid valve's upper and lower stops, as explained in Sect. 2.1. This is shown in Fig. 15. The limited mass flow through the valve enables the pressure in chamber ① is not reduced to a level that activates the quick-exhaust function (a), as explained in Sect. 2.3.

4 Conclusion

This study provides a comprehensive analysis and innovative approach to optimizing quick-exhaust systems in electro-pneumatic applications, particularly in commercial vehicles. By focusing on the magnetic force-stroke characteristics of electromagnetic solenoid valves, it becomes evident that there are considerable opportunities for enhancing control and reducing costs in quick-exhaust systems. To illustrate, a standard quick-exhaust system would benefit from a cost reduction of approximately 20 %. The introduction of a quasi-proportional control

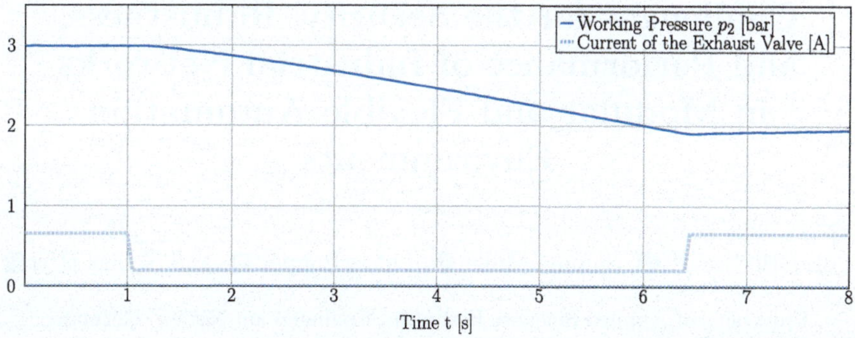

Fig. 15. Measurement of the slow pressure exhaust and the current levels of the exhaust solenoid valve

method for solenoid valves, featuring a bypass mechanism, represents a significant advancement in system design. The empirical validation of the simulation model highlights the practical applicability of the proposed method, with discrepancies between the model and real-world data being thoroughly examined and explained.

In conclusion, this research makes a significant contribution to the field of electro-pneumatic systems by introducing a novel cost-effective, reliable, and controllable quick-exhaust concept. The insights gained from this study are expected to drive further innovations and optimizations in the design and control of electro-pneumatic systems, ultimately benefiting industries reliant on large-volume air management systems. Further work could explore the scalability of the proposed method to other applications and further refine the control strategies for even greater efficiency and performance.

References

1. Sachs, C., Wassermann, M., Dederer, A.: Control Air System for a Hydrodynamic Retarder. World Intellectual Property Organization, Geneva (2023)
2. ISO 8778:2003: Pneumatic Fluid Power—Standard Reference Atmosphere. International Organization for Standardization, Geneva (2003)
3. Schnorr, F.W., Jr.: Design of Pneumatic Pressure Regulating and Quick Exhaust Valve. General Electric Co., Aircraft Nuclear Propulsion Department, Cincinnati (1957)
4. Rannow, M.B., Li, P.Y.: Soft switching approach to reducing transition losses in an on/off hydraulic valve. J. Dyn. Syst-T. Asme. **134**(6), 064501 (2012)
5. Kallenbach, E., Eick, R., Ströhla, T., Feindt, K., Kallenbach, M., Radler, O.: Elektromagnete: Grundlagen, Berechnung. Entwurf und Anwendung. Springer, Berlin (2018)
6. Surek, D., Stempin, S.: Technische Strömungsmechanik. Springer, Berlin (2014)
7. Balint, D.S., Bordas, S.P.A.: Adv. Appl. Mech. Acad. Press **53**, 34–86 (2020)

Challenges for the Security, Robustness, and Performance of Industrial Networks in Modular and Flexible Automation Environments

Lukas Bechtel(✉), Samuel Müller, Lukas Popperl, and Tobias Heer

University of Applied Sciences Esslingen, Esslingen am Neckar, Germany
{lukas.bechtel,samuel.mueller,lukas.popperl,tobias.heer}@hs-esslingen.de

Abstract. Industrial networks must meet stringent requirements for *security*, *robustness*, and *performance* to ensure stable and reliable process control in automation environments. These networks achieve the requirements by using isolated and independent special-purpose segments. However, the transformation towards more flexible and modular (SFs) introduces significant challenges. We identify three major transformation factors demanding novel solutions to maintain industrial standards: *growing interconnectivity*, *expanding multi-tenant capability*, and *increasing application dynamics*. This work analyzes the tension between the fundamental requirements and the transformation factors driving the evolution of industrial networks. We also examine existing solutions, highlighting their inadequacies in addressing the comprehensive needs of SFs. The aim is to identify open areas of research necessary for developing equally secure, robust, and high-performing networks in modern industrial settings.

Keywords: Smart factories · Industrial networks · Robustness · Performance · Security · Transformation

1 Introduction

(IACS), such as factories or process facilities, have the main task of providing stable and reliable process control. Process control, such as supervising machinery or monitoring real-time production metrics, requires secure and timely communication between controllers, sensors, and actuators. IACS require all components and processes to operate (a) unaffected by cyberattacks, (b) with high availability, i.e., without device failures, and (c) optimal performance, i.e., low latency and minimal jitter, even during extreme network loads. Hence, the fundamental requirements to maintain the integrity and efficiency of industrial networks and processes are (a) *security*, (b) *robustness*, and (c) *performance* [1,2]:

A. Öchsner (ed.), *Collaborative Research Advancing Engineering Solutions for Real-World Challenges 2*, Proceedings in Technology Transfer,
https://doi.org/10.1007/978-3-031-79093-5_8

Security involves protecting the communication channels and devices from unauthorized access and cyberattacks. Security measures must ensure the confidentiality, integrity, and availability of the network. For example, current industrial networks use firewalls to segment the network into different zones and, thus, prevent intruders from gaining unauthorized access.

Robustness refers to the ability of a network to maintain stability and recover from failures or disruptions. Robustness measures include redundancy, fault tolerance, and recovery mechanisms. For example, industrial networks utilize hot standby mechanisms and redundant network paths to continue operating in case of device or link failures.

Performance describes the ability of the network to fulfill the real-time communication requirements for synchronized industrial processes. Important performance factors include low latency and minimal jitter. Industrial networks achieve the required performance by isolating critical traffic from best-effort traffic.

Today, industrial networks achieve these requirements through isolated and independent special-purpose network segments. This isolation ensures a controlled and secure environment for each machine inside the factory to operate in. However, applications in IACS evolve from specialized and isolated applications towards interconnected and modular applications. This evolution enables the implementation of novel manufacturing techniques and increases the productivity as well as the operational flexibility of industrial processes [3]. One of these novel manufacturing techniques defines production tools to be more flexible. For example, heavy production material, e.g., a vehicle, resides static on the production floor while production tools move around. This flexibility of tools is achieved with flying drones and (AGVs), transporting the tools and parts to be mounted on the vehicle. The changes in manufacturing are driven by the growing demand for customized and on-demand production, resulting in the development of (SFs). We characterize the resulting network transformation of these SFs through three main factors: (1) *Growing interconnectivity*, (2) *expanding multi-tenant capability*, and (3) *increasing application dynamics*.

Growing interconnectivity refers to network devices being more interconnected in SFs than in current networks. Current networks often utilize (PLCs) as gateways and (FWs) between individual networks to block or filter non-time-critical control traffic. To enable seamless data exchange between devices in a network, as required for SF, this strict block and filter behavior must be adjusted.

Expanding multi-tenant capability refers to the integration of leasing processes into an SF and enables a single network to be shared by multiple users. Current networks do not support this feature, requiring novel network topologies and configurations.

Increasing application dynamics refers to the increasing physical and digital dynamic of SFs in comparison to current networks. Current networks already use mobile devices, such as AGVs, but SFs will increase the diversity of mobile devices, e.g., by introducing flying drones, and the scope of mobile actions, e.g., the time-criticality of communication between mobile devices. Additionally, in SFs, devices continuously perform different tasks over their lifetime, with code and configuration being dynamically adapted during runtime. This dynamic adaptation is not supported in current networks and must be specifically implemented for SFs.

In general, all three transformation factors challenge the requirements of industrial networks and necessitate the development and deployment of new technologies compared to current industrial networks. Figure 1 presents the requirement and transformation matrix of SFs. The vertical lines display the industrial requirements, i.e., *security*, *robustness*, and *performance*. The horizontal lines highlight the identified transformation factors we identified, i.e., *growing interconnectivity*, *expanding multi-tenant capability*, and *increasing application dynamics*. The overlap between the requirements and the transformation factors highlights the required change in multiple areas, as all requirements are intertwined with the outlined transformation factors. Hence, improving a single requirement without considering all transformation factors may negatively impact the other requirements.

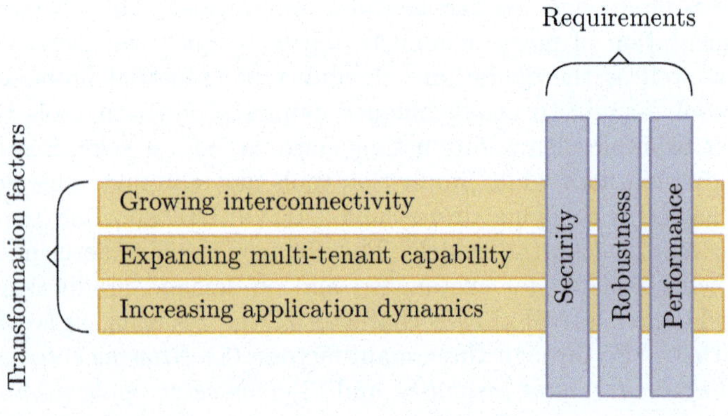

Fig. 1. Requirement and transformation matrix of smart factories

The remainder of the paper is structured as follows. In Chap. 2, we present the three transformation factors for industrial networks in detail. Therefore, for each transformation factor, we highlight its enabling technologies and the challenges caused by its application. Next, in Chap. 3, we present and classify existing solutions for the highlighted challenges and discuss their applicability in SFs. Lastly, we summarize and conclude the paper in Chap. 4.

2 Industrial Transformation

Today, (IACSs) follow strict rules regarding their network architecture and the deployed applications. These networks are guided by industry standards and guidelines, such as the IEC 62443 [4] and the purdue enterprise reference architecture [5]. As mentioned before, the requirements for the deployment and functionality of networks transform towards SFs through the three factors: (1) *Growing interconnectivity*, (2) *expanding multi-tenant capability*, and (3) *increasing application dynamics*. In the following sections, we present these transformation factors and elaborate on the challenges they pose.

2.1 Application of Growing Interconnectivity

Current industrial networks follow a strictly hierarchical approach, i.e., combining connected devices in smaller network segments. These network segments form static and specialized machines, aggregated into production lines. Communication follows a hierarchical pattern by using controllers as gateways between network segments at different hierarchical levels. Figure 2a presents an exemplary network topology for this hierarchical approach. The red arrows mark the communication flow between devices. The communication between segments is limited through controllers acting as gateways. Therefore, the controller must be programmed to expose device information for communication across network segments. For network segments without a gateway, specific traffic is forwarded by a firewall, which only allows communication between controllers.

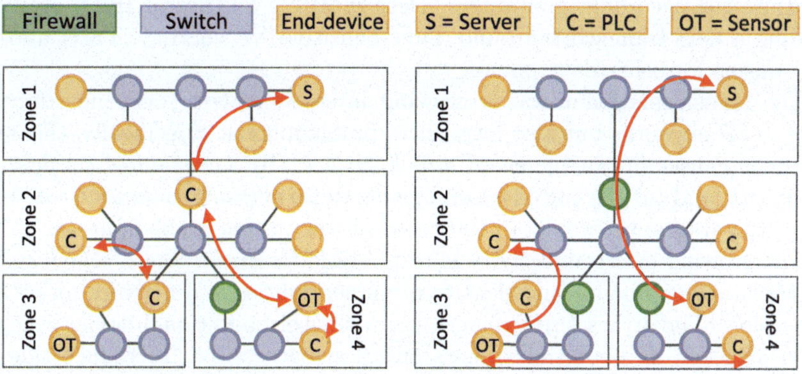

(a) Specialized and isolated network segments.

(b) Interconnected and open network segments.

Fig. 2. Growing interconnectivity

In contrast, SFs require seamless access to all sensor values from either the cloud or from controllers of neighboring machines. This access requires the servers to communicate with OT devices, e.g., sensors, across multiple network segments. Hence, SF networks must be interconnected and less restrictive to allow communication between network zones. Therefore, SF networks cannot use controllers as gateways between network segments but open the network for direct access to all devices. To ensure secure operations, SFs will use firewalls between network segments to filter non-essential traffic. Figure 2b shows an example topology of such an open, interconnected SF network. In this example, the server of the highest network segment can directly access the sensor data of any layer, even the lowest layer, without a controller in between. However, the increasing communication distance from the local data center to the field level adds potential traffic latency. Hence, the network must verify the compliance of the (QoS) requirements.

The QoS requirements of an application define the required bandwidth and latency between controllers and end devices. These QoS requirements do not change during this transformation, as they depend on the physical process. However, with increasing interconnectivity, the distances between devices and the amount of traffic in machine networks increases. The network configuration must ensure stable operation for the novel deployment of time-critical applications on a server distant from the process. Without additional configuration, these long communication paths potentially harm the QoS requirements. To ensure converged networks and the required level of QoS requirements, the IEEE 802.1 develops a set of Time-Sensitive Networking (TSN) standards to offer solutions for networks sharing multiple and different purposes in industrial environments [6], such as control processes, video surveillance, and best-effort applications. Mechanisms, such as Time-Aware Shaper (TAS) [7], Frame Preemption [8], and Per-Stream Filtering and Policing [9] offer the capability to ensure the according guarantees. DetNet [10] is a series of IETF standards building a QoS framework around Time-Sensitive Networking (TSN) and QoS mechanisms for other data planes.

The transformation towards growing interconnectivity does not only add value to SF networks but also introduces problem areas. Specifically, this transformation causes (a) an *increased attack surface*, (b) a *convoluted configuration complexity*, and (c) the *introduction of non-static network configurations*. In the following paragraphs, we detail the cause of these three problem areas.

The growing interconnectivity is responsible for an *increased attack surface*, i.e., the growing amount and density of vulnerabilities in a network an attacker can exploit. The attack surface increases with every additional device reachable by an attacker, as potential vulnerabilities are accessible. To reduce this risk, SFs introduce firewalls between network segments to enable the filtering of non-accessible services. Still, allowed traffic can pass through the firewalls. Specifically, the communication between the sensors and actuators through monitoring and control protocols is allowed. Hence, attackers can interact with the available devices and exploit the vulnerabilities in the accessible services and protocols.

With a growing number of mechanisms to configure on network devices, e.g., (VLANs), TAS, and rate limiting, the number of feature combinations and configuration parameters in the network increases. The *convoluted configuration complexity* stems from the interdependency of mechanisms on one node, as well as the interdependency of configurations across multiple devices. For example, the configuration of a VLAN defines the forwarding path for all traffic within that VLAN and the TAS mechanism the time-based access for the transmission of all traffic classes. Both mechanisms require the configuration of the other as input. Hence, they depend on each other. In meshed networks, the VLAN mechanism enables the reduction of path choices to increase the performance of redundancy protocols. From a security point of view, the VLAN mechanism reduces the reachability of devices in a network. Hence, using VLANs in SF networks is favorable. For the protection of network resources and to achieve high QoS performance, the TSN mechanism TAS [7] allows to use time division multiple access to network resources. However, scheduling the transmission of critical traffic to avoid congested network resources is an NP-hard problem [11–13]. Additionally, the configuration of the TAS mechanism or rate limiting is dependent on the VLAN configuration and requires the exact forwarding path of the critical traffic as input and hence, is dependent on the VLAN configuration. Therefore, the application of such new mechanisms increases the configuration complexity. In network segments of current industrial networks, typically, only one controller communicates with sensors and actuators in an isolated network. Due to the lack of additional low-priority communication, the resources of a network segment, e.g., bandwidth, can exclusively be used for critical traffic. Hence, the VLAN or TAS mechanisms are redundant, as no resources must be protected.

Applications require the network to be configured for their applicational needs, such as timing constraints and bandwidth requirements, before they start communicating. Typically, in small, isolated networks, an application runs on a single controller. Hence, the application announces the requirements to enable the network administrator or the administration tool to configure the network before this single application starts. However, in interconnected networks, not all machines start simultaneously, as they might communicate across multiple network segments and are independent of each other. This delayed startup of independent machines results in additional application needs, which were unavailable during the network configuration. Hence, the growing interconnectivity necessitates the *introduction of non-static network configurations*, i.e., the reconfiguration of network configurations during runtime, which is a problem area with unique challenges. For example, a network configuration for an application with strict timing requirements needs modifications to its TAS configuration. If an application sends data without adjusted network configurations, it competes with all other applications for the configured resources, causing application failures. However, all applications require reconfiguration by the configuration entity during the operational phase of the network without disturbing existing traffic. Specifically, there are two kinds of possible disturbances: (1) a configuration change to a feature that requires the switch to reload the complete configuration, limiting the capability to forward traffic in this time, and (2) a new application requiring resources in a network segment without any resources left. Hence, the network must reroute existing traffic to other network segments to ensure enough resources for the new application.

2.2 Application of Expanding Multi-tenant Capability

SFs offer flexible production environments for novel manufacturing techniques and service models. Specifically, an SF focuses on the modular approach and flexible use of production capacities and capabilities. These production-environment changes enable the use of new network architectures to cope with the arising challenges. Figure 3a presents a state-of-the-art industrial network with four isolated network segments owned by a single user. Each network segment performs its specialized task. The transition towards SFs decouples network architecture from specialized functionality. A novel service model, achievable by SFs, is the leasing of capacities for customized and on-demand production. With this leasing concept, multiple users access the industrial network simultaneously. In parallel, new network architectures break the hierarchical approach and connect devices to a central industrial backbone based on physical location and not based on association with a machine. For example, Fig. 3b highlights this possible expanding multi-tenant capability. Multiple users share the network where OT devices are connected to the nearest switches. The focus of this novel connection strategy is the reduction of cabling and active network infrastructure. In comparison, the existing hierarchical infrastructure use parallel network infrastructures for specific purposes to ensure functional safety and security. The convergence of the network for multiple purposes in SFs requires mechanisms, such as TSN, to protect the different traffic types and users on the wire. As introduced in Sect. 2.1, TSN achieves this convergence by allocating the necessary network resources exclusively for dedicated traffic classes.

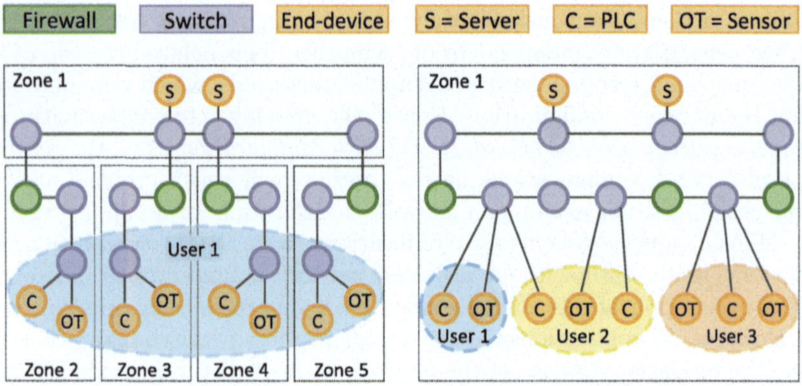

(a) Isolated network segments with one user

(b) Shared network infrastructure with multiple users.

Fig. 3. Expanding multi-tenant capability

One enabling concept for the transformation factor *expanding multi-tenant capability* are virtualized controllers due to their support of flexible and modular applications [14]. This flexible deployment of applications is defined by distributed control standards, such as IEC 61499[15]. With a flexible deployment of control functionality, the assignment of sensors and actuators to this control functionality is more flexible. Hence, sensors and end devices are not bound to a special-purpose machine anymore: Virtual controllers running on servers can integrate the required end devices into their control algorithm. Combined with a flexible leasing model for the SF, the control traffic changes over time. This dynamic application deployment introduces an *increased congestion risk* on switches and firewalls in the network. Previously, network infrastructure was reserved exclusively for a special-purpose machine without additional management of resources. In SFs, network-infrastructure resources are not guaranteed for all applications and must be reserved before executing control functionality in shared network infrastructure. This dynamic reservation requires flexible reconfiguration of TSN mechanisms, but also security configurations, such as firewall rulesets.

Additionally, *expanding multi-tenant capability* increases the configuration complexity. Specifically, the extensive use of TSN mechanisms, especially TAS, introduces a complex technology with strict timing requirements during the operation, as described in Sect. 2.1. Devices communicating regularly with each other are no longer necessarily located in the same network segment, as the non-hierarchical topology of SFs enables broader access to end devices. Hence, by corrupting a central network segment, attackers can easily gain access to all other systems, which increases the *attack surface* (see Sect. 2.1).

2.3 Application of Increasing Application Dynamics

Current industrial networks use dynamic applications that adapt to the changing production needs of a factory during operation. These dynamic applications enable the production of customized products, such as customer-specific cars with special features. Dedicated machines install such features with a dedicated PLC and OT devices in a separate network. Exemplary, Fig. 4a shows two machine networks, i.e., zone 2 and zone 3. Each network contains an access point, enabling wireless communication between the corresponding machine and the machine-dependant mobile OT devices, e.g., a barcode scanner. The transparent OT device with dashed lines symbolizes the potential movement of the OT device within the range of the machine without losing the connection. Compared to current networks, SFs enable new production techniques, where the product is fixed and tools move around. Hence, tools can move from one machine network to another. Figure 4b shows this *increasing application dynamic* where an OT device moves from the left network segment (zone 2) to the right network segment (zone 3). For example, a static production material, such as a vehicle, stays within the zone, but all the tools for the different manufacturing steps move between the zones. Compared to the use-case of AGVs or drones, these tools are controlled by the controllers in different machines. AGVs do not change the controller during operation but are either self-driving or have a fixed centralized controller.

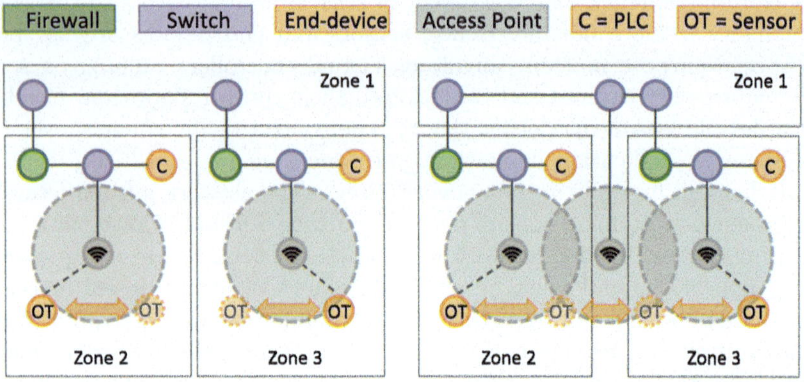

(a) Static network participants and applications.

(b) Mobile devices and dynamic activation of applications.

Fig. 4. Increasing application dynamics

The *increasing application dynamics* require network reconfigurations, depending on the current association of clients and access points. As described in Sect. 2.1, this transition challenges SFs to reconfigure network resources seamlessly when mobile devices transition to another access point. The *increasing application dynamics* make it impossible to configure the network statically. Hence, the network must adapt to the changing communication paths, e.g., by dynamically adjusting the configuration on the path through the network, which introduces the problem area of non-static network configurations. Similarly, the transition to other access points increases the risk of congestion, as described in Sect. 2.2.

2.4 Overview of Challenges Due to the Transformation Factors

The previous three sections on the transformation factors present four different problem areas of current mechanisms in industrial networks that arise during the transformation towards SFs while meeting the requirements of security, robustness, and performance: (1) *Increased attack surface*, (2) *growing complexity of configuration*, (3) *increased congestion risk*, and (4) *introduction of non-static network configurations*. Table 1 highlights these problem areas, emphasizing the transformation factors causing them and the industrial requirements these problem areas affect. Specifically, all identified problem areas challenge multiple requirements simultaneously. In the following, we summarize the causes and impacts of each problem area based on Table 1.

Table 1. Problem areas for industrial requirements caused by SF transformation factors; (R:S), robustness (R:R), performance (R:P), growing interconnectivity (T:C); expanding multi-tenant capability (T:T), increased application dynamics (T:D), "●": affected, "○": unaffected, "→": introduce, "→": violate

Transformation factors				Problem area		Industrial requirements		
T:C	T:T	T:D				R:S	R:R	R:P
●	●	○	→	Increased attack surface	→	●	●	○
●	●	○	→	Convoluted configuration complexity	→	●	○	●
○	●	●	→	Increased congestion risk	→	●	●	●
●	○	●	→	Introduction of non-static network configurations	→	○	●	●

Increased attack surface originates due to expanding multi-tenant capability and growing interconnectivity of SFs. The expanding multi-tenant capability introduces multiple actors accessing and sharing the same underlying network, cf. Sect. 2.2. The growing interconnectivity results in more reachable and potential attack targets exploitable by infected or compromised devices, cf. Sect. 2.1. Consequently, both transformation factors increase the attack surface, providing more possibilities for malicious actors to attack and disrupt the communication, which challenges the security and robustness requirement.

Convoluted configuration complexity is the result of growing interconnectivity and expanding multi-tenant capability in SFs. The number of communication paths between end devices grows, and the number of complex mechanisms on those devices increases. Hence, interdependencies of features occur, as features of a singular device may impact each other and features of other connected devices, cf. Sect. 2.1. Additionally, the expanding multi-tenant capability introduces complex technologies like TSN with strict timing requirements, cf. Sect. 2.2. Both transformation factors introduce new, complex mechanisms or features into industrial networks with partially non-obvious implications. These implications challenge security and performance, as networks with high configuration complexity are prone to misconfiguration causing security gaps and bad performance.

Increased congestion risk is caused by the increasing application dynamics and expanding the multi-tenant capability of SFs. Due to increasing application dynamics, more mobile devices share the same access points, increasing the load on these access points, cf. Sect. 2.3. Similarly, switches and firewalls run at risk of congestion due to expanding multi-tenant capability, as a lot of traffic related to specific applications passes the same nodes, cf. Sect. 2.2. Thus, both transformation factors increase the congestion risk in a network, which is a challenge for all industrial requirements.

Introduction of non-static network configurations is the last problem area we identified. This problem area is introduced because of growing interconnectivity and increasing application dynamics. Due to growing interconnectivity, there is a higher probability of time deviation at the operation start of different applications. However, no configuration entity may disturb existing traffic, requiring a non-static network configuration, cf. Sect. 2.1. Similarly, due to increasing application dynamics, streams frequently change their paths, which also necessitates a non-static network configuration, cf. Sect. 2.3. If carried out wrongly, non-static network configuration may cause latency or even loss of important time-sensitive packets. Hence, the introduction of non-static network configuration challenges the robustness and performance of a network.

3 Existing Solutions for Identified Problem Areas

SFs transform industrial networks. As detailed in Sect. 2, this transformation leads to different problem areas. In this section, we present and discuss existing solution approaches for these four problem areas. Table 2 provides an overview of the four identified problem areas and solutional approaches to addressing them. The table also indicates the negative and positive impacts of each solution on the industrial requirements security, robustness, and performance.

Table 2. Impact of existing solutions for identified problem areas of industrial requirements depicted in Table 1 due to the outlined smart factory transformation factors; security (R:S), robustness (R:R), performance (R:P); "↑": positive impact, "↓": negative impact, "–": no impact

Solutional approaches for identified problem areas	R:S	R:R	R:P
Increased attack surface			
Use of additional firewalls in industrial control systems [4,16]	↑	↓	↓
Extending intrusion detection systems to control traffic [17,18]	↑	↓	↓
Convoluted configuration complexity			
Optimization of FW configurations [19–22]	–	–	↑
Traffic schedule calculations [11–13,23]	↓	–	↑
Increased congestion risk			
Next-generation wireless standards [24–27]	–	↓	↓
On-the-fly resource allocation [28–31]	↓	–	↑
Introduction of non-static network configurations			
Seamless configuration rollout [32,33]	–	↑	↑
Intent-based networking [34–36]	↑	↓	↑

3.1 Increased Attack Surface

Due to the transformation towards SFs, the attack surface of industrial networks increases. However, there exist several solutions to minimize the attack surface, e.g., (a) the placement of additional firewalls inside the network to limit network access and prevent malicious actors, and (b) the utilization of intrusion detection systems to detect intruders inside the network. In the following sections, we outline these two solutions and their shortcomings, i.e., their negative impact on the previously outlined industrial requirements.

3.1.1 Use of Additional Firewalls in Industrial Control Systems As described in Sect. 2.1, PLC will not act as gateways between network zones anymore. Instead, networks will be interconnected with devices forwarding traffic, e.g., switches or routers, thus opening the network for more interconnectivity of devices in SFs. However, this approach increases the potential attack surface for attackers. Hence, researchers suggest replacing gateways with FW as conduits between network zones instead of switches and routers. These conduits define all allowed traffic exchanged between different network zones [4] and enable the controlled opening of networks. Hence, minimizing the overall attack surface.

Critically, due to the reorganization of network topologies in SFs, this negatively affects nearly all time-sensitive traffic, e.g., control data for robots from virtualized PLC, as nearly all traffic now passes at least one FW, which usually leads to additional latency on forwarded traffic [16]. Hence, finding cost-efficient FW that meets performance requirements is crucial to successfully transform towards SFs.

3.1.2 Extending Intrusion Detection Systems to Control Traffic Intrusion detection systems (IDSs) are an essential part of securing a network and helping to fulfill the security requirement of industrial networks [18]. (IDSs) increase the security of the network by monitoring incoming traffic for suspicious activities, e.g., a device suddenly receives more traffic or receives traffic from different addresses. IDSs collect the required information input on the basis of strategically placed devices, i.e., sensors, inside the network. These sensors passively collect the required information, e.g., network packets, and either store or forward it to a host where they can be further analyzed. However, depending on the timing of the analysis, i.e., after forwarding or before forwarding a packet, or whether deep packet inspection is utilized, latency may arise on passing traffic [17]. Hence, similar to FW, IDSs may lower the performance of a network.

Additionally, the growing interconnectivity and expanding multi-tenant capability pose problems for existing IDSs. Both result in additional traffic, requiring extra bandwidth or network routes to forward the traffic to the analysis station. Consequently, lowering the performance of the network due to additional network traffic and less available bandwidth, which may result in congestion.

Furthermore, IDSs often rely on a baseline phase to determine the known good behavior of a network. Thereafter, deviations to the established default behavior, i.e., the baseline, may indicate an attack. However, the increasing application dynamics of SF prevent the establishment of a reliable baseline. Differentiating malicious traffic from valid traffic becomes increasingly difficult, which increases the number of false positives. These false alarms decrease the robustness of the network as valid traffic might become blocked by wrongly applied prevention mechanisms. Additionally, detecting effective sensor positions inside the network becomes more challenging due to the increasing application dynamics. Thus, suboptimally placed sensors may lead to missed attacks.

3.2 Convoluted Configuration Complexity

SF networks require administrators to configure a multitude of features with interdependencies. However, several procedures and tools exist, easing this process by lowering the complexity of necessary calculations, e.g., the (a) optimization of firewall configurations and (b) traffic schedule calculations. In this section, we inspect these solutions and examine their weaknesses in relation to SFs.

3.2.1 Optimization of Firewall Configurations As discussed in Sect. 3.1, researchers currently plan to integrate more firewalls into networks during the transformation to SFs in order to minimize the existing attack surface. A central concern in relation to this approach is the poor performance of FW, i.e., the latency caused by FW on forwarded traffic. To mitigate this latency, researchers already proposed a variety of different optimization approaches. For example, Katic et al. [22] present an optimization algorithm to minimize the number of configurations in a firewall to reduce the necessary checks before a packet is accepted or dropped. The optimizations in [19–21] additionally prioritize specific configurations with high hit counts, i.e., a high number of matching packets to a firewall configuration. In general, these optimization strategies minimize latency without negatively impacting security or robustness. However, as measured by Wüsteney et al. [16], most latency does not occur due to the number of predefined firewall configurations but due to the operation of a firewall application itself. Consequently, the presented firewall-optimization solutions are insufficient for SF implementation as latency requirements remain unfulfilled.

3.2.2 Traffic Schedule Calculations Traffic schedule calculation solves the convoluted configuration complexity problem of specific QoS features arising due to the transformation towards SFs. Traffic schedule calculation for TSN traffic provides suitable configurations, e.g., TAS schedules, and provides the necessary QoS guarantee for time-critical traffic. However, traffic schedule calculation is an NP-hard problem [37]. Therefore, finding suitable schedules in efficient time is an active research area tackled by multiple research groups, some of which include [11–13]. Current research can calculate and model configurations for heterogeneous networks with a unified feature set. Since we expect time-critical traffic across network segments with heterogeneous TSN features in SFs, we contribute a model combining network segments with different TSN-mechanisms and configurations in [23].

Even though existing research provides suitable traffic schedules, it fails to include security mechanisms such as firewalls or (VPNs) in addition to calculating QoS guarantees. Hence, existing research creates large and unprotected networks, which negatively impacts the security of the network. The traffic schedule calculation must consider these security mechanisms to prevent a negative impact on the security of the network.

3.3 Increased Congestion Risk

SFs originate from new and existing technologies, creating highly flexible networks. However, at certain points in SFs, congestions occur more easily than in current networks. There are several approaches to resolve these congestions, e.g., (a) next-generation wireless technology and (b) on-the-fly resource allocation, of which we consider their advantages and disadvantages in this section.

3.3.1 Next-Generation Wireless Technology As wireless technology uses a shared medium, its robustness and performance decrease compared to wired technology. Especially in SFs, deploying more mobile devices, such as AGVs or drones, increases communication density on the wireless medium, leading to a decreasing performance. Next-generation wireless technologies such as WiFi 6 [27] provide new features like orthogonal frequency division multiple access, new modulation, and additional frequency bands. These novel features reduce the increased congestion risk in dynamic and dense mobile applications. Similarly, the new cellular standard 5G [24] enhances cellular communication to differentiate multiple traffic types, implement service level agreements, and schedule transmissions to achieve ultra-low latency. Ambrosy et al. [25] evaluate the performance and reliability of 5G for SFs. For WiFi 6, Mozaffariahrar et al. [26] present these new features and explain their capabilities regarding security, robustness, and performance.

3.3.2 On-the-Fly Resource Allocation Networks for SFs need to serve several traffic classes of multiple users simultaneously. For example, two customers run a robot control and a predictive maintenance application in the same SF network. None of these applications shall fail, requiring knowledge on available and reserved resources. This configuration requires the end devices to announce their QoS requirements to the network for appropriate reconfiguration, as described in Sect. 2. The Resource Allocation Protocol (RAP) serves this purpose in a distributed manner and is in the process of standardization in the draft IEEE P802.1Qdd [28]. Similarly, IEEE Std. 802.1Qdj [29] specifies this mechanism in a centralized manner. For dynamic and open networks, the distributed approach offers more flexibility despite the lack of computational power for the resulting configuration. Osswald et al. [31] proposes a mechanism to connect centralized and hybrid devices. The hybrid RAP protocol does not enhance or limit the reliability of the network but strengthens the performance of the network. Still, any operation of RAP is insecure, as the protocol does not implement security mechanisms to protect against integrity and confidentially attacks. Hence, in Osswald et al. [30], we propose to extend the RAP protocol with security measures called Secure RAP, or SecRAP.

3.4 Introduction of Non-static Network Configurations

SFs cause traffic behaviours requiring on-the-fly adjustments to configurations during runtime. Some solutions for this behavior already exist, e.g., (a) seamless configuration rollout and (b) traffic schedule calculations. In this section, we inspect these solutions and examine their strengths and weaknesses.

3.4.1 Seamless Configuration Rollout Changes in network configuration on multiple devices require caution during rollout. For example, if configuration changes affect device reachability, the configuration sequence is important to finish the rollout successfully. Currently, Simple Network Management Protocol (SNMP) [38] is the leading protocol for network management in the industry. Nevertheless, SNMP has known issues, specifically regarding the security and operations of larger networks. Consequently, the new profile for industrial networks, IEC/IEEE 60802 [6], defines the use of Network Configuration Protocol (NETCONF) [32], which is already commonly used in the IT, as the successor of SNMP. This transition adds better transport security and authentication in network management and solves the problem of configuration rollout order. With SNMP, the management tool or administrator decide on the order to reconfigure devices such that all devices remain reachable. NETCONF introduces a transaction model across multiple network devices, which reduces the number of unreachable devices after reconfiguration and is quicker for larger networks. However, many *yang* models, which are necessary for the data modeling in NETCONF, are still missing for the industry to apply NETCONF to all devices. Additionally, legacy devices either do not receive updates at all or do not support the more resource-consuming NETCONF stack. Hence, Yu et al. [33] present the idea of a NETCONF-gateway, which mediates between legacy devices and a NETCONF-manager. This approach reintroduces the problem of defining a sequence for the device configuration rollout solely on the gateway and not the manager. In summary, if all *yang* models are available for industrial needs, NETCONF can increase the security, robustness, and performance of industrial networks, but until then, NETCONF is an incomplete solution.

3.4.2 Intent-Based Networking Stream and traffic configuration is a complex task for administrators in SFs due to necessary reconfigurations during runtime caused by, e.g., frequently changing positions of stream endpoints and stream requirement adjustments. For example, a particularly critical stream s requires guaranteed bandwidth. Hence, an administrator configures a shaping and policing mechanism for stream s on all the switches on the path of s and prevents other traffic from consuming the required resources through rate limiting. Hence, if the bandwidth requirements of stream s change, the administrator must also adapt all reserved and allowed bandwidth configurations of stream s on all switches and routers along the path of stream s. This scenario becomes even more complex if one or more switches cannot configure bandwidths per stream. In such a case, an administrator must aggregate streams. Then, if a single stream s must be reconfigured, each aggregation of stream s with other streams must be adjusted as well.

To ease such processes for administrators, intent-based networking (IBN) enables a reduction of configuration complexity by expressing intents end-to-end. Tools implementing IBN cover the complexity and track configuration options. One of the most dominant solution spaces in IBN is software-defined networking (SDN) [34]. SDN defines an architecture with a centralized controller handling all switch configurations. These configurations are often more detailed than in typical Ethernet networks, as the network does not learn addresses independently. Instead, the controller programs all forwarding decisions and is aware of all configurations in the network. Other approaches to IBN do not require such a detailed form of switch hardware configuration [35,36]. However, the approach of centralized entities controlling all configurations is a risk for robustness. Nevertheless, a tool-based end-to-end configuration increases security as fewer errors due to forgotten configuration options occur.

4 Conclusion

In this paper, we explore the transformation from traditional industrial networks to SFs, emphasizing the fundamental industrial requirements *security*, *robustness*, and *performance*. Our analysis highlights the transformation factors *growing interconnectivity*, *expanding multi-tenant capability*, and *increasing application dynamics* of SFs revealing significant challenges in maintaining the requirements amidst evolving network conditions. We identify four problem areas challenging multiple requirements simultaneously: (1) *Increased attack surface*, (2) *convoluted configuration complexity*, (3) *increased congestion risk*, and (4) *introduction of non-static network configurations*. For all of these problem areas, we examine existing solutions and their capability to solve the complete transformation towards SFs. Our examination reveals that current solutions insufficiently address the problem areas and often negatively impact the *security*, *robustness*, and *performance*. This gap expresses the need for continuous research and innovative approaches to develop robust, secure, and high-performing industrial networks able to seamlessly adapt to the demands of modern, flexible automation environments. Future work should focus on integrated solutions that holistically consider all critical factors to facilitate a successful transition towards advanced industrial network paradigms.

Acknowledgments. This work has been funded by the Deutsche Forschungsgemeinschaft (DFG, German Research Foundation)—Project-ID 528745080—FIP 68. Additionally, this paper was created through research supported by the project "FHP: Qualifizierung und Entwicklung des professoralen Personals der Hochschule Esslingen fr zukunftsweisende Themen" (FKZ: 03FHP115). The project is part of the Federal States Program "FH-Personal", funded by the BMBF and MWK Baden-Wrttemberg. The authors alone are responsible for the content of the paper.

References

1. Carugi, M., Lou, Z.: Future industrial networks: requirements, challenges, research and standardization needs. In: ITU Kaleidoscope: Connecting Physical and Virtual Worlds (ITU K), IEEE (2021)
2. Mabkhot, M.M., Al-Ahmari, A.M., Salah, B., Alkhalefah, H.: Requirements of the smart factory system: a survey and perspective. Machines **6**(2) (2018)
3. David R. Sjödin, Vinit Parida, M.L., Petrovic, A.: Smart factory implementation and process innovation. Res. Technol. Manage. **61**(5) (2018)
4. International Electrotechnical Commission (IEC): IEC 62443-3-2. Security for Industrial Automation and Control Systems—Part 3-2: Security Risk Assessment for System Design (2020)
5. Williams, T.: The purdue enterprise reference architecture. Proceedings of the Triennal Wold Congress of the International Federation of Automatic Control (IFAC) **26**(2), 559–564 (1993)
6. IEEE 802.1: IEC/IEEE 60802 TSN Profile for Industrial Automation. IEC/IEEE 60802 D2.4 (2024)
7. Jeffree, T.: IEEE Standard for Local and Metropolitan Area Networks—Bridges and Bridged Networks—Amendment 25: Enhancements for Scheduled Traffic. IEEE Std. 802.1Qbv-2015 (2016)
8. Jeffree, T.: IEEE Standard for Local and Metropolitan Area Networks—Bridges and Bridged Networks—Amendment 26: Frame Preemption. IEEE Std. 802.1Qbu-2016 (2016)
9. Jeffree, T.: IEEE Standard for Local and Metropolitan Area Networks—Bridges and Bridged Networks—Amendment 28: Per-Stream Filtering and Policing. IEEE Std. 802.1Qci-2017 (2017)
10. Finn, N., Thubert, P., Varga, B., Farkas, J.: Deterministic Networking Architecture. RFC 8655 (2019)
11. Hellmanns, D., Glavackij, A., Falk, J., Hummen, R., Kehrer, S., Dürr, F.: Scaling TSN scheduling for factory automation networks. In: IEEE International Conference on Factory Communication Systems (WFCS), Porto, Portugal (2020)
12. Schweissguth, E.B., Danielis, P., Timmermann, D., Parzyjegla, H., Mühl, G.: ILP-based joint routing and scheduling for time-triggered networks. In: Proceedings of the International Conference on Real-Time Networks and Systems (RTNS), Grenoble, France (2017)
13. Stüber, T., Eppler, M., Osswald, L., Menth, M.: Performance comparison of offline scheduling algorithms for the time-aware shaper (TAS). IEEE Trans. Ind. Inf. (2024)
14. Cruz, T.J., Queiroz, R., Simões, P., Monteiro, E.: Security implications of SCADA ICS virtualization: survey and future trends. In: European Conference on Cyber Warfare and Security (ECCWS), Munich, Germany (2016)
15. International Electrotechnical Commission (IEC): IEC 61499-1. Function Blocks—Part 1: Architecture (2012)
16. Wüsteney, L., Menth, M., Hummen, R., Heer, T.: Impact of packet filtering on time-sensitive networking traffic. In: IEEE International Conference on Factory Communication Systems (WFCS), Linz, Austria (2021)
17. Hu, Q., Yu, S.Y., Asghar, M.R.: Analysing performance issues of open-source intrusion detection systems in high-speed networks. J. Inf. Security Appl. **51** (2020)
18. Wu, M., Moon, Y.B.: Intrusion detection system for cyber-manufacturing system. J. Manuf. Sci. Eng. **141**(3) (2019)

19. Acharya, S., Wang, J., Ge, Z., Znati, T.F., Greenberg, A.: Traffic-aware firewall optimization strategies. In: IEEE International Conference on Communications, Istanbul, Turkey (2006)
20. Fulp, E.W.: Optimization of network firewall policies using directed acyclical graphs. In: Proceedings of the IEEE Internet Management Conference (2005)
21. Hamed, H., Al-Shaer, E.: Dynamic rule-ordering optimization for high-speed firewall filtering. In: ACM Symposium on Information, Computer and Communications Security, Taipei, Taiwan (2006)
22. Katic, T., Pale, P.: Optimization of firewall rules. In: International Conference on Information Technology Interfaces, Cavtat, Croatia (2007)
23. Wüsteney, L., Hellmanns, D., Schramm, M., Osswald, L., Hummen, R., Menth, M., Heer, T.: Analyzing and modeling the latency and jitter behavior of mixed industrial TSN and DetNet networks. In: Proceedings of the International Conference on emerging Networking EXperiments and Technologies (CoNEXT), Roma, Italy (2022)
24. Release 15: 3rd Generation Partnership Project (3GPP) (2019)
25. Ambrosy, N., Kampa, T., Jumar, U., Großmann, D.: 5G and DetNet: towards holistic determinism in industrial networks. In: IEEE International Conference on Industrial Technology (ICIT), Shanghai, China (2022)
26. Mozaffariahrar, E., Theoleyre, F., Menth, M.: A survey of Wi-Fi 6: technologies, advances, and challenges. Future Internet **14**(10), 293 (2022)
27. Stacey, R.: IEEE Standard for Information Technology—Telecommunications and Information Exchange between Systems Local and Metropolitan Area Networks—Specific Requirements Part 11: Wireless LAN Medium Access Control (MAC) and Physical Layer (PHY) Specifications Amendment 1: Enhancements for High-Efficiency WLAN. IEEE Std. 802.11ax-2021 (2021)
28. Chen, F.: IEEE Standard for Local and Metropolitan Area Networks—Bridges and Bridged Networks—Amendment: Resource Allocation Protocol. IEEE P802.1Qdd Draft 0.9 (2024)
29. Kehrer, S.: IEEE Standard for Local and Metropolitan Area Networks—Bridges and Bridged Networks—Amendment 38: Configuration Enhancements for Time-Sensitive Networking. IEEE Std. 802.1Qdj-2024 (2024)
30. Osswald, L., Lindner, S., Bechtel, L., Heer, T., Michael, M.: Secure Resource Allocation Protocol (SecRAP) for time-sensitive networking. In: IEEE International Conference on Emerging Technologies and Factory Automation (ETFA), Padova, Italy (2024)
31. Osswald, L., Lindner, S., Wüsteney, L., Michael, M.: RAP extensions for the hybrid configuration model. In: IEEE International Conference on Emerging Technologies and Factory Automation (ETFA), Västerås, Sweden (2021)
32. Enns, R., Björklund, M., Bierman, A., Schönwälder, J.: Network Configuration Protocol (NETCONF), RFC 6241 (2011)
33. Yu, J., Al Ajarmeh, I.: An empirical study of the NETCONF protocol. In: IEEE International Conference on Networking and Services (ICNS), Cancun, Mexico (2010)
34. Haleplidis, E., Pentikousis, K., Denazis, S., Salim, J.H., Meyer, D., Koufopavlou, O.: Software-Defined Networking (SDN): Layers and Architecture Terminology. RFC 7426 (2015). https://doi.org/10.17487/RFC7426, https://www.rfc-editor.org/info/rfc7426
35. Leivadeas, A., Falkner, M.: A survey on intent-based networking. IEEE Commun. Surv. Tutorials **25**(1), 625–655 (2022)

36. Zeydan, E., Turk, Y.: Recent advances in intent-based networking: a survey. In: IEEE Vehicular Technology Conference, Antwerp, Belgium (2020)
37. Dürr, F., Nayak, N.G.: No-wait packet scheduling for IEEE time-sensitive networks (TSN). In: Proceedings of the International Conference on Real-Time Networks and Systems (RTNS), New York, USA (2016)
38. Case, D.J.D., Mundy, R., Partain, D., Stewart, B.: Introduction and Applicability Statements for Internet Standard Management Framework, RFC 3410 (2002)

Methods of Laser Beam Shaping and Its Advantages in Metal Laser Additive Manufacturing: A Review

Florian Markowski and Lukas Löber(✉)

Faculty of Mechanical and Systems Engineering, Esslingen University of Applied Sciences, 73728 Esslingen, Germany
{florian.markowski,lukas.loeber}@hs-esslingen.de

Abstract. Industry is increasingly using additively manufactured components in the field of laser powderbed fusion (L-PBF). When it is necessary to manufacture complex component geometries in small quantities, common processes like L-PBF come into play. Often, the costs of subtractive, conventional manufacturing processes and the expensive post-processing of the workpiece are higher than when using a generative build-up. However, the use of these conventional manufacturing processes remains common. This is due to the extensive knowledge and research on the mechanical properties and material behavior during processing. In contrast, the material behavior of the metal powder is significantly and seemingly uncontrollably influenced by high heating and cooling rates, dynamic melt pool behavior and spatter formation. The standard Gaussian energy density distribution of a laser beam has certain physical limits when processing powder material. This is where beam shaping comes into play. Specially manipulated laser beam intensity distributions have a direct influence on the behavior of the molten pool, making the mechanical properties of the material more controllable. This paper, therefore, deals with various methods of beam shaping. This paper summarizes potential defects in an additively manufactured component and demonstrates how shaped laser beams can minimize these defects. The last chapter shows a schematic test setup that should lay the foundation for future research on this topic.

Keywords: Laser additive manufacturing · Beam shaping · Melt pool behavior

1 Introduction

In production technology, additive manufacturing has ushered in a revolutionary development. Specifically, it enables the production of intricate components within the metal processing industry. One example of this is conformal, very fine cooling channels that can be integrated directly into the component without complicated reworking using conventional manufacturing methods [1]. There are now over 20 different processes for additive processing of metallic materials [2]. Among them, the L-PBF process is the most established. It is based on the layer-by-layer buildup of metal parts formed by

A. Öchsner (ed.), *Collaborative Research Advancing Engineering Solutions for Real-World Challenges 2*, Proceedings in Technology Transfer,
https://doi.org/10.1007/978-3-031-79093-5_9

selectively melting fine metal powders. The laser radiation's energy input ensures melting. An uncoated build platform serves as the starting point, allowing for layer-by-layer lowering during the processing time. A so-called squeegee pushes the first layer of metal powder over the build platform. The laser beam precisely melts the areas that correspond to the cross-section of the manufactured part by hitting the powder at the points where the first layer of the component needs to solidify. This process is repeated layer by layer until the complete component has been created. Despite the rapid technological development of the process, there are still challenges in terms of surface quality, process efficiency and the handling of the molten pool under the influence of laser radiation. The standard Gaussian energy density distribution of a laser beam has certain physical limits when processing powder material. One disadvantage, for example, is the low energy utilization of such a beam. This can lead to defects such as spatter formation or porosity [3]. This is where beam shaping comes in. Specially manipulated intensity distributions of the laser beam have a direct influence on the behavior of the molten pool and thus make the mechanical properties of the material more controllable. There are two approaches to manipulating the intensity distribution of a beam: static beam shaping and dynamic beam shaping. In static beam shaping, physically fixed refractive or diffractive optical elements are used to shape a beam. This method is robust and cost-effective, but it has disadvantages in terms of flexibility. In contrast, dynamic beam shaping enables real-time beam adjustment during the ongoing process by using adaptive optics. Adaptive optics include deformable mirrors or spatial light modulators that can change the amplitude, phase or polarization of the incident light. These properties can be adjusted almost freely, depending on the desired process requirements. The integration of such beam shaping technologies into the L-PBF technology opens up new dimensions in additive manufacturing process optimization. In the molten pool, precise control of the energy distribution can significantly improve both the surface quality in the first step and the mechanical properties in the second step. The targeted control of the energy distribution minimizes thermal gradients and thus reduces internal stresses in the component. This reduces the risk of cracking. Additionally, by optimizing the energy input at specific points of the component, we can optimize productivity and energy utilization. This leads to a decrease in energy consumption and an increase in the buildup rate.

As a result, the first two sections of this paper deal with beam-shaping technology and its advantages for additive manufacturing of metallic components. These sections are to be understood as a review and lead to the fourth section, Experimental setup for future investigations and outlook. This chapter deals with the schematic experimental setup, which forms the basis for future investigations on metal samples.

2 Methods of Beam Shaping

As laser technologies continue to evolve, the development of advanced beam shaping techniques will undoubtedly contribute to more efficient, precise, and versatile laser-based systems. The following section addresses various beam shaping technologies and distinguishes between static and dynamic beam shaping methods.

Laser material processing is used in various manufacturing processes. Particularly in the fields of cutting (laser cutting) and joining (laser welding), it is important to be able to precisely control the laser depending on the application. A key technology is to manipulate the intensity profile of the laser by static or dynamic beam shaping. Particularly in the field of powder bed-based laser melting, the spatial distribution of intensity leads to increased productivity and flexibility. When used correctly, the quality of the processed components is also optimized.

Laser beams normally have a Gaussian energy distribution. Due to the high energy density, this intensity distribution is particularly suitable for applications in the fields of laser cutting, welding or drilling. In the additive processing of metal powder, melt pool control plays an important role with regard to the mechanical properties of the component. The shape of the laser spot can have a significant influence on the melt pool behavior. Compared to a Gaussian beam, specially shaped beam profiles can make the energy input more uniform, reduce thermal gradients, influence spatter formation and minimize defects such as porosity and cracking. At the same time, the surface quality can be influenced. These aspects mean that the possibilities in this type of application can be expanded. Bakhtari et al. [4] have summarized a large selection of scientific work in this area in their publication. It can be seen that technical innovations in the field of beam shaping technology have made it increasingly interesting for scientists to integrate new technologies into the field of additive manufacturing. The following section therefore deals with static and dynamic beam shaping. This is followed by a discussion of different intensity profiles and their advantages and disadvantages in the application of additive manufacturing.

2.1 Static Beam Shaping

Static beam shaping is a fundamental technique in optics that involves manipulating the spatial distribution of the intensity of a laser beam using static optical elements such as lenses, mirrors, and other optical components. This process is crucial for applications that require precise control over the beam profile to optimize performance and efficiency. Static beam shaping can generally be divided into three different techniques [5]:

- Physical-static spatial filters (beam masks) that change the amplitude of the incoming laser beam and thus control the beam profile.
- Field mapping manipulates the phase of the incoming laser beam to achieve a specific intensity distribution.
- Beam integrators split the incoming laser beam into a multitude of individual beams that overlap at the output and form a desired shape.

The beam masks mentioned in the previous section represent a proven and practical approach to manipulating an incoming laser beam. One can use the mask to block specific parts of the beam, thereby generating a desired beam profile. The design of a beam mask can vary depending on the application. The simplest version can only ensure an on/off transmission of the light. However, some masks can even be programmed [6]. A major advantage of this technique is the high accuracy of the beam shapes it generates. However, the high loss of incoming laser energy is a disadvantage. In order to achieve a high-quality output shape, the incoming beam must be highly homogeneous. If the input laser has a Gaussian intensity distribution, the beam must be expanded before entering the mask in order to use the flat, central part of the Gaussian beam. Consequently, the Gaussian flanks lose their energy.

Field mapping devices shape the beam by refracting and reflecting (e.g. parabolic mirrors) the incoming laser beam. The input intensity can be transferred into a new form without significant losses. The basic principle of field mapping is to divide the laser beam into an array of smaller partial beams and to control their phase and amplitude independently. By adjusting the relative phases and amplitudes, the partial beams interfere constructively or destructively with each other, resulting in the desired beam shape in the output plane. This is usually achieved by refractive elements with free-form surfaces, for example by collimation or focusing lenses. In the meantime, however, diffractive optical elements (DOEs) have also become established in this technology, as they produce robust results, especially at high laser powers. DOEs manipulate the phase, amplitude and direction of the incoming laser beam by using microstructured surfaces. The function of DOEs is based on the diffraction of light that occurs when a wave encounters an obstacle or an opening. By designing the DOE surface with fine, periodic structures on the scale of the light wavelength, specific optical functions such as focusing, beam splitting or the generation of symmetrical intensity distributions can be achieved. Examples of beam profiles include inverse Gaussian, Bessel and top-hat beams [7]. A major advantage over beam masks is the high energy efficiency, which can be over 90% [4].

The integration or homogenization of a laser beam describes the breaking of a mostly Gaussian input beam into several smaller partial beams (with a desired characteristic intensity distribution). After exiting the beam-shaping optics, these are combined into an output form that can either be a closed intensity distribution or a multiple-point distribution—a so-called beam-splitter profile. This is done by redistributing the energy of the laser beam across its cross-section using a series of optical components and techniques that manipulate the spatial properties of the beam [8]. The choice of these techniques depends largely on the desired energy distribution, beam size, divergence, wavelength and process requirements. This technology also offers advantages in terms of high energy utilization. In addition, it is insensitive to changes in the input beam. However, one disadvantage is that phase displacements in the beam shaping area lead to contrast modulation. This is characterized by a grainy intensity modulation [9].

2.2 Dynamic Beam Shaping

Dynamic beam shaping differs from static beam shaping in that it uses real-time control of the direction, shape and intensity of laser beams. Whereas static beam shaping methods use fixed optical components to manipulate the beam, dynamic beam shaping relies on adaptive optics such as spatial light modulators (SLMs) [10], acousto-optical deflectors (AODs) [11] or deformable mirrors [12]. These techniques allow the phase, amplitude or polarization of the laser beam to be dynamically modified. For example, complex beam shapes of any form can be realized and uneven intensity distributions can be created with dynamic reconfigurations. This leads to a significant increase in the flexibility and adaptability of the process, especially in the field of L-PBF. The best-known SLM systems are the so-called liquid crystal on silicon SLMs (LCoS-SLMs). They are comparable to a light projector that shows a movie. They are programmable displays that can switch between different phase holograms. This allows them to determine an optimal beam shape and continuously vary it. The core component of an LCoS-SLM is a liquid crystal layer. An electric field can align these molecules in different directions. This changes the optical properties, such as the refractive index, of the liquid crystal. Two other layers, each serving as an electrode, sandwich the layer of liquid crystals. These electrodes are substrate plates coated with a transparent, conductive material (usually indium tin oxide (ITO)). Often, the entire assembly is sealed to protect the liquid crystal layer. Changing the refractive index modulates the phase of the incoming beam. By varying the voltage in different areas of the LCoS-SLM, different phase patterns can be generated, enabling precise control of the light wavefront. One advantage of this technology is the high spatial resolution of the modified beam. In addition, complex and dynamic light patterns can be modulated by separately controlling individual pixels [13].

Another dynamic beam shaping technique are microelectromechanical systems spatial light modulators (MEMS-SLMs). MEMS-SLMs consist of small movable micromirrors that can individually tilt or deform in response to electrical signals. The movement of these mirrors changes the angle of reflection, thus modulating the intensity, phase or polarization of the light reflected from the mirror surface. The accuracy of the mirror movement can thus create high-resolution and complex light patterns. A typical MEMS-SLM consists of a mirror array, electrostatic actuators that are responsible for moving the mirrors, and an optical system that can focus and guide the modulated light. One advantage over LCOs-SLMs is the faster response time of the mechanical components compared to the liquid crystal. Furthermore, MEMS-SLMs have a higher optical efficiency, which also leads to better image quality [14, 15].

Another important aspect that should be taken into account when designing the beam shaping technology is the limits of the technical system in terms of cost, efficiency and flexibility. The optical elements mentioned above are cheap to produce. However, their resistance to heat must be taken into account, especially when high laser power is used. In some cases, beam shaping optics have to be recalibrated after each construction process in order to ensure good beam quality [16]. This limitation increases, for example, the maintenance costs and must be taken into account when calculating the initial investment. In the case of a complex beam guidance, several optical components are often required for the beam manipulation in several steps. If SLMs are used, their damage threshold must also be considered. This inevitably leads to power losses, which must be taken into consideration when designing a beam source. This limitation primarily affects manufacturing processes in the field of powder-bed laser melting, as these usually involve powers of over 300 W (at the workpiece).

3 Advantages of Shaped Laser Beams in Metal Laser Additive Manufacturing

3.1 Typical Defects in AM Metal Components and Their Causes

The quality of a metal sample produced by powder bed-based laser melting can be divided into two main aspects: the geometric accuracy of the built sample and its mechanical properties [17]. The following sections characterize specific defects based on these two aspects and address which process parameters can be adjusted to avoid such defects.

The mechanical properties are determined by the porosity or density of the built material. These properties are directly influenced by the powder characteristics in the pre-process stage. Particularly, the homogeneity (uniform size of the powder particles) and the prevailing density of the powder play a decisive role. A common defect is the inclusion of gas pores within the structure after processing the material. These occur due to insufficient fusion of the material, also known as "lack of fusion" (LOF).

In additively manufactured metals, such LOF areas, unmelted particles, or other gas porosities frequently occur. These defects arise due to the layer-by-layer application, local heating, rapid cooling, and the use of powder material and are typically the cause of fatigue fractures. The defects usually result from either insufficient or excessive energy input (LOF defects or porosities) during the manufacturing process. The characteristics of defects, including their type, location, shape, size, orientation, and density, are significantly determined by the process parameters. A reduction of these defects can be achieved by using high-quality powder material and carefully adjusting the process parameters. These parameters include layer thickness [18–20], energy input [21, 22], build direction and scan strategy/speed including the hatch distance [23–32]. By optimizing these factors during the manufacturing process, the occurrence of defects can be minimized. Defects can be reduced by post-processing the manufactured parts, such as through hot isostatic pressing (HIP), but cannot be completely eliminated [33–37]. Other methods, such as preheating and maintaining the base plate at a specific temperature during the process and maintaining appropriate cooling rates, can also influence the defect content of the final product. Gas porosities, which typically appear spherical or ellipsoidal, mainly arise from gas trapped in the metal powder and the gas atmosphere in the AM machine. The energy input by the laser into the melt pool should be optimized so that the previous layer only partially melts to ensure a connection between successive layers. Deviations from this optimal energy amount can lead to insufficient melting or an unstable melt pool. Excessive energy causes strong convection currents and significant recoil forces in the vapor. These phenomena result in the formation of voids and/or gas bubbles in the melt pool [17].

Another example of a defect caused by insufficient energy input is known as "balling". Balling occurs due to capillary effects, which are due to the interaction of surface tension and interfacial tension between the melt and the substrate. Wettability plays a crucial role here. The wetting behavior can be described by Young's equation, where θ is the contact angle, $\sigma_{s,g}$ is the surface tension of the solid, $\sigma_{1,g}$ is the surface tension of the liquid, and $\gamma_{1,s}$ is the interfacial tension between the substrate and fluid

[38]:

$$\cos\theta = \frac{\sigma_{s,g} - \gamma_{1,s}}{\sigma_{1,g}} \tag{1}$$

Complete wetting occurs at a contact angle of $\theta \leq 90°$. If the contact angle is $\theta = 180°$, the substrate is not wetted at all. In the SLM process, balling appears as the formation of fluid spheres or the segmentation of the melt track into separate, consolidated melt sections [39].

Studies on process parameters show that these instabilities occur at insufficient laser power and high scan speeds [40]. A study on optimized process parameters was conducted by Kasperovich et al. [41]. The study aimed to minimize defects in selective laser melting of Ti–6Al–4V. Two different causes for pore formation were identified. On one hand, the powder particles were incompletely melted due to insufficient energy input. On the other hand, excessive energy input led to vaporization within the melt pool. The energy density (E_v) plays a crucial role in the porosity of the resulting component. It is defined as the ratio between laser power (P), scan speed (v), hatch distance (h), and layer thickness (δ):

$$E_v = \frac{P}{v \times \delta \times h} \tag{2}$$

Low energy density leads to incomplete melting of the powder, causing spherical and irregular pores. These pores arise from the balling effect and unmelted particles. High-energy density causes overheating and vaporization, leading to strong convection currents and recoil forces of the vapor. This results in material ejection and instability, which in turn create voids and gas bubbles in the melt pool.

The use of AI to investigate the mechanical properties and potential defects within an additively manufactured component is also the subject of current scientific research. Funcke et al. [42] studied the use of hybrid neural networks (HNN) to predict the tensile strength properties of additively manufactured (LPBF) samples made of AlSi10Mg. They combined a dense neural network (DNN) with a convolutional neural network (CNN) to analyze microstructure images. The investigations showed that the neural network achieves a high level of accuracy in predicting tensile strength and yield strength despite the limited amount of data. Nudelis et al. [43] used neural networks that draw conclusions about the underlying process parameters based on the pore distribution in the finished workpiece. The results show that the classification of pores by the AI model is sufficient to successfully determine the adjusted parameters. The researchers postulated that the use of AI offers a promising way to reduce the experimental effort required to determine optimal process parameters in the future.

3.2 Melt Pool Behavior Under the Influence of Shaped Laser Beams

So far, this paper has presented various methods for static and dynamic beam shaping. However, this section is concerned with the various energy distributions of a shaped laser beam and the effects that shaped beams can have on the behavior of the molten pool. Based on a Gaussian energy distribution, this section will take a closer look at the effects of an elliptical, inverse Gaussian, top hat and Bessel beam on the molten pool.

The energy distribution of the laser beam has a significant influence on the quality of the printed parts. The Gaussian beam is characterized by a narrow process window, low energy efficiency and a high energy gradient. This means that the intensity of the laser beam is highly concentrated in a specific area. This, in turn, has an influence on the behavior of the molten pool. The high local energy input causes instabilities in the molten pool, which can lead to thermocapillary flows [44], spatter formation [45] and, ultimately, porosity in the structure [3]. In order to prevent component defects that arise due to the use of a Gaussian energy distribution, alternative spot shapes were investigated.

In a study, Grünewald et al. [46] investigated various properties of hot spatters during a "laser-based powder bed fusion of metals" process using 316L. They used ring-shaped beam profiles with different diameters. They found that the number of hot spatters in relation to the fused area mainly depends on the beam profile. Beam profiles with high peak intensities lead to a significantly higher number of spatters per fused area than beam profiles with lower peak intensities. The spot size and beam profile also have an effect on the size of the spatter. Larger spot sizes generally lead to larger spatter, while beam profiles with lower peak intensities tend to produce smaller spatter. Due to the lower peak intensity, beam profiles that have less power in the center and more power in the ring area show a significant reduction in spatter from the molten bath.

With regard to the microstructure and the resulting mechanical properties of the produced component, Matthews et al. [47] investigated the use of an elliptical intensity profile in the LPBF process. They concluded that such a spot shape has a strong influence on the solidification microstructure. They postulated that the grain morphology can be specifically modified by varying the spatial beam intensity profile. This creates the potential to optimize the mechanical properties in a targeted manner. In addition, residual stresses could be reduced if more than one laser is used. A similar approach was taken by Shi et al. [48] in their work on the effects of the elliptical laser beam shape on the development of the melt pool geometry.

Tumkur et al. [49] compared the effects of a Bessel beam on the molten pool with a classic Gaussian beam. They found that the lower thermal gradient of the Bessel beam produces a more stable molten pool. This leads to a more uniform solidification of the melt. Furthermore, keyhole formation is reduced. Workpieces processed with the Bessel beam have a higher density and better surface quality. In addition, Bessel beams, due to their non-diffractive character, are more forgiving of any positioning errors in the focal plane during the process.

In their work, Belay et al. [50] compared a top-hat intensity profile with a Gaussian profile. The material used was stainless steel 316L. They found that the size of the molten pool increased significantly when using a top-hat profile. They attributed this finding to the fact that this alternative spot shape results in a more uniform energy distribution in the molten pool. In addition, there is the possibility that flexibility can be increased by dynamically adjusting the top-hat intensity profile. Depending on the process requirements, for example, spot shapes with a smaller diameter can be used to produce finer details and more precise contours. On the other hand, the use of a large profile diameter increases productivity and is used primarily when larger areas are processed in a single step. The dynamic combination of small and large beam profiles increases the efficiency and quality of the processed components.

4 Experimental Setup for Future Investigations and Outlook

This chapter is about future investigation methods to explore the potential of dynamically shaped beams on a metallic powder bed. At the time of writing this paper, the focus was on the initial phase of familiarization with the topic of dynamic beam shaping. This phase includes an in-depth literature review on beam shaping, modularization of additive manufacturing systems, and simulation approaches. The next step involves designing an experimental setup, which will be discussed in detail later in this work. This includes the design and procurement of all necessary components to ensure the operation of the experimental system. At the beginning of the study, two hypotheses are proposed, which will be derived in the following sections.

Bi et al. [3] stated in their review paper that static beam shaping has unlocked the potential of laser bed-based laser melting to a certain extent. At the same time, it is challenging for a single beam with a specific shape to meet all the requirements of the process. The ability to adapt a beam in real time to specific printing properties during the process increases efficiency and component quality. For this reason, part of the future work will focus on the following questions:

- How do dynamically shaped laser beams affect the efficiency of the printing process?
- Can dynamic beam shaping positively influence the component quality in terms of the mechanical properties of a predetermined material?

As part of the orientation period for this work, a concept for a test setup is being developed to print single tracks on 8 × 8 mm sample elements. The concept will include a 2 kW fiber laser as a beam source, a process chamber, a laser-safe enclosure and an optical table on which the individual components can be mounted. All these components still have to be designed, constructed or ordered as purchased parts. Figure 1 therefore only shows the principle structure of the future trial setup. In addition to the components mentioned, a camera system is planned to enable the process to be observed in real time. The core of the system will be a prototype beam-shaping optic. Those optics will integrate a collimator, a 2D scanning system and focusing optics (Fig. 1). The collimator focuses and aligns the radiation of the laser source to improve the beam quality before it enters the beam-shaping optics. The 2D scanning system is used to guide the laser beam precisely over the two-dimensional processing surface. This should also enable complex scanning strategies with a high level of precision and speed. The focusing optic concentrate and bundle the outgoing beam to a specific point or a specific surface. This is particularly important for shaped or split beam paths. Using two integrated spatial light modulators, this setup can shape an incoming laser beam and divide it into any number of individual laser spots. The technology's greatest potential is probably the high dynamic change rate of the beam shape during the process. The technology is based on several patents that have been submitted, which cover both process- [51–55] and system-technology [56–64].

Furthermore, simulation approaches aim to reduce the experimental scope to only what is necessary. Due to the numerous interactions between the laser and the melt pool, process modeling aims to enhance understanding of the process. This also allows for pre-process optimization, enabling the selection of appropriate exposure parameters

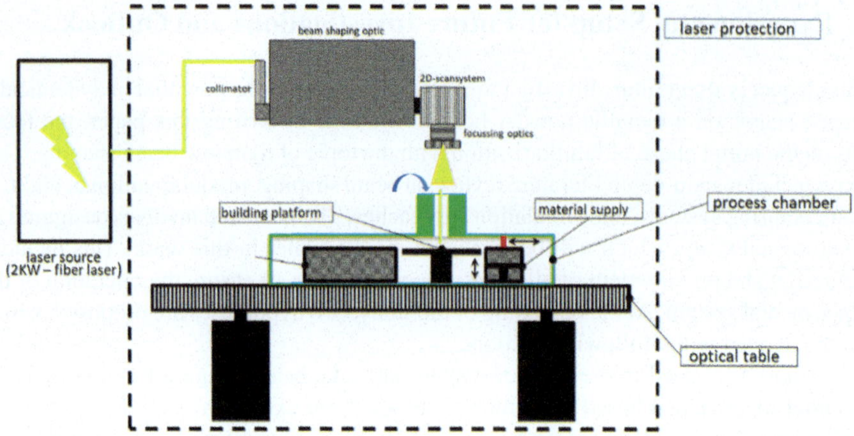

Fig. 1 Expected basic structure of the test stand. The main component is the beam-shaping optic. Along with the process chamber, it is mounted on an optical table to avoid vibrations during processing. A radiation-proof enclosure will cover the entire structure. However, a door will provide access to the test stand, enabling manual sample removal

and predictions regarding part behavior. When planning the experiment, it is important to choose the right simulation approach and, therefore, to become familiar with the topic. Generally, such simulations can be categorized into three abstraction levels: macrostructure Simulation [65–67], mesostructure Simulation [68–70] and microstructure Simulation [71–74]. In addition to the test setup, analysis and familiarization with the process simulation play a decisive role. This also represents an important milestone in project planning.

Acknowledgements. This work has been funded by the Deutsche Forschungsgemeinschaft (DFG, German Research Foundation)—Project-ID 528745080—FIP 68. The authors alone are responsible for the content of the paper.

Furthermore, I would like to thank Prof. Lukas Löber for his support in writing this paper. I would also like to thank Philippe du Maire[0000-0003-4844-6290], who has always helped me with organizational aspects.

References

1. Tan, C., Wang, D., Ma, W., Chen, Y., Chen, S., Yang, Y., Zhou, K.: Design and additive manufacturing of novel conformal cooling molds. Mater. Design **196**, 109147 (2020). https://doi.org/10.1016/j.matdes.2020.109147
2. matzesl: Metal Additive Manufacturing and Metal 3D printing. https://ampower.eu/infographics/metal-additive-manufacturing/. Last accessed 14 Aug 2024
3. Bi, J., Wu, L., Li, S., Yang, Z., Jia, X., Starostenkov, M.D., Dong, G.: Beam shaping technology and its application in metal laser additive manufacturing: a review. J. Mater. Res. Technol. **26**, 4606–4628 (2023). https://doi.org/10.1016/j.jmrt.2023.08.037
4. Bakhtari, A.R., Sezer, H.K., Canyurt, O.E., Eren, O., Shah, M., Marimuthu, S.: A review on laser beam shaping application in laser-powder bed fusion. Adv. Eng. Mater. **26**, 2302013 (2024). https://doi.org/10.1002/adem.202302013

5. Dickey, F.M.: Laser beam shaping techniques. In: Proceedings of SPIE. pp. 338–348. SPIE, Santa Fe, NM, USA (2000). https://doi.org/10.1117/12.407361
6. Matthews, M.J., Guss, G., Drachenberg, D.R., Demuth, J.A., Heebner, J.E., Duoss, E.B., Kuntz, J.D., Spadaccini, C.M.: Diode-based additive manufacturing of metals using an optically-addressable light valve. Opt. Express **25**, 11788 (2017). https://doi.org/10.1364/OE.25.011788
7. Laskin, A., Laskin, V.: Variable beam shaping with using the same field mapping refractive beam shaper. Presented at the SPIE LASE, San Francisco, California, USA February 9 (2012). https://doi.org/10.1117/12.903606
8. Bich, A., Rieck, J., Dumouchel, C., Roth, S., Weible, K.J., Eisner, M., Voelkel, R., Zimmermann, M., Rank, M., Schmidt, M., Bitterli, R., Ramanan, N., Ruffieux, P., Scharf, T., Noell, W., Herzig, H.-P., De Rooij, N.: Multifunctional micro-optical elements for laser beam homogenizing and beam shaping. Presented at the Lasers and Applications in Science and Engineering, San Jose, CA, 7 Feb 2008. https://doi.org/10.1117/12.762923
9. Mikhaylov, D.: Ultrashort Pulse Laser Ablation of Bulk Materials using shaped Laser Beams (2020)
10. Häfner, T., Strauß, J., Roider, C., Heberle, J., Schmidt, M.: Tailored laser beam shaping for efficient and accurate microstructuring. Appl. Phys. A **124**, 111 (2018). https://doi.org/10.1007/s00339-017-1530-0
11. Akemann, W., Léger, J.-F., Ventalon, C., Mathieu, B., Dieudonné, S., Bourdieu, L.: Fast spatial beam shaping by acousto-optic diffraction for 3D non-linear microscopy. Opt. Express **23**, 28191 (2015). https://doi.org/10.1364/OE.23.028191
12. Mi, Y., Mahade, S., Sikström, F., Choquet, I., Joshi, S., Ancona, A.: Conduction mode laser welding with beam shaping using a deformable mirror. Opt. Laser Technol. **148**, 107718 (2022). https://doi.org/10.1016/j.optlastec.2021.107718
13. Zhang, Z., You, Z., Chu, D.: Fundamentals of phase-only liquid crystal on silicon (LCOS) devices. Light. Sci. Appl. **3**, e213–e213 (2014). https://doi.org/10.1038/lsa.2014.94
14. Becker, T., Bifano, T.G., Lee, H., Miller, M., Bierden, P.A., Cornelissen, S.: MEMS spatial light modulators with integrated electronics. Presented at the Micromachining and Microfabrication, San Jose, CA (2003). https://doi.org/10.1117/12.477931
15. Dunayevsky, Y., Marom, D.M.: MEMS spatial light modulator for phase and amplitude modulation of spectrally dispersed light. In: 2010 IEEE 26-th Convention of Electrical and Electronics Engineers in Israel, pp. 000836–000838. IEEE, Eilat, Israel (2010). https://doi.org/10.1109/EEEI.2010.5662090
16. Wischeropp, T.M.: Advancement of Selective Laser Melting by Laser Beam Shaping. Springer, Berlin, Heidelberg (2021). https://doi.org/10.1007/978-3-662-64585-7
17. Sanaei, N., Fatemi, A.: Defects in additive manufactured metals and their effect on fatigue performance: a state-of-the-art review. Prog. Mater. Sci. **117**, 100724 (2021). https://doi.org/10.1016/j.pmatsci.2020.100724
18. Dixit, T., Sahu, P.K., Jonnalagadda, K., Prasad, K.E.: Effect of powder layer thickness and scan orientation on the deformation and failure of selectively laser melted Ti-6Al-4V alloy over six decades of strain rates. Mater. Sci. Eng. **822**, 141656 (2021). https://doi.org/10.1016/j.msea.2021.141656
19. Cutolo, A., Neirinck, B., Lietaert, K., De Formanoir, C., Van Hooreweder, B.: Influence of layer thickness and post-process treatments on the fatigue properties of CoCr scaffolds produced by laser powder bed fusion. Addit. Manuf. **23**, 498–504 (2018). https://doi.org/10.1016/j.addma.2018.07.008
20. Ma, M., Wang, Z., Gao, M., Zeng, X.: Layer thickness dependence of performance in high-power selective laser melting of 1Cr18Ni9Ti stainless steel. J. Mater. Process. Technol. **215**, 142–150 (2015). https://doi.org/10.1016/j.jmatprotec.2014.07.034

21. Montero-Sistiaga, M.L., Godino-Martinez, M., Boschmans, K., Kruth, J.-P., Van Humbeeck, J., Vanmeensel, K.: Microstructure evolution of 316L produced by HP-SLM (high power selective laser melting). Addit. Manuf. **23**, 402–410 (2018). https://doi.org/10.1016/j.addma.2018.08.028
22. Fu, J., Qu, S., Ding, J., Song, X., Fu, M.W.: Comparison of the microstructure, mechanical properties and distortion of stainless steel 316 L fabricated by micro and conventional laser powder bed fusion. Addit. Manuf. **44**, 102067 (2021). https://doi.org/10.1016/j.addma.2021.102067
23. Pant, P., Salvemini, F., Proper, S., Luzin, V., Simonsson, K., Sjöström, S., Hosseini, S., Peng, R.L., Moverare, J.: A study of the influence of novel scan strategies on residual stress and microstructure of L-shaped LPBF IN718 samples. Mater. Design **214**, 110386 (2022). https://doi.org/10.1016/j.matdes.2022.110386
24. Rashid, R., Masood, S.H., Ruan, D., Palanisamy, S., Rahman Rashid, R.A., Brandt, M.: Effect of scan strategy on density and metallurgical properties of 17-4PH parts printed by Selective Laser Melting (SLM). J. Mater. Process. Technol. **249**, 502–511 (2017). https://doi.org/10.1016/j.jmatprotec.2017.06.023
25. Cheng, B., Shrestha, S., Chou, K.: Stress and deformation evaluations of scanning strategy effect in selective laser melting. Addit. Manuf. **12**, 240–251 (2016). https://doi.org/10.1016/j.addma.2016.05.007
26. Ramos, D., Belblidia, F., Sienz, J.: New scanning strategy to reduce warpage in additive manufacturing. Addit. Manuf. **28**, 554–564 (2019). https://doi.org/10.1016/j.addma.2019.05.016
27. Haines, M., List, F., Carver, K., Leonard, D., Plotkowski, A., Fancher, C., Dehoff, R., Babu, S.: Role of scan strategies and heat treatment on grain structure evolution in Fe-Si soft magnetic alloys made by laser-powder bed fusion. Addit. Manuf. **50**, 102578 (2022). https://doi.org/10.1016/j.addma.2021.102578
28. Ni, C., Zhu, L., Zheng, Z., Zhang, J., Yang, Y., Hong, R., Bai, Y., Lu, W.F., Wang, H.: Effects of machining surface and laser beam scanning strategy on machinability of selective laser melted Ti6Al4V alloy in milling. Mater. Design **194**, 108880 (2020). https://doi.org/10.1016/j.matdes.2020.108880
29. Xu, J., Ding, Y., Gao, Y., Wang, H., Hu, Y., Zhang, D.: Grain refinement and crack inhibition of hard-to-weld Inconel 738 alloy by altering the scanning strategy during selective laser melting. Mater. Design **209**, 109940 (2021). https://doi.org/10.1016/j.matdes.2021.109940
30. Zhao, C., Bai, Y., Zhang, Y., Wang, X., Xue, J.M., Wang, H.: Influence of scanning strategy and building direction on microstructure and corrosion behaviour of selective laser melted 316L stainless steel. Mater. Design **209**, 109999 (2021). https://doi.org/10.1016/j.matdes.2021.109999
31. Spierings, A.B., Dawson, K., Uggowitzer, P.J., Wegener, K.: Influence of SLM scan-speed on microstructure, precipitation of Al3Sc particles and mechanical properties in Sc- and Zr-modified Al-Mg alloys. Mater. Design **140**, 134–143 (2018). https://doi.org/10.1016/j.matdes.2017.11.053
32. Wang, Y., Shi, J.: Influence of laser scan speed on micro-segregation in selective laser melting of an iron-carbon alloy: a multi-scale simulation study. Procedia Manuf. **26**, 941–951 (2018). https://doi.org/10.1016/j.promfg.2018.07.121
33. Li, G., Chi, W., Wang, W., Liu, X., Tu, H., Long, X.: High cycle fatigue behavior of additively manufactured Ti-6Al-4V alloy with HIP treatment at elevated temperatures. Int. J. Fatigue **184**, 108287 (2024). https://doi.org/10.1016/j.ijfatigue.2024.108287
34. Günther, J., Krewerth, D., Lippmann, T., Leuders, S., Tröster, T., Weidner, A., Biermann, H., Niendorf, T.: Fatigue life of additively manufactured Ti–6Al–4V in the very high cycle fatigue regime. Int. J. Fatigue **94**, 236–245 (2017). https://doi.org/10.1016/j.ijfatigue.2016.05.018

35. Edwards, P., Ramulu, M.: Fatigue performance evaluation of selective laser melted Ti–6Al–4V. Mater. Sci. Eng. **598**, 327–337 (2014). https://doi.org/10.1016/j.msea.2014.01.041
36. Wycisk, E., Siddique, S., Herzog, D., Walther, F., Emmelmann, C.: Fatigue performance of laser additive manufactured Ti–6Al–4V in very high cycle fatigue regime up to 109 cycles. Front. Mater. **2**, (2015). https://doi.org/10.3389/fmats.2015.00072
37. Leuders, S., Thöne, M., Riemer, A., Niendorf, T., Tröster, T., Richard, H.A., Maier, H.J.: On the mechanical behaviour of titanium alloy TiAl6V4 manufactured by selective laser melting: Fatigue resistance and crack growth performance. Int. J. Fatigue **48**, 300–307 (2013). https://doi.org/10.1016/j.ijfatigue.2012.11.011
38. Kruth, J.P., Froyen, L., Van Vaerenbergh, J., Mercelis, P., Rombouts, M., Lauwers, B.: Selective laser melting of iron-based powder. J. Mater. Process. Technol. **149**, 616–622 (2004). https://doi.org/10.1016/j.jmatprotec.2003.11.051
39. Gunenthiram, V., Peyre, P., Schneider, M., Dal, M., Coste, F., Fabbro, R.: Analysis of laser–melt pool–powder bed interaction during the selective laser melting of a stainless steel. J. Laser Appl. **29**, 022303 (2017). https://doi.org/10.2351/1.4983259
40. Yadroitsev, I., Gusarov, A., Yadroitsava, I., Smurov, I.: Single track formation in selective laser melting of metal powders. J. Mater. Process. Technol. **210**, 1624–1631 (2010). https://doi.org/10.1016/j.jmatprotec.2010.05.010
41. Kasperovich, G., Haubrich, J., Gussone, J., Requena, G.: Correlation between porosity and processing parameters in TiAl6V4 produced by selective laser melting. Mater. Design **105**, 160–170 (2016). https://doi.org/10.1016/j.matdes.2016.05.070
42. Funcke, F., Forster, T., Mayr, P.: Advantages of hybrid neural network architectures to enhance prediction of tensile properties in laser powder bed fusion. KEM **964**, 65–71 (2023). https://doi.org/10.4028/p-0tCaMF
43. Nudelis, N., Mayr, P.: Process parameter prediction in laser powder bed fusion using an artificial neural network. KEM **964**, 59–64 (2023). https://doi.org/10.4028/p-rl51NI
44. Khairallah, S.A., Anderson, A.T., Rubenchik, A., King, W.E.: Laser powder-bed fusion additive manufacturing: physics of complex melt flow and formation mechanisms of pores, spatter, and denudation zones. Acta Mater. **108**, 36–45 (2016). https://doi.org/10.1016/j.actamat.2016.02.014
45. Matthews, M.J., Guss, G., Khairallah, S.A., Rubenchik, A.M., Depond, P.J., King, W.E.: Denudation of metal powder layers in laser powder bed fusion processes. Acta Mater. **114**, 33–42 (2016). https://doi.org/10.1016/j.actamat.2016.05.017
46. Grünewald, J., Reimann, J., Wudy, K.: Influence of ring-shaped beam profiles on spatter characteristics in laser-based powder bed fusion of metals. J. Laser Appl. **35**, 042009 (2023). https://doi.org/10.2351/7.0001153
47. Matthews, M.J., Roehling, T.T., Khairallah, S.A., Tumkur, T.U., Guss, G., Shi, R., Roehling, J.D., Smith, W.L., Vrancken, B.K., Ganeriwala, R.K., McKeown, J.T.: Controlling melt pool shape, microstructure and residual stress in additively manufactured metals using modified laser beam profiles. Procedia CIRP **94**, 200–204 (2020). https://doi.org/10.1016/j.procir.2020.09.038
48. Shi, R., Khairallah, S.A., Roehling, T.T., Heo, T.W., McKeown, J.T., Matthews, M.J.: Microstructural control in metal laser powder bed fusion additive manufacturing using laser beam shaping strategy. Acta Mater. **184**, 284–305 (2020). https://doi.org/10.1016/j.actamat.2019.11.053
49. Tumkur, T.U., Voisin, T., Shi, R., Depond, P.J., Roehling, T.T., Wu, S., Crumb, M.F., Roehling, J.D., Guss, G., Khairallah, S.A., Matthews, M.J.: Nondiffractive beam shaping for enhanced optothermal control in metal additive manufacturing. Sci. Adv. **7**, eabg9358 (2021). https://doi.org/10.1126/sciadv.abg9358

50. Belay, G.Y., Kinds, Y., Goossens, L., Gurung, K., Bosmans, N., Diltoer, R., Eraly, J., Vervaeke, M., Thienpont, H., Van Hooreweder, B., Van Erps, J.: Dynamic optical beam shaping system to generate Gaussian and top-hat laser beams of various sizes with circular and square footprint for Additive Manufacturing applications. Procedia CIRP **111**, 75–80 (2022). https://doi.org/10.1016/j.procir.2022.08.134
51. Loeber, L., Heeling, T., Kiedrowski, T.: Multi-Spot-Laservorrichtung und Verfahren zum Erzeugen eines Werkstücks (2023)
52. Ilin, A., Kiedrowski, T.: Verfahren und Vorrichtung zum generativen Herstellen eines Bauteils (2018)
53. Loeber, L., Ilin, A., Slodczyk, M., Hartel, U., Heeling, T., Schaenzel, M.: Verfahren und Vorrichtung zum generativen Herstellen eines Bauteils und Verwendung des Verfahrens (2021)
54. Ilin, A., Slodczyk, M.: Verfahren zum Herstellen eines Bauteils (2021)
55. Loeber, L., Slodczyk, M., Kiedrowski, T.: Verfahren zur additiven Fertigung eines Bauteils und Laserbearbeitungsvorrichtung zur Fertigung eines Bauteils (2022)
56. Slodczyk, M., Kiedrowski, T., Graf, T., Mikhaylov, D.: Räumliche Lichtmoduliervorrichtung und Verfahren zum Betreiben einer räumlichen Lichtmoduliervorrichtung (2023)
57. Slodczyk, M., Kiedrowski, T., Graf, T., Mikhaylov, D.: Räumliche Lichtmoduliervorrichtung und Verfahren zum Betreiben eines räumlichen Lichtmodulators (2023)
58. Slodczyk, M.: Steuervorrichtung und Verfahren zum Betreiben eines räumlichen Lichtmodulators (2021)
59. Mikhaylov, D., Shakirov, D.: Verfahren und Vorrichtung zur Formung kohärenter Strahlung (2018)
60. Mikhaylov, D., Gauch, R., Graf, T.: Verfahren und Vorrichtung zur Formung von Strahlung für die Laserbearbeitung (2018)
61. Mikhaylov, D., Gauch, R., Graf, T.: Verfahren und Vorrichtung zur Formung von Strahlung für die Laserbearbeitung (2019)
62. Mikhaylov, D., Kroschel, A., Ramsayer, R., Shakirov, D., Ilin, A., Kiedrowski, T.: Verfahren und Vorrichtung zur prozessorientierten Strahlformanpassung und Strahlorientierung (2018)
63. Mikhaylov, D., Shakirov, D.: Vorrichtung und Verfahren zur Formung eines Laserstrahls durch einen programmierbaren Strahlformer (2018)
64. Mikhaylov, D., Graf, T.: Vorrichtung zur Strahlformung für einen zur Materialbearbeitung eingesetzten Laserstrahl (2020)
65. Branner, G.: Modellierung transienter Effekte in der Struktursimulation von Schichtbauverfahren. Herbert Utz Verlag, München (2011)
66. Keller, N., Ploshikhin, V.: New Method for fast predictions of residual stress and distortion of AM parts. Presented at the Solid Freeform Fabrication Symposium, Austin, Texas August (2024)
67. Schänzel, M., Shakirov, D., Ilin, A., Ploshikhin, V.: Coupled thermo-mechanical process simulation method for selective laser melting considering phase transformation steels. Comput. Math. Appl. **78**, 2230–2246 (2019). https://doi.org/10.1016/j.camwa.2019.01.019
68. Gusarov, A.V., Smurov, I.: Modeling the interaction of laser radiation with powder bed at selective laser melting. Phys. Procedia **5**, 381–394 (2010). https://doi.org/10.1016/j.phpro.2010.08.065
69. Khairallah, S.A., Anderson, A.: Mesoscopic simulation model of selective laser melting of stainless steel powder. J. Mater. Process. Technol. **214**, 2627–2636 (2014). https://doi.org/10.1016/j.jmatprotec.2014.06.001
70. King, W.E., Anderson, A.T., Ferencz, R.M., Hodge, N.E., Kamath, C., Khairallah, S.A., Rubenchik, A.M.: Laser powder bed fusion additive manufacturing of metals; physics, computational, and materials challenges. Appl. Phys. Rev. **2**, 041304 (2015). https://doi.org/10.1063/1.4937809

71. Mohebbi, M.S., Ploshikhin, V.: Implementation of nucleation in cellular automaton simulation of microstructural evolution during additive manufacturing of Al alloys. Addit. Manuf. **36**, 101726 (2020). https://doi.org/10.1016/j.addma.2020.101726
72. Zinoviev, A., Zinovieva, O., Ploshikhin, V., Romanova, V., Balokhonov, R.: Evolution of grain structure during laser additive manufacturing. Simulation by a cellular automata method. Mater Design **106**, 321–329 (2016). https://doi.org/10.1016/j.matdes.2016.05.125
73. Zinovieva, O., Zinoviev, A., Ploshikhin, V.: Three-dimensional modeling of the microstructure evolution during metal additive manufacturing. Comput. Mater. Sci. **141**, 207–220 (2018). https://doi.org/10.1016/j.commatsci.2017.09.018
74. Zhang, J., Liou, F., Seufzer, W., Taminger, K.: A coupled finite element cellular automaton model to predict thermal history and grain morphology of Ti-6Al-4V during direct metal deposition (DMD). Addit. Manuf. **11**, 32–39 (2016). https://doi.org/10.1016/j.addma.2016.04.004

Fluid Mechanical Investigation of the Outgoing Free Jet of Nozzels

Philippe du Maire[1(✉)], Michael Johlitz[2], and Andreas Öchsner[1]

[1] Faculty of Mechanical and Systems Engineering, Esslingen University of Applied Sciences, Kanalstrasse 33, 73728 Esslingen, Germany
{philippe.du-maire,andreas.oechsner}@hs-esslingen.de

[2] Institute of Mechanics, Universität der Bundeswehr München, Faculty of Aerospace Engineering, Werner-Heisenberg-Weg 39, 85577 Neubiberg, Germany
michael.johlitz@unibw.de

Abstract. This study investigates the fluid mechanical behaviour of outgoing free jets from circular nozzles, focusing on jet angle and velocity profile analysis. Two nozzles with outlet diameters of 8 and 12 mm were examined using schlieren photography and Pitot tube measurements. Schlieren imaging revealed a consistent jet angle of approximately 22° unaffected by variations in pressure or nozzle diameter. Additionally, the schlieren method revealed that turbulence within the jet boundary increased at lower pressures due to greater mixing with the ambient air. The velocity profiles, obtained through dynamic pressure measurements, followed a Gaussian distribution within the core and transition regions, with the larger 12 mm nozzle producing a more stable and extended core region compared to the 8 mm nozzle. Deviations were observed near the jet boundaries, particularly at lower velocities, highlighting the challenges in measuring low dynamic pressures. The study's findings contribute to the understanding of free jet dynamics, offering insights for optimising industrial cooling and drying systems.

Keywords: Subsonic free jet · Schlieren photography · Pitot tube measurements

1 Introduction

Fluid mechanics encompasses a wide range of phenomena that are crucial for understanding and optimising systems that involve fluid flow. Among these phenomena, the behaviour of free jets expelled from a circular opening or nozzle into an ambient medium can be found in various engineering applications and is thus of particular interest. Free jets are encountered for example in aerospace engineering, automotive exhaust systems or industrial cooling processes.

A. Öchsner (ed.), *Collaborative Research Advancing Engineering Solutions for Real-World Challenges 2*, Proceedings in Technology Transfer,
https://doi.org/10.1007/978-3-031-79093-5_10

Especially, in light of climate change and the imminent need for a holistic rethink, sustainable development is becoming an increasingly crucial aspect in engineering. The objective is to develop and construct machines and systems with optimal efficiency, defined as the highest possible degree of effectiveness. A high degree of efficiency not only optimises the utilisation of resources, making processes more sustainable but also often results in significant financial savings, making it a sound business decision.

This paper presents an experimental investigation on the characteristics of the outgoing free jets of circular nozzles from an industrial cooling and drying system. The primary objectives of the study are to determine the jet angle and analyze the velocity profile, i.e. the effective area of the free jet. To this end, two nozzles with different outlet diameters were analysed. The investigation employs schlieren photography to visualise the outgoing free jet and Pitot tube measurements to assess the velocity distribution. The schlieren photography allows the determination of the jet angle and provides qualitative insights into the flow characteristics. By capturing images at various pressures, the schlieren method reveals how pressure influences jet formation and turbulence.

In addition to visualising the jet, the velocity profile is measured using a Pitot tube, determining the dynamic pressure within the jet. The velocity distribution is derived from these measurements, providing a quantitative analysis of the jet's behaviour and the determination of the area of effect.

This investigation aims to gain a better understanding of the outgoing free jet behaviour of nozzles by investigating the interaction between nozzle geometry, jet angle, and velocity distribution. By comparing results from different nozzles and pressures, the study aims to contribute knowledge on free jet dynamics and provide practical recommendations for the area of effect of the free jet.

2 Theoretical Background

2.1 Turbulent Free Shear Flow

The turbulent free shear flow is a special case among turbulent flows. Turbulent free shear flows are flows in which no walls are in the immediate vicinity. They occur, for example, in free jets, lift jets or wake flows, the so-called slipstream [1].

Similar to laminar flows, momentum transfer in the immediate vicinity of the wall also occurs in turbulent flows due to the viscosity of the flow medium (viscous wall layer). As there are no walls in free shear flows, the viscosity effects can be neglected in these flows, as the turbulent friction is always much greater than the viscous friction [1]. Viscosity only has an influence on the mean movement of the free turbulent shear layers at their edges. Here, similar to the outer edge of turbulent boundary layers, the so-called viscous overlayer prevails.

The viscous overlayer is the area between non-turbulent, rotation-free external flow and fully turbulent and thus rotational flow at the boundary layer edge of turbulent flows [1]. According to Schlichting, this area is subject to strong temporal and spatial fluctuations. The width of the viscous overlayer is proportional to the Kolmogorov length and thus proportional to $\nu^{3/4}$ or $Re^{-3/4}$, where ν is the kinematic viscosity and *Re* the Reynolds number. Due to this relationship, the viscous overlayer can be neglected for high Reynolds numbers [1].

According to Schlichting, the slenderness of the turbulent free shear flow was proven by the experimental findings [1]. This means that the flow has only small widening angles. For this reason, the boundary layer equation reduced by the viscous term can be used instead of the Navier-Stokes equations to calculate turbulent free shear flows. As a result, turbulent free shear flows are also referred to as free boundary layers or free shear layers.

2.2 Rotationally Symmetrical Subsonic Free Jet

Free jets are an example of a turbulent free shear flow. They are confined flow fields and can be categorised into supersonic and subsonic free jets, among others. The topic of free jets covers an extensive and specialised subject area with many special and individual cases. For reasons of clarity, only the rotationally symmetrical subsonic free jet from circular longitudinal nozzles is considered in more detail below, as this is an essential part of this work.

Rotationally symmetric subsonic free jets occur when a fluid flows out of a circular longitudinal nozzle into a limited or unlimited medium. Longitudinal nozzles are nozzles whose openings lie in the same plane as the pipe axis, as shown in Fig. 1a [2]. The basic prerequisites described in Sect. 2.1 apply to the formation of free jets, in particular that there are no boundary walls in the immediate vicinity of the free jet.

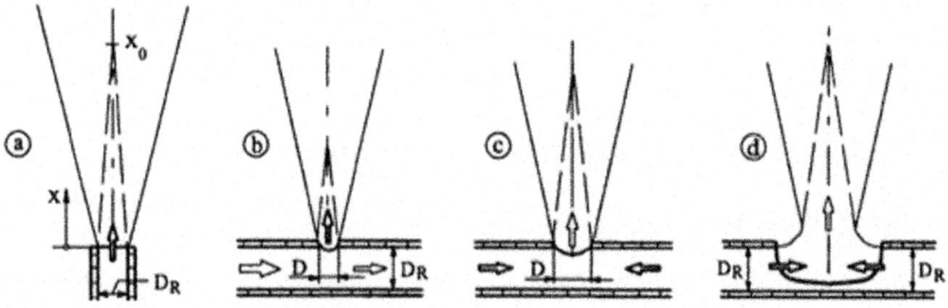

Fig. 1. Definition of longitudinal and transverse nozzle. **a** Common nozzle and pipe axis (longitudinal nozzle), **b–d** nozzle axis perpendicular to pipe axis (transverse nozzle) [2]. Reprinted with permission from SpringerNature publishers

When the fluid jet exits, it mixes with the stationary ambient medium due to the turbulent and the momentum exchange [4]. As a result, the volume transported in the jet increases downstream and the diameter of the free jet increases with increasing distance from the nozzle, i.e. the jet expands.

Assuming that the fluid and flow medium are the same and the ambient medium is stationary, as is the case with air nozzles in ambient air, for example, the free jet no longer has any velocity under idealised conditions until it reaches infinity. In reality, frictional losses during the momentum exchange between the fluid jet and the ambient medium mean that the jet centre velocity decreases with increasing nozzle distance, but the transported volume increases. For example, at a relative nozzle distance of $x/D \approx 70$, the jet centre velocity is only 0.1% of the velocity at the nozzle outlet, where x corresponds to the distance to the nozzle and D to the diameter of the nozzle outlet [2].

This decrease in velocity, which results from the momentum exchange and mass mixing, can be calculated realistically assuming a constant total momentum flow [2]. Therefore, the following applies to the momentum flux of a rotationally symmetrical free jet:

$$i_x = \frac{\mathrm{d}I_x}{\mathrm{d}t} = 2 \cdot \pi \cdot \int_{y=0}^{\frac{D_s}{2}} \rho \cdot w^2 \cdot y \cdot \mathrm{d}y \approx \text{const.} \tag{1}$$

If the impulse flow at the nozzle outlet and the increase in the jet diameter D_s and the transverse distributions of the velocity w/w_M and the density ρ/ρ_M are known, the decrease in the centre velocity of the jet in the x-direction can be calculated.

If a swirl-free, quasi-parallel jet flow has a small Reynolds number ($Re \approx 1000$), a laminar jet boundary layer forms at the beginning of the free jet, which only becomes turbulent at some distance. However, small Reynolds numbers are rarely the case; the Reynolds number of the flow is usually larger ($Re > 6000$) and the free jet is therefore turbulent right from the start. Before the fluid jet expands, subsonic free jets always begin with an adapted expansion.

Adapted expansion is the expansion of the gas emerging from the nozzle to a pressure equal to the ambient pressure. The static pressure of the subsonic free jet is therefore equal to the ambient pressure. This is not the case with supersonic free jets. Here, the free jet usually has a higher static pressure than the ambient pressure at the outlet, the so-called underexpansion [5].

Another special characteristic of the rotationally symmetric subsonic free jet is its propagation angle. As described in Sect. 2.1, turbulent free shear flows are slender and have small propagation angles. Various experimental investigations have shown that the propagation angles of different circular free jets are similar. The propagation angle of a circular free jet is therefore $\delta \approx 24°$ [6].

The question of the exact reasons why turbulent shear flows are slender and the free jet propagates at approximately the same angles has not yet been conclusively clarified [6].

The flow field of a subsonic free jet can be divided into three areas. At the beginning, at the nozzle outlet, there is the core region, followed by the transition region and the fully developed region. The region of the fully developed flow can further be subdivided into a near and a far field. Figure 2 shows the flow field of a rotationally symmetrical subsonic free jet and its various flow field areas.

2.2.1 Core Region of the Free Jet In subsonic free jets, the core region, also known as the jet core, is located directly at the nozzle outlet (see Fig. 2). In this area, the value of the centre or central velocity corresponds to the value of the velocity at the nozzle outlet.

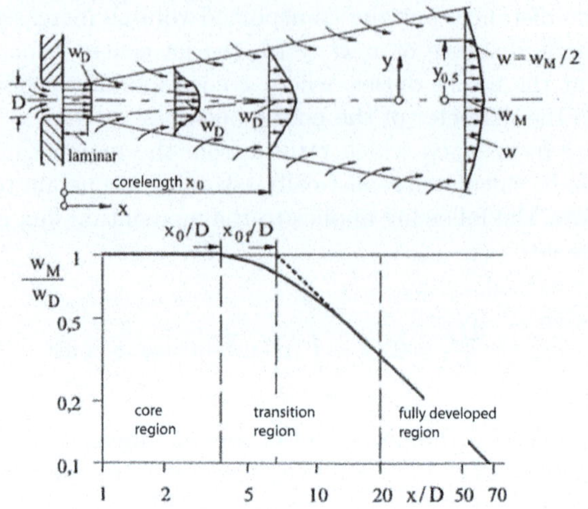

Fig. 2. Flow field areas of a subsonic free jet [2]. Reprinted with permission from SpringerNature publishers

The core length x_0 shown in Fig. 2 above is the distance to the nozzle up to which the velocity profile present in the nozzle outlet has changed due to the interaction with the ambient medium in such a way that the original nozzle velocity w_D only prevails on the jet center axis. This means that all fluid particles on the jet centre line have the nozzle outlet velocity w_D over the entire length of the core area.

In theory, the jet centre velocity can be represented via the nozzle spacing, as shown in Fig. 2 below, with the aid of two straight lines on a double-logarithmic scale. Their intersection provides the fictitious core length x_{0f}. The real change in velocity is represented by a curve in the transition region, which is why the real core length is significantly shorter than the fictitious core length. Experimental measurements have shown that the relative core length $x_0/D \approx 2$ to 8, whereas the relative fictitious core length assumes values of $x_{0f}/D \approx 4.4$ to 7.3 [2]. The core length is influenced by various parameters. These include the following:

- The shape of the nozzle and the resulting velocity profile at the nozzle outlet (standard nozzle, long pipe, etc.).
- Density and temperature ratio between flowing fluid and ambient medium.
- Degree of turbulence.
- Reynolds number.

2.2.2 Transition Region of the Free Jet The transition region follows on from the core region; in it, the velocity profile of the free jet gradually changes to a Gaussian bell shape as a result of mixing with the surrounding medium. Only at the boundary between the transition region and the fully developed region is the transformation complete and corresponds to the Gaussian bell shape. Due to the variable core length x_0, the length of the transition region is also not fixed and depends on various parameters. Some of these are listed in Sect. 2.2.1. In the transition area, the decrease in the centre velocity w_{M} of a rotationally symmetrical free jet can be described with the following equation [2].

$$\frac{w_{\mathrm{M}}}{w_{\mathrm{D}}} \approx f(1/\sqrt{x/D}) \tag{2}$$

2.2.3 Fully Developed Region of the Free Jet After the transition region, the fully developed region follows at a relative nozzle spacing of $x/D \approx 20$ (see Fig. 2). In this region, the turbulent free jet is fully developed. This means that the velocity profile in all cross-sections can be represented by the same function; in this case, this is the Gaussian bell curve. Due to this fact, the velocity profiles are fluid mechanically similar, which is why the fully developed region is also referred to as the similarity region. If the local velocity $w(y)$ is now related to the jet centre velocity w_{M} and the distance in the y-direction (transverse distance) y to $y_{0.5}$, all velocity profiles can be represented by a single curve. The transverse distance at which the local velocity $w(y)$ is half the centre velocity of the free jet is referred to as $y_{0.5}$ (see Eq. (3)) [2].

$$w(y_{0.5}) = 0.5 \cdot w_{\mathrm{M}} \tag{3}$$

The calculation of the velocity profiles is discussed in more detail in Sect. 2.3. Furthermore, the fully developed region of a rotationally symmetrical subsonic free jet can be divided into a near field and a far field. This clear subdivision is only possible if it is an undisturbed free jet and if it is not influenced by walls, interfering bodies or similar [2].

The near field is the area of the fully developed region of undisturbed free jets that directly follows the transition region. In this area, the nozzle configuration, the properties of the free jet medium and the thermodynamic state of the jet at the nozzle outlet are largely responsible for the formation of the free jet. The thermodynamic state and the state of motion of the surrounding medium also influence the formation of the jet in the near field [2].

After the near field comes the far field. In this area, the velocity profiles of the free jet can no longer be used to draw conclusions about the geometric shape of the nozzle or the initial jet structure. Due to the reduced kinetic energy or the low specific impulse, buoyancy forces and gravitational forces as well as other environmental influences become more important for the further jet development [2].

2.3 Calculation of the Velocity Profiles in the Free Jet

Different approaches exist for calculating the velocity profiles in the subsonic free jet. Free turbulence was first described by Ludwig Prandtl's mixing path approach for the exchange of turbulent work processes. The mixing length is the length along which a fluid element loses its individuality due to turbulent mixing with the surrounding medium [3]. According to Prandtl, the mixing length l in each cross-section of a free turbulent flow is proportional to the jet width b at that point and can be described using Eq. (4).

$$l = \alpha \cdot b \tag{4}$$

Here, b corresponds to the semi-diameter of the parabolic velocity distribution and α to a proportionality constant that must be determined experimentally. This approach forms the basis for today's approximation equations. As already mentioned, there are several approaches for the mathematical representation of the velocity distribution, including the Tollmien approximation, and the Reichardt or Schlichting approaches. Each of these approaches has advantages and disadvantages, although all are valid. The Reichardt approximation is discussed below [2,3].
Reichardt uses the Gaussian curve approach as an approximation equation for the velocity profile in the free jet [2]. He describes the velocity distribution with the following equation.

$$\frac{w_{\mathrm{M}}}{w_{\mathrm{D}}} \approx \exp[-0.693 \cdot (y/y_{0.5})^2] \tag{5}$$

Here $y_{0.5}$ corresponds to the transverse distance at which the local velocity w is half as great as the beam centre velocity w_{M}.

$$y_{0.5} = y(w/w_{\mathrm{M}} = 0.5) \tag{6}$$

2.4 Flow and Pressure Losses

This section deals with the basics of calculating flows, pressure losses and loss coefficients. The mathematical derivation and fluid mechanical background, the formulae and equations required for analysing the experiments are explained in more detail.
The Bernoulli equation forms the basis for calculating stationary, incompressible flows. This is obtained by integrating Euler's flow equation (Eq. 7) [7].

$$g \cdot z + \frac{p}{\rho} + \frac{w^2}{2} = \text{const.} \tag{7}$$

The formula describes the law of conservation of energy that the sum of potential energy and kinetic energy within a system always remains constant, meaning that energy cannot be lost. The potential energy can be divided into position energy and pressure energy. If the Bernoulli equation is extended by the specific dissipation ϕ, the total energy balance of a system can now be calculated, taking friction into account. This results in the following formula (Eq. 8) [7].

$$g \cdot z_2 + \frac{p_2}{\rho} + \frac{w_2^2}{2} = g \cdot z_1 + \frac{p_1}{\rho} + \frac{w_1^2}{2} - \phi_{12} \tag{8}$$

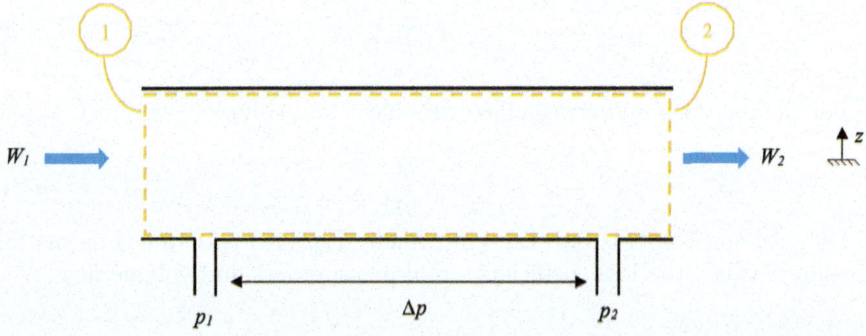

Fig. 3. Calculation model specific dissipation of a pipe flow (stationary, incompressible)

To calculate the specific dissipation of a pipe flow, the model shown in Fig. 3 can be used as the basis for the mathematical derivation.

From Fig. 3 can be seen that the positional energy is zero, as the flow does not cover any height difference. The height at the inlet z_1 and the height at the outlet z_2 is zero, which is why these terms from Eq. (8) can be eliminated. As the velocity does not change due to the constant cross-section, this can also be eliminated. The following Eq. (9) therefore results for the specific dissipation ϕ of a stationary and incompressible flow [7].

$$\begin{aligned} g \cdot z_2 + \frac{p_2}{\rho} + \frac{w_2^2}{2} &= g \cdot z_1 + \frac{p_1}{\rho} + \frac{w_1^2}{2} - \phi_{12} \\ \frac{p_2}{\rho} &= \frac{p_1}{\rho} + \frac{w_1^2}{2} - \phi_{12} \\ \phi_{12} &= \frac{p_1 - p_2}{\rho} \end{aligned} \tag{9}$$

With

$$p_{\text{ges}} = p_1 - p_2 \quad , \tag{10}$$

the following relationship is obtained for the total pressure loss.

$$\Delta p_{\text{ges}} = \rho \cdot \phi_{12} \tag{11}$$

According to Wagner, the specific dissipation is defined by Eq. (12). It represents the relationship between specific dissipation ϕ, flow velocity w and loss coefficient of the pipe flow ξ [7].

$$\phi = \xi \cdot \frac{1}{2} \cdot w^2 \tag{12}$$

If Eq. (12) is solved for the loss coefficient ξ and the specific dissipation ϕ is replaced by the relationship described in Eq. (11), the following formula is obtained for calculating the loss coefficient ξ of a flow.

$$\xi = \frac{2 \cdot \Delta p_{\text{ges}}}{\rho \cdot w^2} \tag{13}$$

For the relationship between flow velocity w and volume flow rate $\dot{V}$ applies.

$$w = \frac{\dot{V}}{A} \tag{14}$$

By inserting Eq. (14) into Eq. (13), finally Eq. (15) is obtained as the relationship between the loss coefficient, total pressure loss and volume flow.

$$\xi = \frac{2 \cdot \Delta p_{\text{ges}}}{\rho} \cdot \left(\frac{A}{\dot{V}}\right)^2 \tag{15}$$

2.5 Schlieren Photography

Schlieren photography belongs to the category of optical measurement methods. The main advantage of optical measurement methods in experimental fluid mechanics is that they work without contact. A negative influence on the measured values due to the introduction of measurement set-ups or measurement equipment such as sensors or impellers into the flow, which could result in disturbances or changes to the flow, is therefore completely ruled out.

The basic principle of the optical methods used in experimental fluid mechanics is based on the fact that density differences or density gradients in a fluid result in a change in the optical refractive index [8]. This is also the case for the schlieren photography, which visualises the optical inhomogeneity of a flow by projecting light and shadow patterns, the so-called schlieren, onto a camera, a ground-glass screen or a screen. Strictly speaking, the schlieren photography does not depict the flow field, but the density field of the flow. This is possible due to the dependency between the density ρ and the refractive index n. The relationship according to Gladson-Dale can be seen in Eq. (16) [9].

$$n - 1 = K \cdot \rho \tag{16}$$

K corresponds to the Gladson-Dale constant [m^3/kg]. This material constant is independent of the wavelength and temperature of the fluid, as it only changes significantly with resonance excitation of atoms and molecules and, according to Fiedler, these areas are not in visible light [9]. The following equation for the relationship between density differences and refractive indices in the flowing fluid results from ratio formation. The material constant K is eliminated due to the ratio formation.

$$\frac{n_0 - 1}{n_1 - 1} = \frac{\rho_0}{\rho_1} \tag{17}$$

A distinction is essentially made between two methods for imaging schlieren: The simple schlieren method, also known as the shadow method, and Toepler's schlieren photography. The latter is used most frequently, which is why it is discussed here in more detail.

2.5.1 Simple Schlieren Photography (Shadow Method) The simple schlieren photography is a subset of Toepler's schlieren photography and is used to visualise flow or density fields. The schematic set-up of the shadow method can be seen in Fig. 4. It essentially consists of a light source and a lense [10]. Parallel light is generated with the aid of the lens or a parabolic mirror; no other optical elements are used. The light rays are guided through the flowing fluid via the lens before they finally image the schlieren and shadows on a ground glass disc, camera or screen. These are images of both the unfractionated and the refracted light rays. For this reason, the shadow method is far less sensitive to scattered light than Toepler's schlieren photography [10].

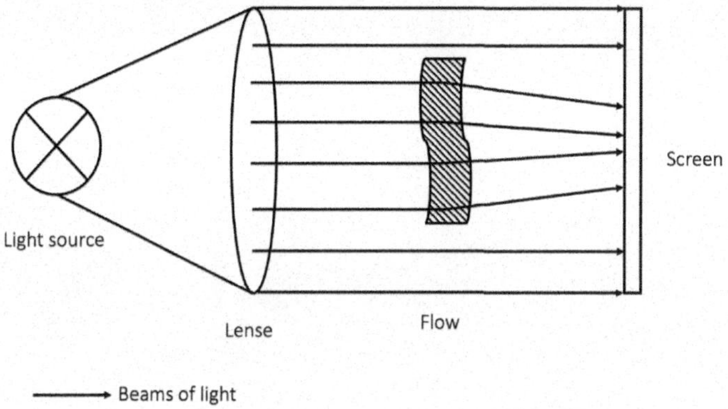

Fig. 4. Schematic principle of the simple schlieren photography (shadow method)

2.5.2 Toeplers's Schlieren Photography In contrast to the simple schlieren method, Toepler's schlieren photography uses additional optical elements, such as apertures, in the beam path of the measuring section. For this reason, the experimental setup is far more complex, but the resolution is better. Furthermore, Toepler's schlieren photography differs from the shadow method in the following points [9]:

- The differences in brightness are proportional to the first derivative of the refractive index, i.e. proportional to the gradient of the light path, whereas the shadow method is proportional to the second derivative.
- The schlieren can also be displayed in light colour on a dark background (dark field method).
- Better brightness gradation possible.
- The direction of the density gradient can be determined (light-dark/dark-light).
- Significantly higher sensitivity.
- Colour visualisation of density differences possible (colour schlieren).

As mentioned above, Toepler's schlieren method has several advantages over the shadow method. However, these also result in a significantly more complex structure. The basic structure of a schlieren system in Toepler's Z arrangement is visualised in Fig. 5. As with the shadow method, the light source is one of the most important components of the system in Toepler's schlieren photography. Ideally, a high-pressure gas discharge lamp with thermal white light is used for this purpose. This has a high intensity with a very small light centre, which offers the advantage of effective focusing. As can be seen in Fig. 5, the light from the light source is focussed in a converging lens before passing through an aperture with a cutting edge on both sides. The prism shown in Fig. 5 is only required for the colour schlieren method and is an extension of Toepler's schlieren photography. This will be discussed in more detail later.

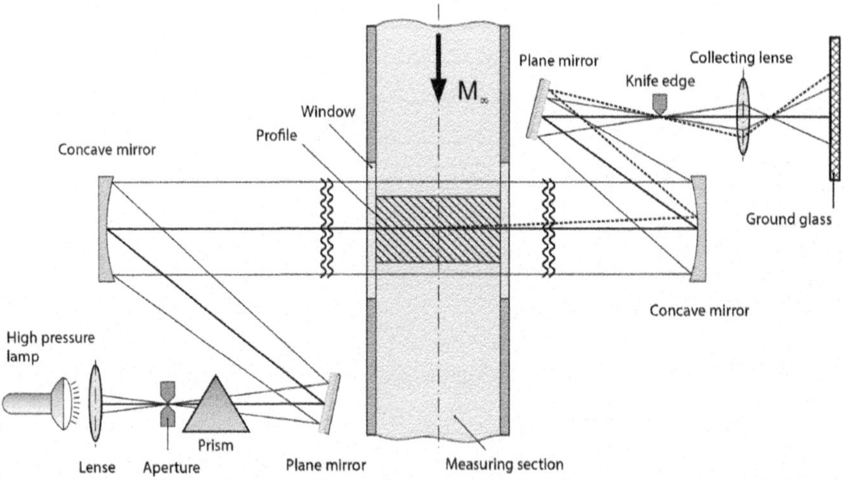

Fig. 5. Schematic principle of schlieren system in Toepler's Z arrangement [8]. Reprinted with permission from SpringerNature publishers

In the image of the flow field, the areas with positive density gradients are lightened and those with negative density gradients are darkened as a result of the separation by the aperture with a one-sided cutting edge. The incline of the density gradient can be defined with the position of the schlieren aperture (aperture with a one-sided cutting edge). The direction of the density gradient, i.e. light-dark or dark-light, can therefore be set. For smaller systems, prisms can also be used instead of concave mirrors to generate parallel light beams [8]. However, as these become larger and thus heavier and more unwieldy with increasing system size, concave mirrors are often the only option.

The structure of Toepler's schlieren system described above produces black and white images of the schlieren with significant differences in brightness, depending on the difference in density of the flow. For the human eye, colour differences are usually more vivid and clearer to see than shades of grey [8].

For this purpose, there are two ways of modifying the arrangement described above to generate colour streaks. One method involves the additional use of a prism. As shown in Fig. 5, this prism is placed behind the slit aperture in the beam path. The installation of the prism means that the white light is split into its spectral colours and the black and white streaks are displayed in colour. By rotating the prism in the beam axis, the colour can be changed according to the spectral distribution [8]. Another option is to use a colour stripe filter instead of the schlieren aperture. This contains a green stripe in the centre and a red and blue stripe on the outside. Such a colour stripe filter can be created using a slide, for example, by photographing the above-mentioned colour configuration and developing a positive film from it.

3 Experimental Setup and Methodology of the Investigation of the Free Jet

Two different nozzles are analysed in the course of this work. The nozzles are made of POM-C and differ only in the cross-section of their outlet opening. One has an outlet diameter of 8 mm and the other one 12 mm. Both have the same flow geometry with a conical reduction and an apex angle of 20°.

3.1 Determination of the Jet Angle of the Nozzles

In order to gain further insights into the effective range of the air jet, the free jet of the nozzles is visualised using the schlieren photography. This is primarily intended to provide information on the jet angle and the characteristics of the free jet. To visualise the free jet, the schlieren optics must be set up as shown schematically in Fig. 6. A converging lens is installed behind the light source to focus the light beams. This is followed by an optical aperture. The light beam is then guided via a concave mirror, a plane mirror and a further concave mirror before passing through a second aperture.

A converging lens is mounted behind this aperture in order to project the light beam onto a reflex camera. In order to visualise the images and align the nozzles, the reflex camera is connected to a screen. The nozzle is clamped in the holding device and attached to a tripod so that the direction of the beam axis points vertically upwards, as shown in Fig. 7. Tripod and nozzle must be placed between the first and second parabolic mirrors. The height of the nozzle is adjusted via the tripod so that the nozzle tip, with the outlet opening, is extended about 5 mm into the image from below. A differential pressure sensor with a resolution of 0 to 0.2 bar is connected to the T-piece in front of the nozzle in order to determine the flow rate or pressure at the nozzle and to be able to adjust it during the experiment. The differential pressure sensor also requires a power supply and a voltmeter to display the pressure at the nozzle (see Fig. 7). CO_2 is used as the fluid for the experiment instead of compressed air. With a density of 1.98 kg/m^3, CO_2 has a higher density than air at 1.29 kg/m^3 [11]. Due to the different density, the CO_2 flowing out of the nozzle stands out even more

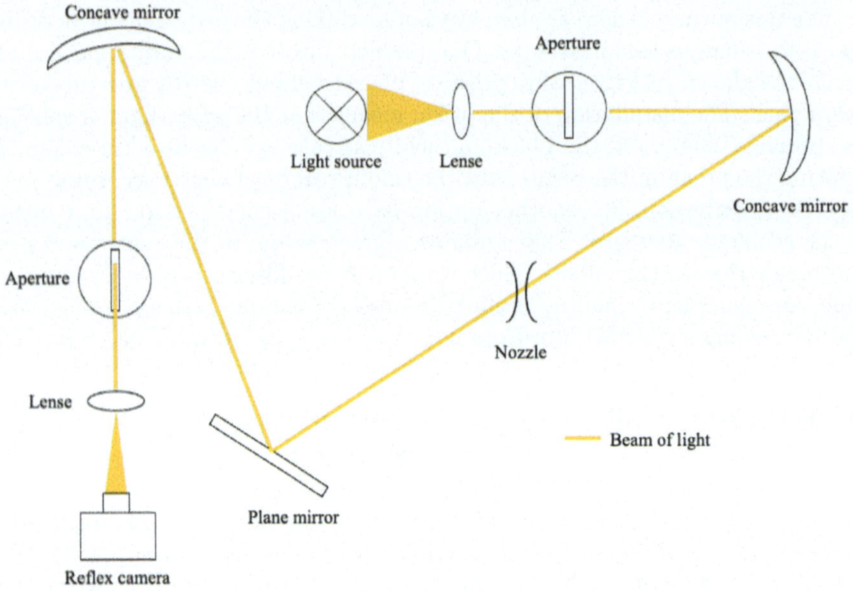

Fig. 6. Schematic principle of the experimental set-up of the schlieren system

clearly from the ambient air in the schlieren optics, ensuring a better imaging of the free jet. Any gas that has a different density to air is suitable for the experiment. However, CO_2 is an inexpensive, non-hazardous and non-explosive gas, which is why it is ideal for the experiments and is used here for this reason.

Ideally, the experiment should be carried out in cloudy weather, before sunrise or after sunset. In this way, the effects of stray light can be avoided. In addition, all interfering light sources should be switched off. Using the pressure reducer and the pressure sensor, different pressures are set at the nozzle and two images are taken at the same pressure level. The pressures to be set are 10 mbar, 20, 40, 80 and 120 mbar. The reflex camera is used in manual photo mode. For optimum photos, select an exposure time of 1/4000 s and a sensitivity of the image sensor of ISO 100.

3.2 Determination of the Velocity Profile of the Free Jet

In order to be able to make more precise statements about the effective range of the nozzles, the velocity profile of the nozzles is analysed in addition to determining the free jet angle. A Pitot tube is used to create a pressure profile of the dynamic pressure of the free jet. The velocity profile of the nozzle can then be derived from this.

A dynamic pressure sensor, also known as a Pitot tube, is required to measure the dynamic pressure of the free jet at different positions. This is produced using two stainless steel tubes with an internal diameter of 1.5 and 2 mm, which

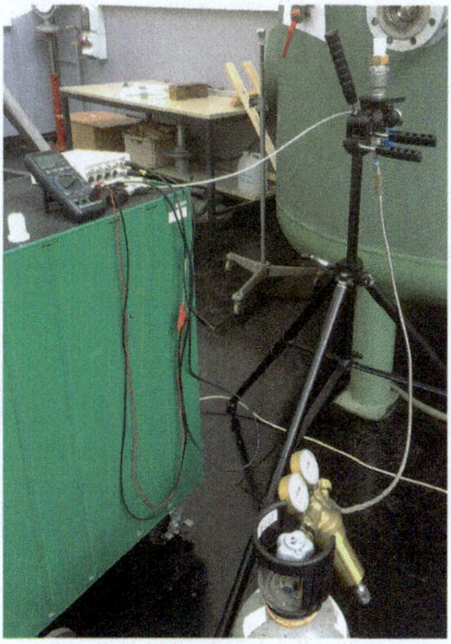

Fig. 7. Experimental set-up of the nozzle in the schlieren system

are bent by hand and inserted into each other, as shown in Fig. 8. The tubes must then be glued airtight with super glue. In order to ensure the accurate measurement of the defined positions, a holder is required to guarantee precise positioning. An aluminium block with the dimensions $60 \times 60 \times 100\ \mathrm{mm}^3$ is employed as the holder. The inlet opening of the Pitot tube is aligned with the front edge of the holder and the centre axis of the tube is positioned 30 mm from the sides of the block. The bore diameter is selected to be 6.2 mm, allowing a pneumatic tube with a diameter of 6 mm to be firmly positioned within the bore. Subsequently, the tube is led out at an angle to one side of the holder with the aid of a second hole. The Pitot tube is inserted into the pneumatic tube on the holder, as illustrated in Fig. 8, ensuring that the inlet opening of the Pitot tube is aligned with the front edge of the block. To prevent the probe from bending, it must also be secured in place with modelling clay. The position of the inlet opening at the base of the holder on the front should then be marked, serving as an adjustment aid for the measuring position.

The experiment also requires a grid with the coordinates of the points to be measured. For this purpose, a grid of 0.5 cm in the y-direction and 3 cm in the x-direction is printed on an A3 sheet (landscape format). The x-axis, and therefore all coordinates with the value $y = 0$, correspond to the centre axis of the free jet in the experiment.

The nozzle is mounted on a stand in its holder and connected to the compressed air supply via a pressure reducer (see Fig. 9). To accurately set the required pressure upstream of the nozzle, a differential pressure sensor is mounted via a T-piece on the holder. The pressure sensor for measuring the upstream pressure at the nozzle has a resolution of 0–600 mbar. As shown in Fig. 9, the Pitot

Fig. 8. Pitot tube used to determine the velocity profile of the free jet

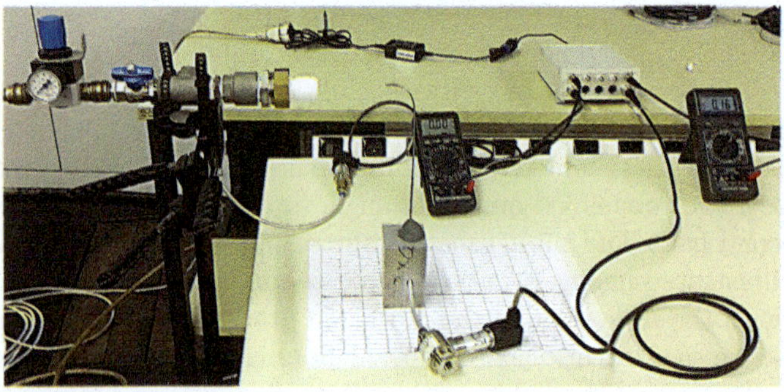

Fig. 9. Test setup for dynamic pressure measurement of the free jet

tube is also connected to differential pressure sensors. For an accurate measurement of the dynamic pressure, two differential pressure sensors with a resolution of 0–200 mbar and 0–100 mbar are used. As shown in Fig. 9, the height of the nozzle is adjusted so that the centre axis of the resulting free jet and the centre axis of the Pitot tube are identical.

The tests were conducted with an upstream pressure of 120 mbar upstream. Subsequently, the probe is employed to ascertain the pressures at the designated measurement coordinates on the grid. It is imperative that, when positioning

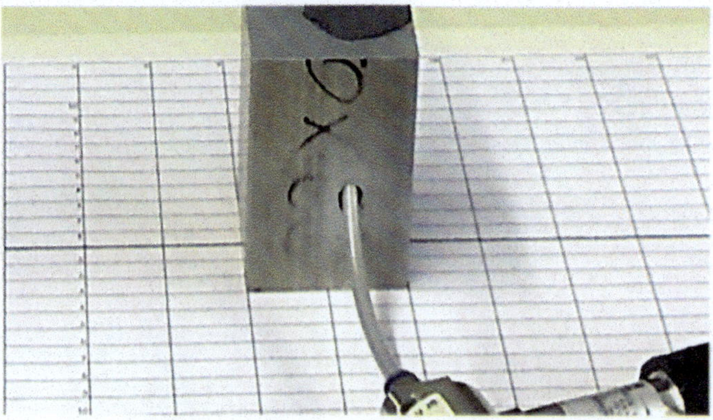

Fig. 10. Positioning of the Pitot tube

the Pitot tube, the centre axis of the inlet opening is maintained in a state of parallelism with the centre axis of the free jet. Parallelism can be achieved by ensuring that the front and side edges of the probe holder are always aligned with the grid lines, as illustrated in Fig. 10.

Upon the occurrence of a pressure reading below 100 mbar, the differential pressure sensor is substituted with one of greater precision, namely one with a resolution of 0–100 mbar. The rotationally symmetrical geometry of the nozzle results in a free jet that is also rotationally symmetrical. Consequently, only half of the jet cone needs to be measured. As the pressure sensors employed are differential pressure sensors, the values obtained for the total pressure correspond to the dynamic pressure. Consequently, it is not necessary to additionally measure the static pressure and subtract this from the dynamic pressure. The velocity profile of the free jet can then be calculated based on the dynamic pressure.

4 Results and Discussion

4.1 Results of the Schlieren Photography

The results of the 8 and 12 mm nozzles, each with the highest and lowest pressure, are shown in the following images as representative of the entire test series. The shape and expansion of the free jet can be clearly seen in Figs. 11 and 12. Now, the beam angle can be determined using the images.

It is striking that the jet angle always remains the same regardless of the pressure and outlet opening; this is confirmed by the literature (see Sect. 2.2) [6]. The jet angle is approximately 22° in each case (see Figs. 11 and 12). It is also noticeable that the turbulence in the edge areas of the free jet is significantly greater at lower pressures than at higher pressures. This is due to the fact that the flow velocity and thus the energy of the free jet are lower at low pressures, resulting in greater mixing with the still ambient air. This can be clearly seen in Fig. 12 when comparing the free jet at 10 mbar and at 120 mbar.

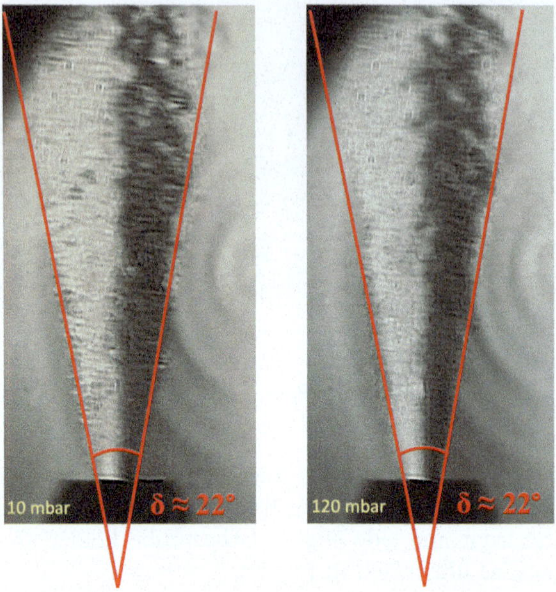

Fig. 11. Schlieren photography of the free jet of the 8 mm nozzle

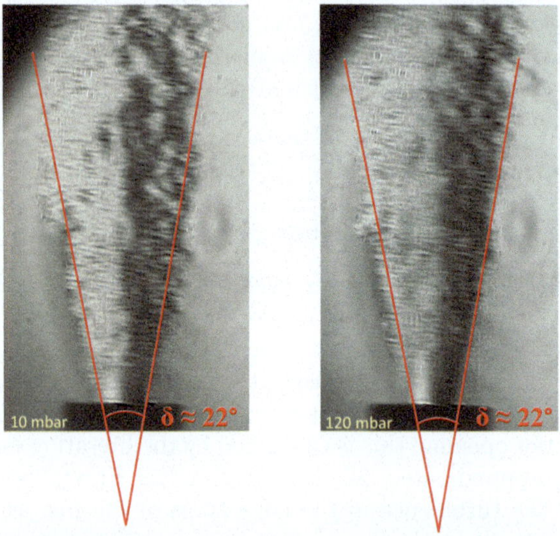

Fig. 12. Schlieren photography of the free jet of the 12 mm nozzle

4.2 Results of Dynamic Pressure Measurements

As the measured pressure corresponds to the local dynamic pressure of the flow, the velocity profile of the free jet can be determined from the definition of the dynamic pressure (Eq. 18).

$$p_{\mathrm{dyn}} = \frac{\rho}{2} \cdot w^2 \tag{18}$$

Rearranging Eq. (18) according to the velocity w results in the following expression to determine the velocity profile of the free jet.

$$w = \sqrt{\frac{2 \cdot p_{\mathrm{dyn}}}{\rho}} \tag{19}$$

Based on the measured values and a density of $\rho = 1.2041\ \mathrm{kg/m^3}$, the velocity distributions shown in Tables 1 and 2 are obtained.

Table 1. Velocity distribution of the half free jet from the 8 mm nozzel in m/s

		Parallel distance x in cm										
		0	3	6	9	12	15	18	21	24	27	30
Radial distance y in cm	6.0											0
	5.5											0
	5.0										0	4.1
	4.5									0	4.1	5.8
	4.0								0	4.1	5.8	7.1
	3.5							0	4.1	5.8	7.1	8.2
	3.0						0	5.8	8.2	9.1	10.0	10.8
	2.5					0	5.8	8.2	12.2	12.9	14.1	13.5
	2.0				0	7.1	12.9	14.7	17.3	18.2	16.8	15.8
	1.5		0	0	8.2	19.1	20.8	21.2	23.8	21.9	20.8	19.1
	1.0		4.1	24.1	32.9	35.5	34.1	31.3	28.8	25.8	23.8	21.6
	0.5	0	85.5	68.8	62.1	52.8	43.3	39.3	33.4	29.7	25.8	24.1
	0	149.2	140.0	120.2	86.8	65.8	51.9	42.9	36.0	32.1	27.9	25.5
		0	3	6	9	12	15	18	21	24	27	30

In order to evaluate the local velocity profiles of the nozzles, a transformation is applied following the methodology described in Sect. 2.3, whereby the velocity profiles are transformed into a dimensionless velocity distribution. The transformation allows the velocity distributions to be superimposed and compared. In order to facilitate evaluation, the local velocities are initially normalised by determining the ratio of the local velocity to the local beam centre velocity for each value.

In the second step, the local transverse distances (y-values) associated with the respective local velocities are transformed with $y/y_{0.5}$ (see Sect. 2.3). In this context, $y_{0.5}$ represents the transverse distance at which the local velocity w is equal to the local yet centre velocity w_{M} (see also Eq. 6). In the absence of an exact measured value at which the local velocity is equal to half the beam centre

Table 2. Velocity distribution of the half free jet from the 12 mm nozzel in m/s

		Parallel distance x in cm										
		0	3	6	9	12	15	18	21	24	27	30
Radial distance y in cm	6.0											0
	5.5										0	4
	5.0									0	4	7.1
	4.5								0	6	7.1	9.1
	4.0							0	4	7.1	10.0	12.9
	3.5						0	6	9.1	11.5	13.5	14.7
	3.0					0	4	11.5	14.7	17.8	18.2	19.5
	2.5				0	6	7.1	18.2	20.8	21.6	22.3	23.1
	2.0			0	12	18.2	25.8	29.1	28.8	28.8	27.9	27.6
	1.5		0	14	31.3	36.5	40.8	40.8	38.7	35.1	32.9	30.2
	1.0		29.9	47.0	60.4	55.0	53.9	49.9	44.6	41.0	37.6	33.6
	0.5	0	128.2	104.7	91.1	76.0	65.2	58.2	50.4	44.5	39.7	35.8
	0.0	147.6	143.4	134.6	110.9	87.9	71.8	60.4	52.5	46.3	40.8	36.9
		0	3	6	9	12	15	18	21	24	27	30

velocity ($w(y) = 0.5$), the value $y_{0.5}$ is determined using linear interpolation. To achieve this, the two pairs of values between which the value $w(y) = 0.5$ is located are taken and the value $y(w = 0.5)$ is calculated using the linear equation $y = mx + b$.

The normalised value pairs w/w_{M} and $y/y_{0.5}$ of the 8 mm nozzle are then transferred to a coordinate system. Subsequently, the measured values can be verified by plotting the Reichardt functions for the velocity distribution. The velocity distribution may be represented in the form of a Gaussian bell curve in accordance with Eq. 5. Ideally, all pairs of values would lie on the curve described. However, only the pairs of values from the core and transition area exhibit this form of velocity distribution, as outlined in Sect. 2.2. Consequently, they may deviate from the curve. For the 8 mm nozzle, the results from evaluating the velocity profiles are illustrated in Fig. 13.

From Fig. 13 can be seen that the measurement outcomes exhibit minimal deviation from the theoretically calculated velocity distribution, which confirms the accuracy of the measurement results. It is noticeable that the deviations near the jet boundary, specifically for values $w/w_{\mathrm{M}} < 0.3$, become more pronounced. This is due, among other factors, to the dynamic pressures measured, which are markedly low in these regions of the free jet, reaching a maximum of 1 mbar. Such low pressures, i.e. flow velocities are highly susceptible to interference from the surrounding environment, which can result in deviations. Additionally, the flow measurement technology employed may also contribute to deviations at such low pressures, due to its measurement tolerances within the wide resolution range of 0–100 mbar. The dimensionless velocity distribution for the 12 mm nozzle is illustrated in Fig. 14. As with the previous example, the measured values exhibit a distinct bell curve and are within the approximation function, with only minimal deviations.

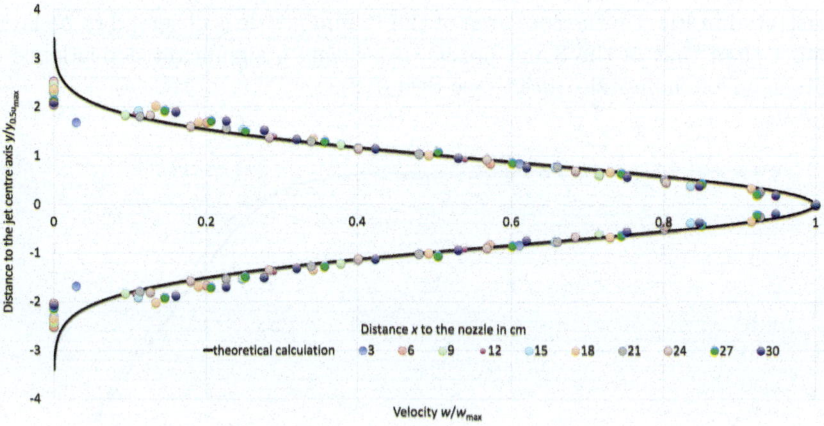

Fig. 13. Dimensionless velocity distribution of the 8 mm nozzle

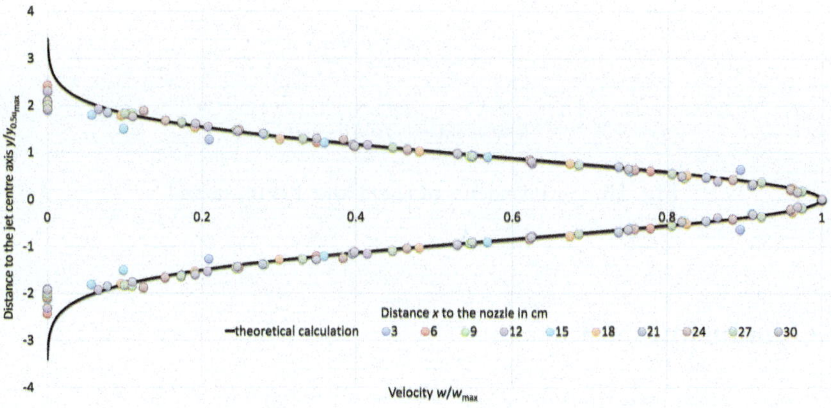

Fig. 14. Dimensionless velocity distribution of the 12 mm nozzle

As with the 8mm nozzle, the deviations between the measured value and the approximation function increase in the area of the jet boundaries. This is also due to the aforementioned deviations resulting from the low flow velocities. Particularly striking in Fig. 14 are the blue measured values of the velocity distribution at a distance of 3 cm from the nozzle outlet. These exhibit a significant deviation from the approximation function, indicating that the measured values originate from the core area of the free jet and that the velocity distribution has not yet assumed the shape of the Gaussian bell curve (see Sect. 2.2).

The obtained results for the beam centre velocity are illustrated in Fig. 15. In case of the 8 mm nozzle the resulting values for the velocity are already in the transition area at a measuring distance of 3 cm from the nozzle, whereas the measured values of the 12 mm nozzle originate from the core area. Furthermore, as illustrated in Fig. 15, the core area of the 12 mm nozzle is observed to be larger or longer than that of the 8 mm nozzle, indicating a significant dependence on the diameter of the nozzle outlet (see Sect. 2.2).

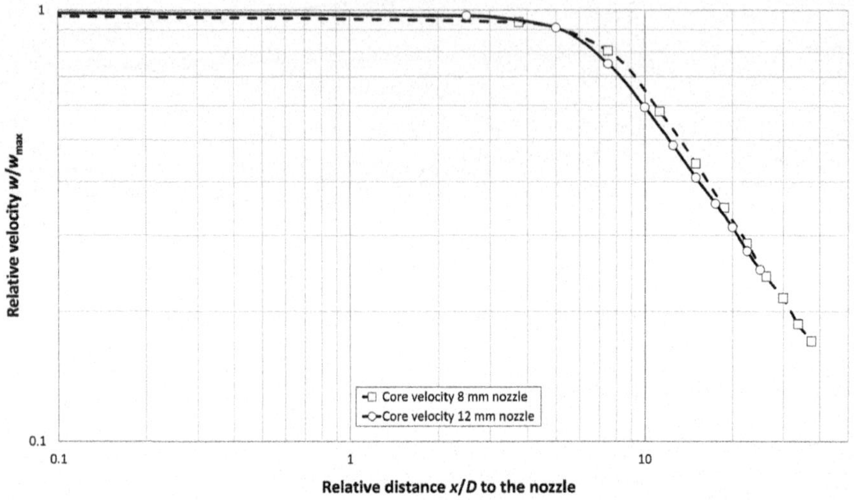

Fig. 15. Core velocity of the 8 and 12 mm nozzel

5 Conclusion

The experimental investigation about the free jet behaviour has provided insights into the relationship between nozzle diameter, pressure, and jet characteristics. Through the use of schlieren photography and Pitot tube measurements, this study has elucidated various aspects of jet formation, including the jet angle and velocity distribution.

One of the most notable findings from the schlieren photography is the consistent jet angle observed across different pressures and nozzle diameters. Despite variations in pressure and nozzle size, the jet angle remained approximately 22°. This result aligns with existing literature, confirming that the jet angle is predominantly influenced by the nozzle design and not significantly affected by the pressure changes within the experimental range.

Furthermore, the schlieren photography revealed that turbulence in the jet's boundary regions increases at lower pressures. At reduced pressures, the flow velocity decreases, leading to greater mixing with the surrounding air and enhanced turbulence. This effect is particularly pronounced in the lower pressure settings, where the jet exhibits more pronounced boundary turbulence compared to higher pressures. This finding highlights the importance of using adequate pressure levels to achieve more stable jet behaviour.

The Pitot tube measurements provided detailed velocity profiles for both the 8 and 12 mm nozzles. The velocity profiles were analysed to determine the distribution of flow velocities within the jet. The results indicated that the velocity profiles for both nozzle sizes followed a Gaussian distribution in the core and transition areas, with deviations observed at the jet boundaries. The 12 mm nozzle, having a larger diameter, exhibited a more pronounced core area compared to the 8 mm nozzle. This suggests that larger nozzles produce jets with more extended and coherent cores, which may be advantageous for applications requiring a stable and high-velocity jet.

The study also identified that deviations in the velocity profiles increased near the jet boundaries, particularly at low velocities. These deviations can be attributed to the susceptibility of low-pressure measurements to environmental interference and the limitations of the measurement technology. Despite these challenges, the core and transition areas of the jet showed good agreement with theoretical predictions, confirming the accuracy of the experimental setup and measurement techniques.

Acknowledgments. The authors acknowledge the Studienstiftung des deutschen Volkes for its comprehensive support of this research work through the doctoral scholarship. Also many thanks to the staff of the laboratory for thermodynamics and fluid dynamics engineering for the kind support to realise the experiments.

References

1. Schlichting, H.: Grenzschicht-Theorie. Springer, Berlin (2006)
2. Rist, D.: Dynamik realer Gase: Grundlagen, Berechnungen und Daten für Thermogasdynamik. Strömungsmechanik und Gastechnik. Springer, Berlin (1996)
3. Oertel, H.: Prandtl - Führer durch die Strömungslehre. Vieweg+Teubner Verlag, Wiesbaden (2008)
4. Schlichting, H.: Grenzschicht-Theorie. G. Braun, Karlsruhe (1965)
5. Ganzer, U.: Gasdynamik. Springer, Berlin (1988)
6. Kümmel, W.: Technische Strömungsmechanik: Theorie und Praxis. B.G. Teubner Verlag, Wiesbaden (2007)
7. Wagner, W.: Strömung und Druckverluste. Vogel Buchverlag, Würzburg (2012)
8. Nitsche, W.: Strömungsmesstechnik. Springer, Heidelberg (2006)
9. Fiedler, O.: Strömungs- und Durchflussmesstechnik. R. Oldenbourg, München (1992)
10. Eckelmann, H.: Einführung in die Strömungsmesstechnik. B. G. Teubner, Stuttgart (1996)
11. Gomeringer, R.: Tabellenbuch Metall. Europa Lehrmittel, Messstetten (2014)

Overview Lecture

A Short Introduction to the Shape Optimization of Structures

Andreas Öchsner(✉)

Esslingen University, Faculty of Mechanical and Systems Engineering, 73728 Esslingen, Germany
andreas.oechsner@gmail.com

Abstract. This chapter provides an introduction to the topic of judging the lightweight potential of structures. Two different concepts, i.e. the so-called lightweight index and the specific energy absorption, are discussed. The basic idea of shape optimization is discussed in the context of one-dimensional structural members, i.e., bars and beams, and one- and two design variables are considered.

Keywords: Lightweight design · Lightweight potential · Lightweight index · Specific energy absorption · One-dimensional elements

1 Introduction

Lightweight design plays a central role in the transportation sector (for example, in aerospace or automotive engineering) because a weight reduction is immediately reflected in a reduced fuel consumption.

As a rule of thumb for the influence of the weight on the fuel costs of airplanes, one can expect for a *reduction of 1% of the weight, a fuel economy—depending on the airplane engines—of 0.75 to 1%* [17]. Considering a total fuel consumption, for example, of the entire Lufthansa fleet in 2015 of 8,947,766 t [10], one can expect—depending on the actual kerosine price [6]—a potential of a few millions of Euro as potential savings.

The crucial question in this context is how engineers can judge and compare the lightweight potential of materials and structures. It is obvious that solely the mass density or the total mass cannot serve for this purpose since mechanical properties (e.g. stiffness or strength) and structural parameters (e.g. geometry and boundary conditions) influence as well this question. The following sections will introduce different concepts to judge the lightweight potential. A common concept, which is based on the so-called lightweight index or the specific energy absorption allows the consideration of the structure, including support and load conditions. In addition, the classical concept to design a structure based on the

A. Öchsner (ed.), *Collaborative Research Advancing Engineering Solutions for Real-World Challenges 2*, Proceedings in Technology Transfer,
https://doi.org/10.1007/978-3-031-79093-5_11

maximum stress value can be included. Thus, we can combine the stress-based design criterion, commonly applied in the first semesters of applied mechanics education, with the weight of a structure. If the specific energy absorption is chosen, a quite powerful index number is available, see [14] for more details (e.g. the consideration of plastic material behavior or complex loading conditions).

2 Lightweight Potential of Structures

Table 1 summarizes some typical material properties. The following calculations are based on the average values of the corresponding intervals.

Table 1. Properties of different materials: E: Young's modulus; ν: Poisson's ratio; $R_{\mathrm{p}0.2}$: initial yield stress; ε_{A}: strain at failure

	E, $\frac{\mathrm{N}}{\mathrm{mm}^2}$	ν, –	ϱ, $10^{-6}\,\frac{\mathrm{kg}}{\mathrm{mm}^3}$	$R_{\mathrm{p}0.2}$, $\frac{\mathrm{N}}{\mathrm{mm}^2}$	ε_{A}, –
Stainless steel	195000	0.30	7.8	393	0.5
(austenitic)	190000 ... 200000		7.5 ... 8.1	286 ... 500	0.45 ... 0.65
Al alloy	74000	0.33	2.75	364	0.75
	69000 ... 79000		2.6 ... 2.9	100 ... 627	0.05 ... 0.3
Ti alloy	105000	0.35	4.7	750	0.8
	80000 ... 130000		4.3 ... 5.1	180 ... 1320	0.06 ... 0.3

Adapted from [2]

2.1 Estimation Using the Lightweight Index

In order to characterize the performance of a lightweight construction, the so-called lightweight index M according to Klein [9] will first be used as defined below:

$$M = \frac{F_0}{F_{\mathrm{G}}}, \tag{1}$$

where the following comments should be done:

- external force F_0 (multiplied with safety factor SF),
- dead weight $F_{\mathrm{G}} = mg = V\varrho g$, for mass $m = \text{const.}$,
- dimensionless performance index; the larger the value, the more efficient a lightweight construction.
- Alternative formulation according to Ashby [3]: mass m and not F_{G}.

The concept of the lightweight index and the associated estimation of lightweight potential can also be applied to other problem areas. For *stability problems*, i.e. for the buckling slender bars or thin-walled tubes or plates, the external force F_0 in Eq. (1) can be replaced by the so-called critical force F_{cr} (buckling force): $M = \frac{F_{\mathrm{cr}}}{F_{\mathrm{G}}}$.

For the example of the tensile bar shown in Fig. 1a, a constant normal force distribution results and thus a constant normal stress along the entire tension bar.

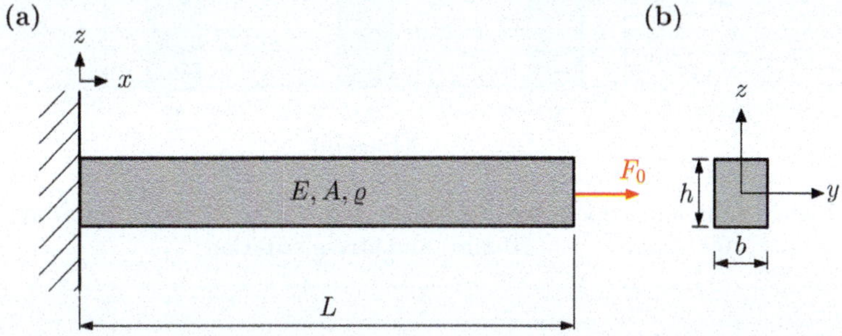

Fig. 1. **a** General configuration of a tensile bar with a single force and **b** cross section

Thus, the *limit of the load-carrying capacity* can be defined by the state where the tensile stress reaches a material characteristic (here the 0.2% strain limit or initial yield stress $R_{\mathrm{p0.2}}$, see Table 1:

$$\sigma_x = \frac{N_x}{A} = \frac{F_0}{A} \stackrel{!}{=} R_{\mathrm{p0.2}}. \tag{2}$$

Therefore, in the definition of the lightweight index according to Eq. (1), the external force F_0 can be replaced by material and geometry parameters:

$$M = \frac{F_0}{F_{\mathrm{G}}} = \frac{AR_{\mathrm{p0.2}}}{AL\varrho g} = \frac{R_{\mathrm{p0.2}}}{\varrho g L} \sim \frac{R_{\mathrm{p0.2}}}{\varrho}. \tag{3}$$

Based on the definition according to Eq. (3) and the characteristic values according to Table 1, the lightweight potential of stainless steel (St), Al and Ti alloys is shown comparatively in Fig. 2. It can be seen that the highest lightweight potential is not achieved by the lightest material (Al), but by the Ti alloy.

2.2 Estimation Using the Specific Energy Absorption

The previous explanations also reveal some disadvantages in connection with the definition of the lightweight index M (see Eq. (1)):

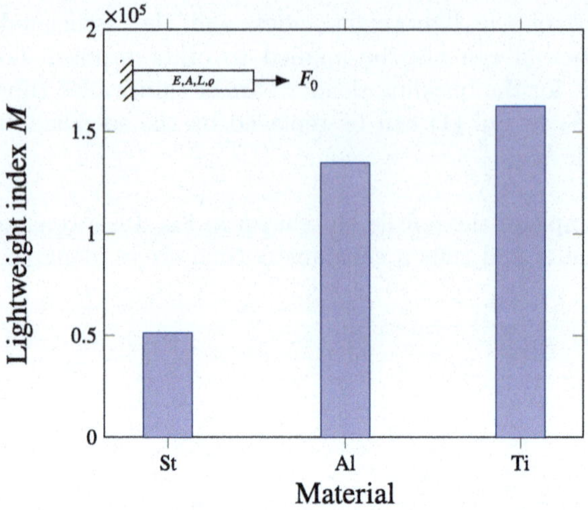

Fig. 2. Lightweight index for a tension bar as a function of the material at constant geometry ($L = 100$ mm, $b = h = 10$ mm) and stress criterion

- Complicated load cases may not be possible to be summarized (vector addition) to a single resulting force;
- the consideration of distributed loads is not directly possible;
- the consideration of volume loads is not directly possible;
- the consideration of external bending moments or torsional moments is not possible;
- The consideration of plastic material behavior is difficult, since, for example, in the case of ideal plasticity, the deformation increases at a constant stress.

Therefore, an alternative approach will be introduced below. The specific energy absorption (SEA) allows for the consideration of the above-mentioned limitations and can be stated as follows:

$$SEA = \frac{\Pi}{m}, \tag{4}$$

where Π is the so-called total strain energy and m is the mass of the member. This index number is often used to assess the crash behavior of lightweight structures (see [8,18]). A descriptive definition of the total strain energy can be based on the tensile test as the area under the force-displacement diagram (see Fig. 3a).

An easily interpreted graphic representation of the total strain energy—area under the external force curve—is

$$\Pi = \frac{1}{2} F_0 u_0, \tag{5}$$

or in an integral approach:

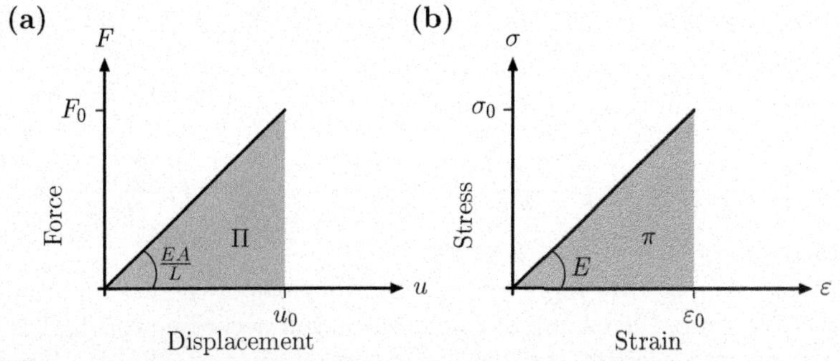

Fig. 3. Experimental data from a tensile test: **a** force-displacement and **b** stress-strain diagram

$$\Pi = \int_0^{u_0} F(u)\mathrm{d}u = \int_0^{u_0} \frac{EA}{L} u\mathrm{d}u = \frac{EA}{2L} u_0^2 = \frac{F_0^2 L}{2EA} = \frac{1}{2} F_0 u_0. \tag{6}$$

Alternatively, the strain energy can be expressed based on the internal reactions, i.e. the normal force distribution. For this purpose it is advisable to integrate the over the specific strain energy, see Fig. 3b:

$$\mathrm{d}\Pi = \mathrm{d}\pi \mathrm{d}V = \sigma \mathrm{d}\varepsilon\, \mathrm{d}V = E\varepsilon \mathrm{d}\varepsilon\, \mathrm{d}V = \frac{E\varepsilon^2}{2}\mathrm{d}V = \frac{\sigma^2}{2E}(A\mathrm{d}x) = \frac{N^2}{2EA}\mathrm{d}x\,. \tag{7}$$

Similar derivations can be written for other simple modes of deformation and the following cases can be distinguished for linear-elastic material behavior, see [12]:

- Tension or compression:

$$\Pi = \int_0^L \frac{N_x(x)^2}{2EA}\mathrm{d}x. \tag{8}$$

- Bending:

$$\Pi = \int_0^L \frac{M_y(x)^2}{2EI_y}\mathrm{d}x. \tag{9}$$

- Shear:

$$\Pi = \int_0^L \frac{Q_z(x)^2}{2GA_\mathrm{s}}\mathrm{d}x = \int_0^L \frac{Q_z(x)^2}{2k_\mathrm{s}GA}\mathrm{d}x. \tag{10}$$

- Torsion:

$$\Pi = \int_0^L \frac{M_x(x)^2}{2GI_\mathrm{p}}\mathrm{d}x. \tag{11}$$

Thus, the total strain energy Π in a bar/beam-like structural member can be expressed as:

$$\Pi = \int_0^L \frac{N_x(x)^2}{2EA} \mathrm{d}x + \int_0^L \frac{M_y(x)^2}{2EI_y} \mathrm{d}x + \int_0^L \frac{Q_z(x)^2}{2GA_s} \mathrm{d}x + \int_0^L \frac{M_x(x)^2}{2GI_p} \mathrm{d}x, \tag{12}$$

where N_x, M_y, Q_z, M_x represent the distributions of the internal reactions.

In the following, the example of the tensile rod according to Fig. 1 is reevaluated, taking into account the specific energy absorption (SEA). The normal force distribution follows from Fig. 4 to $N_x(x) = F_0$.

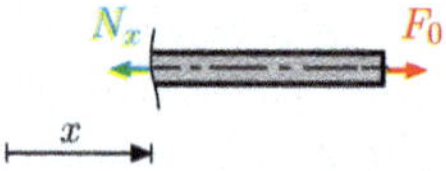

Fig. 4. Internal reactions for tensile bar

Thus, we can calculate the total strain energy as:

$$\Pi = \int_0^L \frac{N_x(x)^2}{2EA} \mathrm{d}x = \frac{F_0^2}{2EA} \int_0^L \mathrm{d}x = \frac{F_0^2 L}{2EA}, \tag{13}$$

whereby the specific energy absorption can be written as:

$$SEA = \frac{\Pi}{m} = \frac{F_0^2 L}{2EA \times AL\varrho} = \frac{F_0^2}{2EA^2\varrho}. \tag{14}$$

Assuming that the limit of the load-carrying capacity can be defined by the state where the tensile stress reaches a material characteristic (0.2% strain limit), the external force is obtained from Eq. (2) as $F_0 = AR_{\mathrm{p}0.2}$. Thus, the maximum specific energy absorption is obtained as:

$$SEA = \frac{A^2 R_{\mathrm{p}0.2}^2}{2EA^2\varrho} = \frac{R_{\mathrm{p}0.2}^2}{2E\varrho} \sim \frac{R_{\mathrm{p}0.2}^2}{E\varrho}. \tag{15}$$

Based on Eq. (15) and the characteristic values according to Table 1, the specific energy absorption of stainless steel (St), Al and Ti alloys is shown in Fig. 5. Again, it is found that the highest energy absorption is achieved not by the lightest material (Al), but by the Ti alloy, thus giving the same trend as when using the lightweight index M, see Fig. 2.

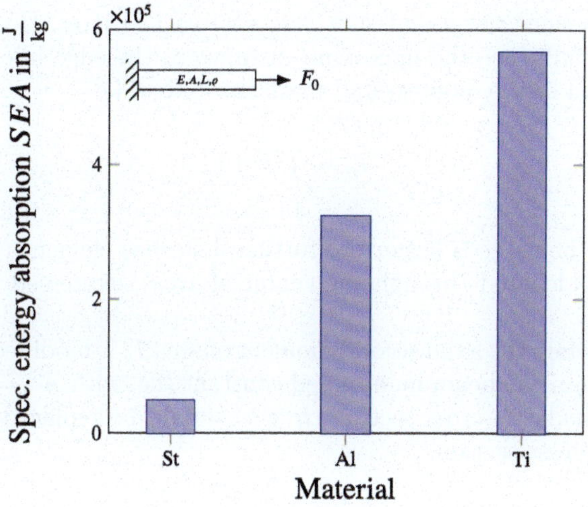

Fig. 5. Specific energy absorption for a tensile bar as a function of the material at constant geometry ($L = 100$ mm, $b = h = 10$ mm) and stress criterion

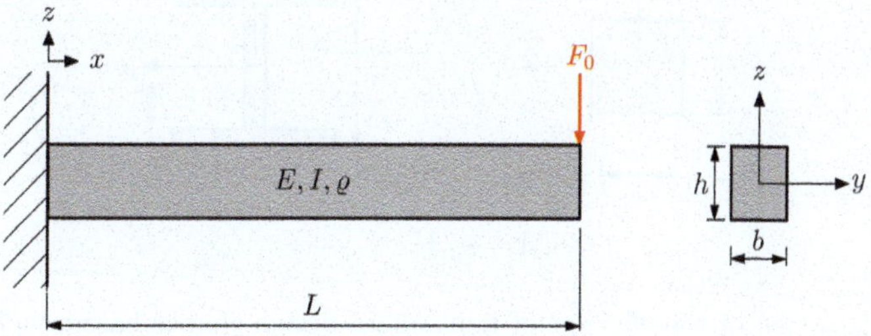

Fig. 6. **a** General configuration of a cantilever beam loaded by a single force and **b** cross section

3 Introduction to Shape Optimization

3.1 Evaluation Using the Lightweight Index

Lightweight design shapes in the context of one-dimensional structural elements is about adapting or optimizing the cross-section to increase the lightweight potential. Thus, it is a purely geometrical problem. Let us consider in the following the example of a cantilever bending beam with a point load at the free end (see Fig. 6) and optimize the structure based on a stress criterion.

In general, the maximum normal stress for the cantilever results in, see [11]:

$$\sigma_{x,\max} = \frac{M_y(x=0)}{I_y} \times z_{\max}, \tag{16}$$

where the cross sectional quantities I_y and $z_{\max}$ influence the maximum stress value. By increasing I_y the stress is reduced and thus the lightweight index should increase since the permissible stress is reached later. However, a change in I_y may also influence the maximum distance to the edge fiber ($z_{\max}$). Based on the definition of the axial second moment of area, i.e.

$$I_y = \int_A z^2 \mathrm{d}A, \tag{17}$$

it follows that the value is larger the further a surface element ($\mathrm{d}A$) is from the y-axis (which is located through the center of area), expressed by the distance (z).

In the following, the axial second moment of area I_y is modified and optimized with the weight of the beam maintained constant (i.e. with $A^{\square} = A^{\mathtt{I}}$), see Fig. 7. The original quadratic cross section ($h \times h$) should be replaced by an I-profile with the same cross-section.

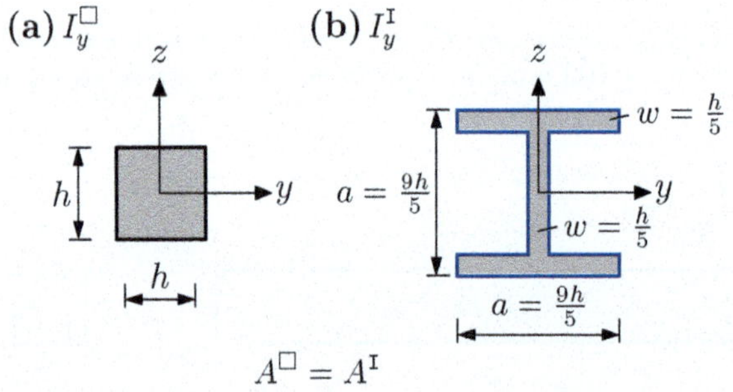

Fig. 7. Realization of the concept of lightweight design shapes based on $I_y^{\mathtt{I}} > I_y^{\square}$ and $A^{\square} = A^{\mathtt{I}}$: **a** Original cross section with $b = h$; **b** modified cross section

The axial second moment of area (with $A^{\square} = A^{\mathtt{I}}$) can be calculated according to [1] as[1]:

$$I_y^{\square} = \frac{1}{12} h h^3, \tag{18}$$

$$I_y^{\mathtt{I}} = \frac{w}{12}(a - 2w)^3 + 2\left(\frac{aw^3}{12} + aw\left(\frac{a}{2} - \frac{w}{2}\right)^2\right). \tag{19}$$

These two equations can be further simplified under consideration of $w = \frac{h}{5}$ and $a = \frac{9h}{5}$:

[1] The axial second moment of area according to Eq. (19) can also alternatively be calculated using the basic shape of a rectangle: $I_y^{\mathtt{I}} = \frac{a^4}{12} - \frac{(a-w)(a-2w)^3}{12}$.

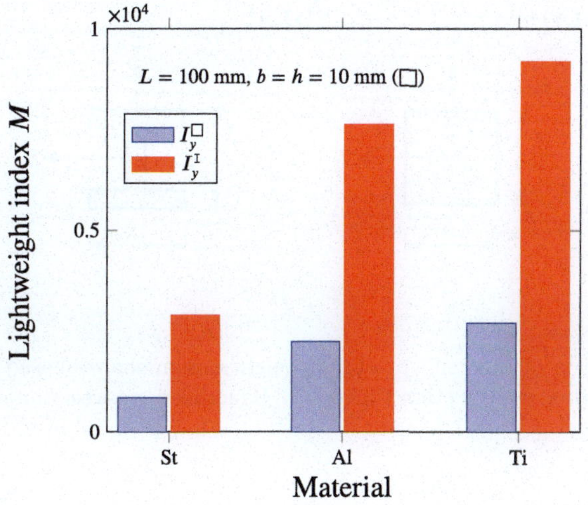

Fig. 8. Lightweight index for cantilever beams (Euler-Bernoulli) as a function of different axial second moments of area ($L = 100$ mm, $h = 10$ mm) and stress criterion. Same cross-sectional area ($A^\square = A^\mathtt{I}$) and thus same mass ($m^\square = m^\mathtt{I}$) assumed

$$I_y^\square = C^\square h^4 \quad \text{mit} \quad C^\square = \tfrac{1}{12}, \tag{20}$$

$$I_y^\mathtt{I} = C^\mathtt{I} h^4 \quad \text{mit} \quad C^\mathtt{I} = \tfrac{3817}{7500}. \tag{21}$$

Thus, the lightweight index is obtained with $z_{\text{max}}^\square = \frac{h}{2}$ and $z_{\text{max}}^\mathtt{I} = \frac{a}{2} = \frac{9h}{10}$ as (see also Fig. 8):

$$M^\square = 2 \times \frac{R_{\text{p0.2}}}{\varrho g \frac{L^2}{h}} \times C^\square, \quad M^\mathtt{I} = \frac{10}{9} \times \frac{R_{\text{p0.2}}}{\varrho g \frac{L^2}{h}} \times C^\mathtt{I}. \tag{22}$$

Alternatively, $I_y^\square \overset{!}{=} I_y^\mathtt{I}$ may be required. With $w = \frac{h}{5}$, we obtain $a = 1.055h$ (see Fig. 9) and the following mass ratio is finally obtained: $m^\mathtt{I} = 0.55m^\square$.

The comparison between the two profile sections shown in Figs. 7 and 9 was based on the assumption that the I-profile has the same height as width (see b) and furthermore the thicknesses of the flanges and the web are identical ($w = \frac{h}{5}$). A further improvement in the lightweight index can be achieved by considering the width-to-height ratio as a design variable (see Fig. 10c). Furthermore, it should be assumed that the cross-sectional areas ($A^\square = A^\mathtt{I}$) and the flanges and web thicknesses ($w = \frac{h}{5}$) remain identical. Figure 10 also contains two limit cases, where either the web height becomes zero (see Fig. 10a) or the flange width is equated with the web thickness (see Fig. 10d).

For equal cross-sectional areas between a square cross section ($h \times h$) and an I-profile in Fig. 10c one gets

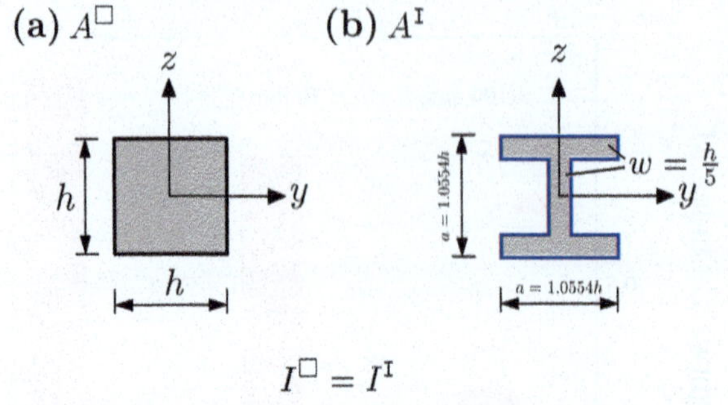

Fig. 9. Realization of the concept of lightweight design shapes based on $I^{\square} = I^{\mathtt{I}}$ and $A_y^{\square} > A_y^{\mathtt{I}}$: **a** Original cross section with $b = h$; **b** modified cross section with $a = 1.0554h$ and $w = \frac{h}{5}$

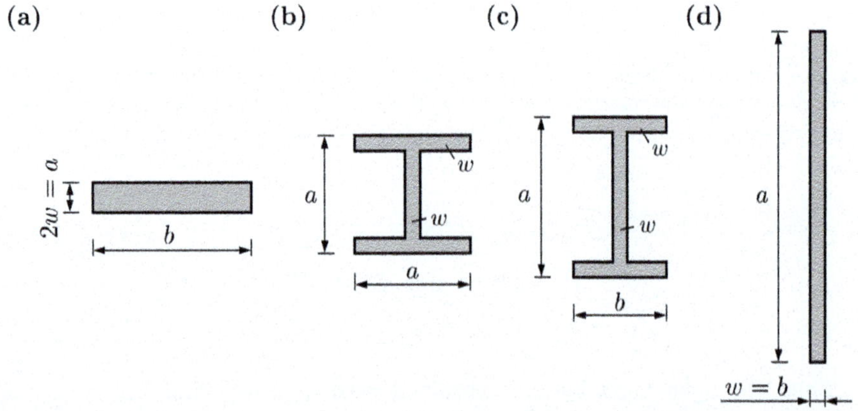

Fig. 10. Profiles with identical cross-sectional areas: **a** degenerated profile (web height → 0), **b** I-profile with the same width and height, **c** I-profile with different width-to-height ratio, **d** degenerated profile (flange width $b \rightarrow w$)

$$A^{\square} = h^2 = 2bw + w(a - 2w) = A^{\mathtt{I}}, \tag{23}$$

or rearranged with $w = \frac{h}{5}$:

$$b = \frac{27}{10}h - \frac{a}{2}. \tag{24}$$

For the I-profile, the axial second moment of area is still obtained as

$$I_y^{\mathtt{I}}(a,b) = \frac{w}{12}(a - 2w)^3 + 2\left(\frac{bw^3}{12} + bw\left(\frac{a}{2} - \frac{w}{2}\right)^2\right), \tag{25}$$

where by means of the degenerated profiles in Fig. 10 the following condition can be derived for the profile height:

$$2w = \frac{2h}{5} \leq a \leq 5h, \tag{26}$$

or as a condition for the profile width:

$$\frac{h}{5} \leq b \leq \frac{5h}{2}. \tag{27}$$

The last two equations can also be combined for the ratio of the design variables:

$$\underbrace{\frac{4}{25}}_{0.16} \leq \frac{a}{b} \leq 25. \tag{28}$$

The axial second moment of area according to Eq. (25) is shown in Fig. 11. It can be seen that there are large changes for small ratios $\frac{a}{b}$, but the functional values are for larger ratios converging to a constant value.

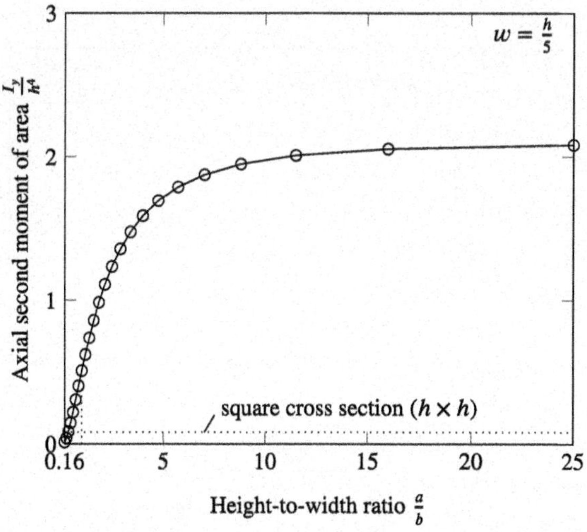

Fig. 11. Axial second moment of area for different I-profiles (see Fig. 10). Same cross sectional area ($A^{\square} = A^{\mathrm{I}}$) and thus same mass ($m^{\square} = m^{\mathrm{I}}$) assumed

To evaluate the lightweight index M, it should be considered that the determination equation contains both, i.e. the axial second moment of area (see Fig. 11) as wells as the distance to the surface $z_{\text{max}} = \frac{a}{2}$ (see Fig. 12) and both parameters change nonlinearly with the $\frac{a}{b}$ ratio:

$$M = \frac{F_0}{F_{\text{G}}} = \frac{R_{\text{p}0.2}}{h^2 L^2 \varrho g} \times \frac{I_y^{\mathrm{I}}}{\frac{a}{2}}. \tag{29}$$

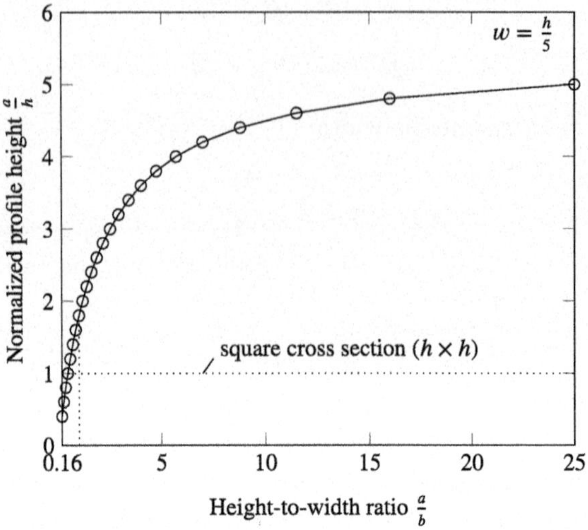

Fig. 12. Profile height for different I-profiles (see Fig. 10). Same cross sectional area ($A^{\square} = A^{\mathrm{I}}$) and thus same mass ($m^{\square} = m^{\mathrm{I}}$) assumed

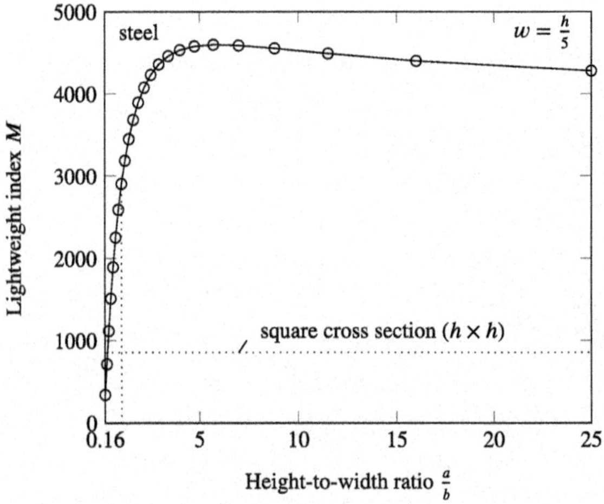

Fig. 13. Lightweight index for different I-profiles (see Fig. 10) and stress criterion ($\sigma_{\max} = R_{\mathrm{p0.2}}$). Same cross sectional area ($A^{\square} = A^{\mathrm{I}}$) and thus same mass ($m^{\square} = m^{\mathrm{I}}$) assumed ($L = 100\,\mathrm{mm}, h = 10\,\mathrm{mm}$)

The graphical representation of this lightweight index is shown in Fig. 13. It can be seen that a maximum value is obtained at $a \approx 5.7b$.

In regards to the diagrams shown in Figs. 11 and 13, it must be considered that there is no closed form $I_y = I_y(\frac{a}{b})$ and $M = M(\frac{a}{b})$. Thus, a tabular representation might be appropriate to generate pairs of values, see Table 2. Introducing in the equation for the second moment according to (25) the geometrical condition $w = \frac{h}{5}$ and replacing the width b based on (24), we obtain after a short calculation the following normalized representation:

$$\frac{I_y}{h^4} = \frac{1}{60}\left(\frac{a}{h} - \frac{2}{5}\right)^3 + \left(\frac{1}{750}\left(\frac{27}{10} - \frac{a}{2h}\right) + \frac{2}{5}\left(\frac{27}{10} - \frac{a}{2h}\right)\left(\frac{a}{2h} - \frac{1}{10}\right)^2\right). \tag{30}$$

The last relation is formulated as a function of the ratio $\frac{a}{h}$, which can be represented based on Eq. (24) as follows:

$$\frac{a}{h} = \left(\frac{h}{a}\right)^{-1} = \left(\frac{10}{27} \times \left(\frac{b}{a} + \frac{1}{2}\right)\right)^{-1}. \tag{31}$$

In the relation for the lightweight index according to Eq. (29), one should use finally for a graphical representation equations (30) and (31). The pairs of values for the generation of the diagrams are summarized in Table 2.

Finally, the classic design of a profile, i.e. taking only a stress criterion into account, is discussed below. Using the values in Table 2, the maximum stress is

$$\sigma_{\text{max}} = \frac{M_{\text{max}}}{I} \times z_{\text{max}} = \frac{M_{\text{max}}\left(\frac{a}{h}\right) \times \frac{h}{2}}{\left(\frac{I}{h^4}\right) h^4}, \tag{32}$$

or in normalized representation as (see Fig. 14):

$$\frac{\sigma_{\text{max}}}{\frac{M_{\text{max}}}{2h^3}} = \frac{\left(\frac{a}{h}\right)}{\left(\frac{I}{h^4}\right)}. \tag{33}$$

Here too, an optimal height-to-width ratio of $a \approx 5.7b$ is obtained, i.e. to achieve a minimum stress. Note that the diagram was created under the assumption of a constant mass. For the square reference cross-section ($h \times h$), the normal stress is:

$$\sigma_{\text{max}} = \frac{M_{\text{max}}}{I} \times \frac{h}{2} = \frac{M_{\text{max}}}{\frac{1}{6}h^3} = \frac{M_{\text{max}}}{\frac{1}{12} \times 2h^3}. \tag{34}$$

According to Eq. (16) it can also be seen that the stress is proportional to the bending moment. Thus, it may be useful to increase the axial second moment of area at the location of the largest bending moment (see Fig. 15a) and to make a reduction at the point of the minimum moment (see Fig. 15b).

Table 2. Pairs of values for the generation of Figs. 11, 12 and 13

$\frac{a}{b}$	$\frac{a}{h}$	$\frac{I_y}{h^4}$	M
0.16	0.4	0.013333	342.403095
0.25	0.6	0.041733	714.481124
0.347826	0.8	0.086933	1116.234089
0.454545	1.0	0.147333	1513.421678
0.571429	1.2	0.221333	1894.630457
0.70	1.4	0.307333	2254.968952
0.842105	1.6	0.403733	2591.991427
1.00	1.8	0.508933	2904.339139
1.176471	2.0	0.621333	3191.196842
1.375	2.2	0.739333	3452.045746
⋮	⋮	⋮	⋮
3.40	3.4	1.475333	4457.282639
4.00	3.6	1.587733	4530.373391
4.75	3.8	1.692933	4576.307467
5.714286	4.0	1.789333	4595.049531
7.00	4.2	1.875333	4586.570978
8.80	4.4	1.949333	4550.848404
11.50	4.6	2.009733	4487.862475
16.00	4.8	2.054933	4397.597080
25.00	5.0	2.083333	4280.038684

According to Fig. 16 this results in a considerable enlargement of the lightweight index. However, it should be noted that in this example the cross section at $x = 0$ has been increased to $\frac{7h}{5}$, resulting in a significantly greater force F_0. Therefore, larger values of the lightweight index M also result.

In the previous example, the beam cross-section has been changed so that at the position of the maximum bending moment ($x = 0$) the stress just reached the limit value. However, in the following, we require that the stress at each point x of the beam just reaches the limit. Thus, the beam cross section along the beam axis is optimized. For comparison, we use the example with constant rectangular cross section ($b = h = 10\,\text{mm}$) according to Fig. 8, i.e., the blue bars for the quadratic cross section. For the sake of simplicity, it is further assumed that the beam width (b) remains unchanged. If the limit condition is again assumed to be the maximum stress, that is,

$$\sigma_{x,\max}(x) = \frac{M_y(x)}{I_y(x)} \times \frac{h(x)}{2} = \frac{F_0 L(1-\frac{x}{L})}{I_y(x)} \times \frac{h(x)}{2} = \frac{6F_0 L(1-\frac{x}{L})}{bh(x)^2} \stackrel{!}{=} R_{\text{p0.2}}, \tag{35}$$

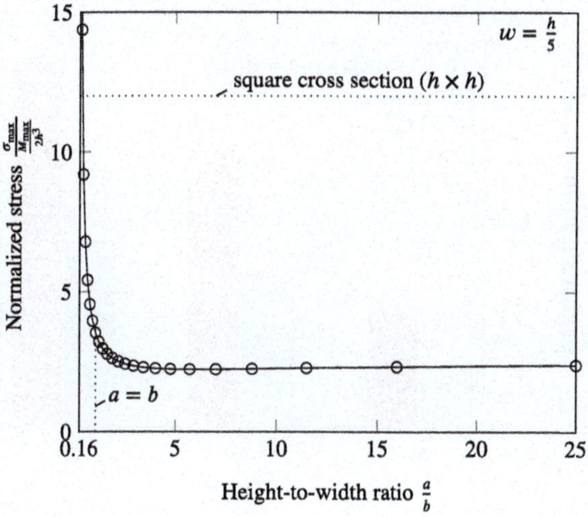

Fig. 14. Normalized stress for different I-profiles (see Fig. 10). Same cross sectional area ($A^{\square} = A^{I}$) and thus same mass ($m^{\square} = m^{I}$) assumed

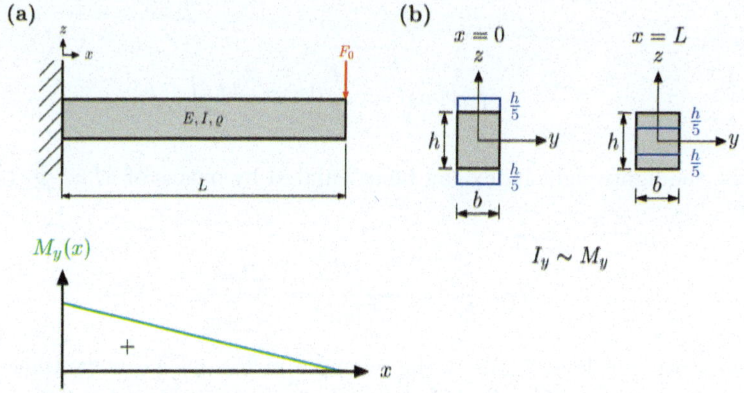

Fig. 15. Realization of the concept of lightweight design shapes based on $I_y \sim M_y$: **a** beam configuration and moment distribution; **b** adaptation of cross section

the variable beam height is obtained (see also Fig. 17):

$$h(x) = \underbrace{\sqrt{\frac{6F_0L}{bR_{p0.2}}}}_{h_0} \times \sqrt{1 - \frac{x}{L}} = h_0 \times \sqrt{1 - \frac{x}{L}}. \tag{36}$$

Figure 18 shows that for the optimized beam the stress limit is reached at each point x.

The volume of the optimized beam is given for a small volume element as $\mathrm{d}V = bh(x)\mathrm{d}x$, or for the entire beam:

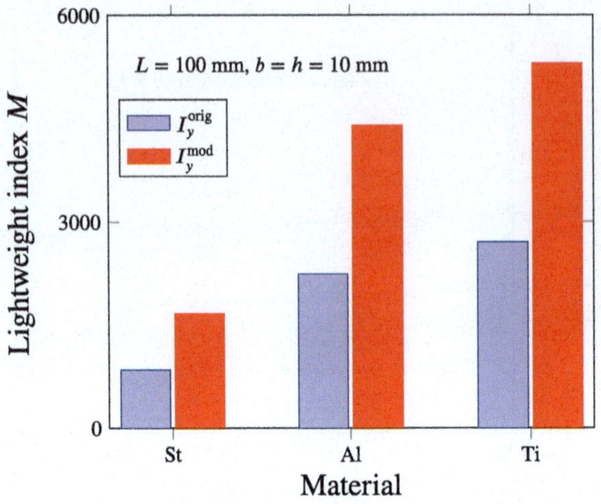

Fig. 16. Lightweight index for a cantilever beam made of different materials for $I_y = I_y(x)$ ($L = 100$ mm, $h = 10$ mm) and stress criterion

$$V = \int_{x=0}^{L} bh_0\sqrt{1-\frac{x}{L}}\,\mathrm{d}x = \frac{2}{3}bh_0L. \tag{37}$$

Thus, the lightweight index can be calculated by means of $M = \frac{F_0}{F_G}$ (see also Fig. 19):

$$M = \frac{F_0}{F_G} = \frac{\frac{R_{p0.2}bh_0^2}{6L}}{\varrho g\frac{2}{3}bh_0L} = \frac{R_{p0.2}}{4\varrho g\frac{L^2}{h_0}}. \tag{38}$$

Finally, we will look again at the example in Fig. 6a, i.e. a cantilever beam clamped on one side with a single force at the free end, taking the stress criterion into account. However, in the following we will consider an I-profile that is to be optimized along the entire beam axis, see Fig. 20. Compare the mass and lightweight index with a beam of constant cross-section according to Fig. 20a.

At the fixed support, i.e. at $x = 0$, the axial second moment of area $I_{y,0}^{\mathrm{I}}$ and the cross-sectional area A_0^{I} are:

$$I_{y,0}^{\mathrm{I}} = \frac{a_0^4}{12} - \frac{1}{12}(a_0 - w_0)(a_0 - 2w_0)^3 \overset{w_0=\frac{a_0}{5}}{=} \frac{517}{7500}a_0^4 = C^{\mathrm{I}}a_0^4, \tag{39}$$

$$A_0^{\mathrm{I}} = a_0^2 - (a_0 - w_0)(a_0 - 2w_0) = \frac{13}{25}a_0^2. \tag{40}$$

At arbitrary location x the same expressions as in Eqs. (39)–(40) result, but without the index '0'. The beam should be optimized by allowing plastic flow to

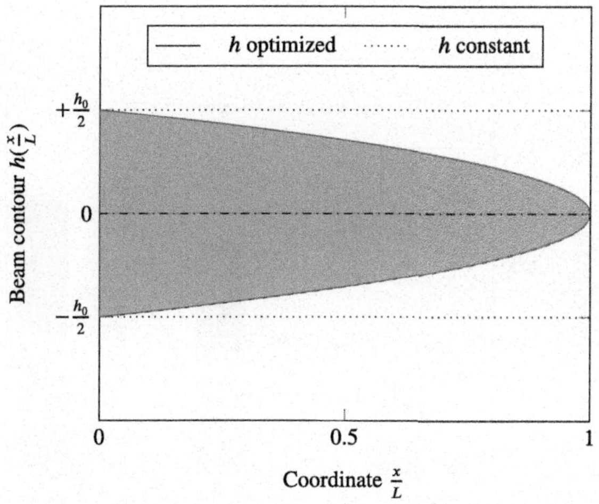

Fig. 17. Beam contour along the beam axis

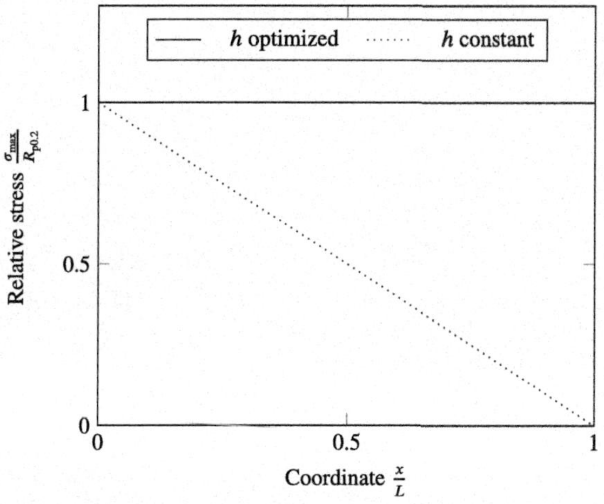

Fig. 18. Relative stress along the beam axis

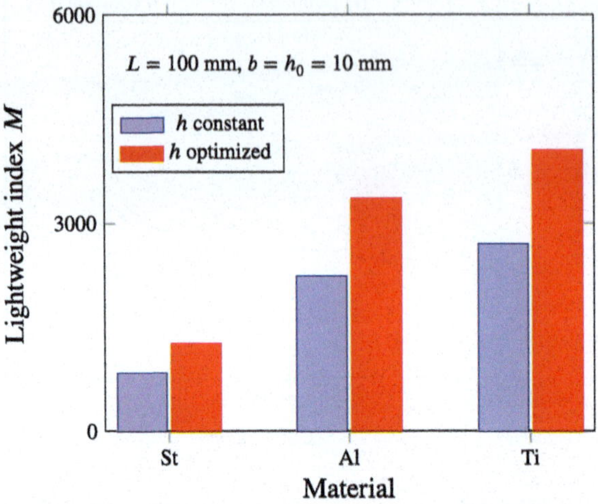

Fig. 19. Lightweight index for cantilever beam made from different materials for $h = h(x)$ ($L = 100$ mm, $h_0 = 10$ mm) and stress criterion

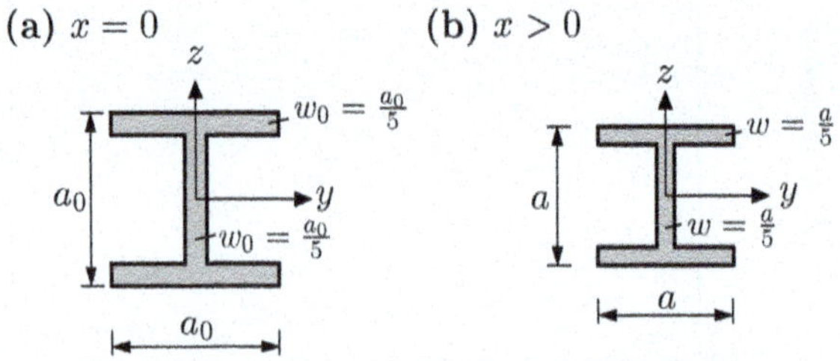

Fig. 20. Beam with changing cross section along the axis: **a** location of support; **b** arbitrary position

occur simultaneously at any point of the beam (and not only at the clamping point). The stress criterion can therefore be formulated as follows:

$$\sigma_{x,\max} = \frac{M_y(x)}{I_y^{\mathrm{I}}(x)} \times \frac{a(x)}{2} = \frac{F_0 L\left(1-\frac{x}{L}\right)}{C^{\mathrm{I}} a(x)^4} \times \frac{a(x)}{2} \stackrel{!}{=} R_{\mathrm{p}0.2}, \tag{41}$$

which results in the function of the side dimension (see Fig. 21):

$$a(x) = \sqrt[3]{\frac{F_0 L}{2C^{\mathrm{I}} R_{\mathrm{p}0.2}}} \times \sqrt[3]{1-\frac{x}{L}} = a_0 \times \sqrt[3]{1-\frac{x}{L}}. \tag{42}$$

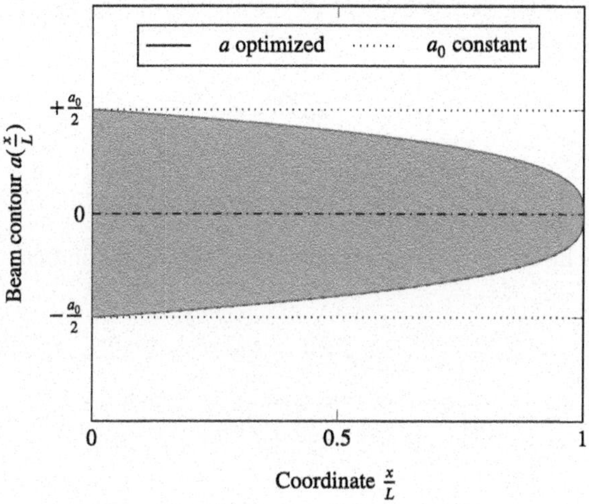

Fig. 21. Beam contour along the axis

The volume of the optimized beam is given for a small volume element as $\mathrm{d}V = A(x)\mathrm{d}x$, or for the entire beam as:

$$V = \int\limits_{x=0}^{L} \frac{13}{25} a(x)^2 \mathrm{d}x = \frac{13}{25} a_0^2 \int\limits_{x=0}^{L} \left(1-\frac{x}{L}\right)^{\frac{2}{3}} \mathrm{d}x = \frac{39}{125} a_0^2 L. \tag{43}$$

Thus, the lightweight index can be calculated according to Eq. (1):

$$M^{\mathrm{I}} = \frac{F_0}{F_{\mathrm{G}}} = \frac{\frac{2C^{\mathrm{I}} R_{\mathrm{p}0.2} a_0^3}{L}}{\varrho g \frac{39}{125} a_0^2 L} = \frac{250}{39} \times \frac{C^{\mathrm{I}} R_{\mathrm{p}0.2}}{\varrho g \frac{L^2}{a_0}}. \tag{44}$$

The lightweight index for a beam with constant cross-section A_0 (see Fig. 20a) can also be determined accordingly:

$$M_0^{\mathrm{I}} = \frac{50}{13} \times \frac{C^{\mathrm{I}} R_{\mathrm{p0.2}}}{\varrho g \frac{L^2}{a_0}}. \tag{45}$$

This gives the ratio of the lightweight indices as:

$$\frac{M^{\mathrm{I}}}{M_0^{\mathrm{I}}} = \frac{5}{3} \approx 1.667. \tag{46}$$

The ratio of the masses is obtained as $\frac{m^{\mathrm{I}}}{m_0^{\mathrm{I}}} = \frac{3}{5} = 0.6$.

A comparison of the beam with the constant I-profile with a beam with a constant square cross-sectional area ($h \times h$) gives for the same mass, i.e. ($A^{\mathrm{I}} = A^{\square}$), the length ratio $h^2 = \frac{13}{25} a_0^2$. This result gives:

$$A^{\square} = h^2 = \frac{13}{25} a_0^2, \tag{47}$$

$$I_y^{\square} = \frac{1}{12} \left(\frac{13}{25} \right)^2 a_0^4 = \frac{169}{7500} a_0^4. \tag{48}$$

This results in the following ratio of the lightweight indices for a stress criterion:

$$\frac{M_0^{\mathrm{I}}}{M^{\square}} = \frac{\frac{50}{13} \times C^{\mathrm{I}} \times \frac{R_{\mathrm{p0.2}}}{\varrho g \frac{L^2}{a_0}}}{\frac{1}{6} \times \sqrt{\frac{13}{25}} \times \frac{R_{\mathrm{p0.2}}}{\varrho g \frac{L^2}{a_0}}} \approx 2.206. \tag{49}$$

3.2 Evaluation Using the Specific Energy Absorption

Let us reconsider the example with $I_y^{\square}$ and I_y^{I} (with $A^{\square} = A^{\mathrm{I}}$) after Fig. 7, i.e. a cantilever beam with a single shear force at the free end, treated under the consideration of a the stress criterion. The definition of specific energy absorption according to Eq. (4) requires that the strain energy is determined. The definition based on the internal reaction, see Eq. (9), yields the following relationship:

$$\Pi = \int_0^L \frac{M_y(x)^2}{2EI_y} \mathrm{d}x = \int_0^L \frac{F_0^2 (L-x)^2}{2EI_y} \mathrm{d}x = \frac{F_0^2}{2EI_y} \int_0^L \left(L^2 - 2Lx + x^2 \right) \mathrm{d}x$$
$$= \frac{F_0^2 L^3}{6EI_y}. \tag{50}$$

In the case of a stress criterion according to Eq. (16), the acting single force can be replaced by:

$$F_0 = \frac{R_{\mathrm{p0.2}} I_y}{L z_{\mathrm{max}}} \tag{51}$$

Considering that the mass is given by $m = \varrho V = \varrho AL = \varrho Lh^2$, the specific energy absorption can be calculated as follows:

$$SEA = \frac{F_0^2 L^3}{6EI_y \times \varrho Lh^2} = \frac{R_{\text{p0.2}}^2 I_y}{6E\varrho h^2 z_{\text{max}}^2} \sim \frac{R_{\text{p0.2}}^2}{E\varrho}. \tag{52}$$

Based on the expressions for the axial second moments of area according to Eqs. (20) and (21) and the maximum distance to the free surface $z_{\text{max}}^{\square} = \frac{h}{2}$ and $z_{\text{max}}^{\text{I}} = \frac{a}{2} = \frac{9h}{10}$ (see also Fig. 8), one obtains:

$$SEA^{\square} = \frac{2}{3} \times \frac{R_{\text{p0.2}}^2}{E\varrho} \times C^{\square}, \quad SEA^{\text{I}} = \frac{50}{243} \times \frac{R_{\text{p0.2}}^2}{E\varrho} \times C^{\text{I}}. \tag{53}$$

The graphical representation of the specific energy absorption in Fig. 22 shows that the same trends are found as in the case of the lightweight index M (see Fig. 8). Using the I-profile with the same cross-sectional area results in a significant increase in the lightweight potential, with the highest value again achieved by the Ti alloy.

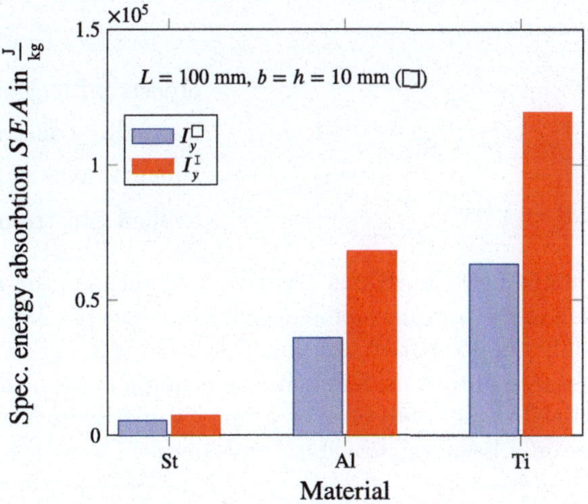

Fig. 22. Specific energy absorption for cantilever beam made of different materials for different second moments of area ($L = 100$ mm, $h = 10$ mm) and stress criterion. Same cross sectional area ($A^{\square} = A^{\text{I}}$) and thus same mass ($m^{\square} = m^{\text{I}}$) assumed

3.3 Consideration as a Classic Optimization Problem

In general, a classical optimization problem can be described (see [19,20,23]) by specifying an objective function F, e.g. the weight (mass) of a mechanical structure or the cost of a project, i.e.

$$F(\boldsymbol{X}) = F(X_1, X_2, X_3, \ldots, X_n), \tag{54}$$

which is to be minimized under the following constraints

$$g_j(\boldsymbol{X}) \leq 0 \qquad j = 1, m \qquad \text{inequality constraints,} \tag{55}$$

$$h_k(\boldsymbol{X}) = 0 \qquad k = 1, l \qquad \text{equality constraints,} \tag{56}$$

$$X_i^{\min} \leq X_i \leq X_i^{\max} \qquad i = 1, n \qquad \text{explicit constraints,} \tag{57}$$

where the column matrix of n design variables, e.g. the geometric dimensions of the mechanical structure, can be represented as follows:

$$\boldsymbol{X} = \left\{ \begin{matrix} X_1 \\ X_2 \\ X_3 \\ \vdots \\ X_n \end{matrix} \right\}. \tag{58}$$

In the case of a one-dimensional objective function, i.e. $\boldsymbol{X} \to X$, the following simplified set of equations results:

$$F(X) \qquad \text{objective function,} \tag{59}$$

$$g_j(X) \leq 0 \qquad j = 1, m \qquad \text{inequality constraints,} \tag{60}$$

$$h_k(X) = 0 \qquad k = 1, l \qquad \text{equality constraints,} \tag{61}$$

$$X^{\min} \leq X \leq X^{\max} \qquad \text{explicit constraints.} \tag{62}$$

In practice, numerical algorithms are used to determine the minimum of the objective function in the n-dimensional design space (see [13, 23]). In this description, however, only one- or two-dimensional design spaces, i.e. with a maximum of two design variables, are considered. In these simpler cases, the minimum can often be determined using analytical or graphical methods. The intention here is also to convey only basic relationships and ideas.

In the following, a simple example[2] with a design variable (X) is considered to explain the basic approach to the optimization process in the case of *one* design variable. Figure 23 shows a generalized beam that is clamped at one side and is subjected to a single force F_0 in tension and a single moment M_0 in bending. The geometric cross-sectional dimensions of the beam, i.e. the width and the height (with a fixed ratio between height and width (α)), are to be optimized for minimal mass. As a restriction (g), it must be noted that no plastic deformations may occur. For this purpose, the beam can be considered thin. The beam length L, the external loads (F_0, M_0), the material properties ϱ and $R_{\text{p}0.2}$, as well as the ratio α are known.

[2] The idea for this problem is taken from [21].

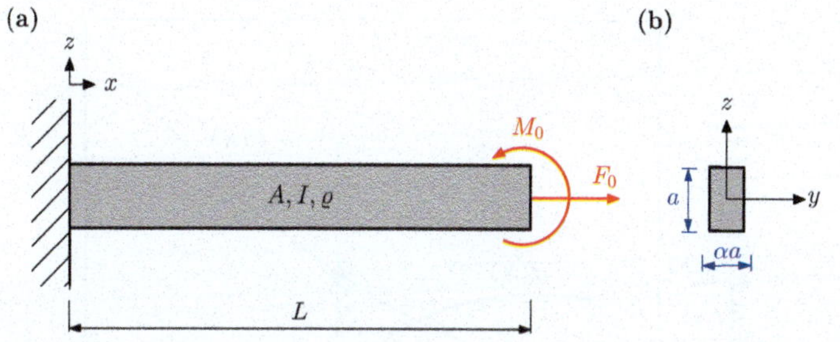

Fig. 23. Generalized beam fixed on one side under tension and moment loading: **a** General configuration; **b** beam cross section

Due to the acting force and the moment, a composed normal stress state is obtained. This results in the condition that no plastic deformations may occur:

$$\sigma_{x,\max} = \frac{M_y}{I_y} \times z_{\max} + \frac{N}{A} \leq R_{\text{p0.2}}, \tag{63}$$

$$= \frac{-M_0}{\frac{1}{12}a(\alpha a)^3} \times \left(-\frac{\alpha a}{2}\right) + \frac{F_0}{a \times \alpha a} \leq R_{\text{p0.2}}, \tag{64}$$

$$\rightarrow \; 6M_0 + \alpha a F_0 - \alpha^2 a^3 R_{\text{p0.2}} \leq 0. \tag{65}$$

The mass of the beam, i.e. the objective function of the optimization problem, can be easily expressed using the following relation:

$$F = m = \varrho AL = \varrho La(\alpha a). \tag{66}$$

If one sets $X = a$ to generalize and normalizes the stress condition with M_0 for better graphical representation, the following set of equations results to describe the classical optimization problem:

$$F(X) = \varrho L \alpha X, \tag{67}$$

$$g(X) = 6 + \frac{F_0}{M_0} \times X - \frac{\alpha}{M_0 R_{\text{p0.2}}} \times X^3 \leq 0. \tag{68}$$

From the graphical representation of Eqs. (67)–(68) in Fig. 24 one can see that the permissible region must be to the right of the root of the inequality restriction. Thus, the optimal design for minimal mass is calculated using the following condition:

$$g(X_{\text{extr}}) \stackrel{!}{=} 0. \tag{69}$$

In the following, a simple example[3] with two design variables (X_1, X_2) is considered to explain the basic approach to the optimization process in the case

[3] The idea for this problem is taken from [19].

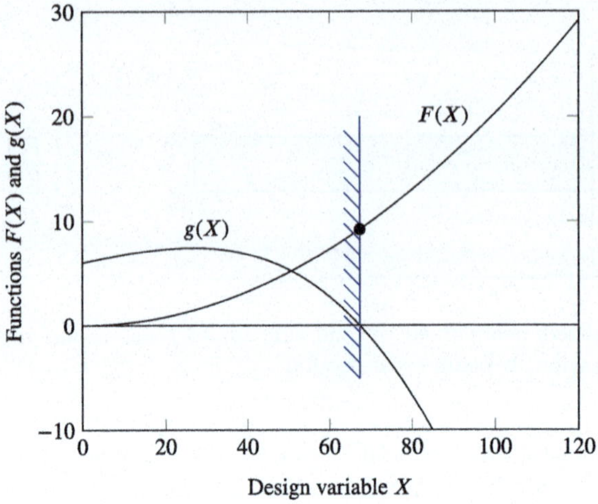

Fig. 24. Optimal design of a generalized beam fixed on one side under tension and moment loading: $L = 2500\,\text{mm}$, $F_0 = 90\,\text{kN}$, $M_0 = 1130\,\text{kNmm}$, $R_{\text{p0.2}} = 140\,\text{MPa}$, $\varrho = 2.7 \times 10^{-6}\,\text{kg/mm}^3$, $\alpha = 0.3$ (original set of equations (67)–(68)). Exact solution for the minimum: $X_{\text{extr}} = 67.372$ mm (indicated by the • marker)

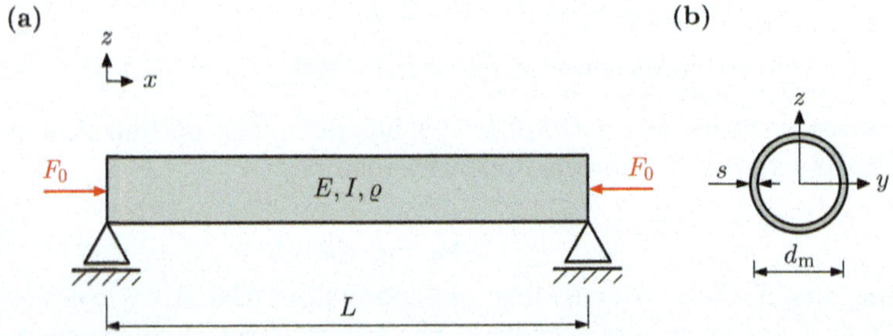

Fig. 25. Simply supported tube under compressive load: **a** General configuration; **b** tube cross section

of *two* design variables. Figure 25 shows a thin tube with a circular cross-section that is simply supported at both sides and is subjected to a compressive load by means of the force F_0. The geometric dimensions of the thin cross-section, i.e. the mean diameter d_{m} and the wall thickness s, are to be optimized for minimal mass. The restrictions (g_i) to be considered are that no global or local instabilities may occur, see Table 3. The tube length L, the external load F_0 and the material properties E, ν and ϱ can be considered as known.

The global instability failure due to buckling can be described for purely elastic material behavior using the buckling formula according to Euler (see [1]) as follows:

Table 3. Instabilities of simply supported thin-walled tubes with a circular cross-section under axial load

	Schematic representation	Description
(a)	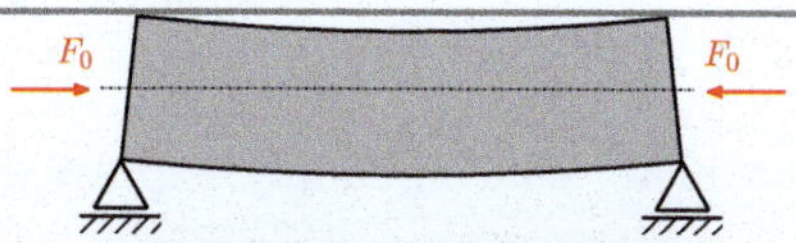	Global instability: buckling
(b)	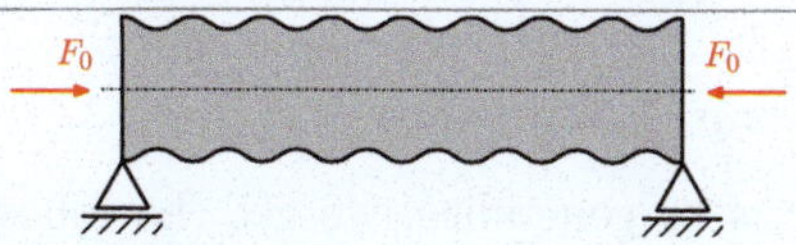	Local instability: wrinkling

$$F_{\rm cr} = \frac{\pi^2 EI}{L^2}. \tag{70}$$

To prevent buckling from occurring, the loading force F_0 must always be smaller than the buckling force $F_{\rm cr}$.

The theoretical wrinkling stress ($\sigma_{\rm cr}$) for a thin-walled tube with circular cross-section under axial pressure is for a simple support on both sides, see [1,5,22]:

$$\sigma_{\rm cr} = \underbrace{\frac{1}{\sqrt{3(1-\nu^2)}}}_{K} \times \frac{Es}{r_{\rm m}}, \tag{71}$$

where the wrinkling coefficient for metallic materials with $\nu = 0.3$ is $K = 0.605$. Since the task involves a thin tube, the axial second moment of area I and the cross-sectional area A can be simplified as follows, see [14]:

$$I \approx \pi r_{\rm m}^3\, s = \frac{\pi d_{\rm m}^3\, s}{8}, \tag{72}$$

$$A \approx 2\pi r_{\rm m} s = \pi d_{\rm m} s. \tag{73}$$

With these approximations and the critical loads according to Eq. (70) and (71) the two inequality constraints result in:

$$g_1 = F_0 - \frac{\pi^2 EI}{L^2} \leq 0, \tag{74}$$

$$= F_0 - \frac{\pi^3 E d_{\rm m}^3\, s}{8L^2} \leq 0, \tag{75}$$

or in relation to local buckling:

$$g_2 = \frac{F_0}{A} - 0.605 \times \frac{Es}{\frac{d_m}{2}} \leq 0, \tag{76}$$

$$= F_0 - 1.21 \times E\pi s^2 \leq 0. \tag{77}$$

The objective function, i.e. the mass, can be simply expressed using the following relationship:

$$F = m = \varrho AL = \varrho L\pi d_{\mathrm{m}} s. \tag{78}$$

If we set $X_1 = d_{\mathrm{m}}$ and $X_2 = s$ to generalize, we get the following set of equations to describe a classical optimization problem:

$$F(X_1, X_2) = \varrho L\pi X_1 X_2, \tag{79}$$

$$g_1(X_1, X_2) = F_0 - \frac{\pi^3 E X_1^3 X_2}{8L^2} \leq 0, \tag{80}$$

$$g_2(X_1, X_2) = F_0 - 1.21 \times E\pi X_2^2 \leq 0. \tag{81}$$

For graphical representation, it is advisable to formulate the three equations in the form $f(x)$ or in the current notation as $X_2(X_1)$:

$$X_2 = \frac{\text{const.}}{\varrho L\pi X_1}, \tag{82}$$

$$X_2 \geq \frac{8F_0L^2}{\pi^3 E X_1^3}, \tag{83}$$

$$X_2 \geq \sqrt{\frac{F_0}{1.21E\pi}}. \tag{84}$$

From the graphical representation of Eqs. (79)–(81) in Fig. 26 one can see that the permissible range is above the red g_2 line and to the right of the blue g_1 curve. Thus, the optimal design for minimal mass is calculated as the intersection of the two inequality constraints:

$$X_{1,\mathrm{extr}} = \left(\frac{8F_0L^2}{\pi^3 E X_2}\right)^{\frac{1}{3}}, \tag{85}$$

$$X_{2,\mathrm{extr}} = \sqrt{\frac{F_0}{1.21E\pi}}. \tag{86}$$

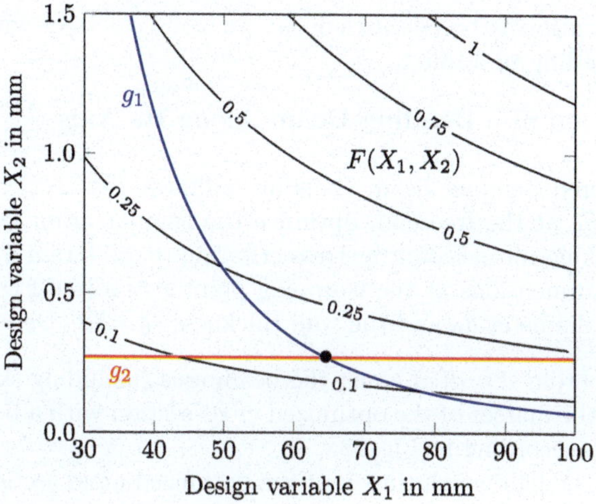

Fig. 26. Optimal design of a tube with a circular cross section: $L = 1000\,\text{mm}$, $F_0 = 20000\,\text{N}$, $E = 70{,}000\,\text{MPa}$, $R_{\text{p0.2}} = 247\,\text{MPa}$, $\varrho = 2.691 \times 10^{-6}\,\text{kg/mm}^3$, (original set of equations (79)–(81)). Exact solution for the minimum: $X_{1,\text{extr}} = 64.544$ mm, $X_{2,\text{extr}} = 0.274$ mm (indicated by the • marker)

3.4 Exercises

1. Implementation of the Concept of Lightweight Design Shapes: Optimization of a Box Profile

For a cantilever beam ($L = 100\,\text{mm}$) with end force F_0, the cross section is to be optimized based on a stress criterion using the lightweight index. The original square cross-section ($h = 10\,\text{mm}$) is to be modified and optimized by a box profile while maintaining the beam weight (i.e. with $A^{\boxtimes} = A^{\square}$), see Fig. 27.

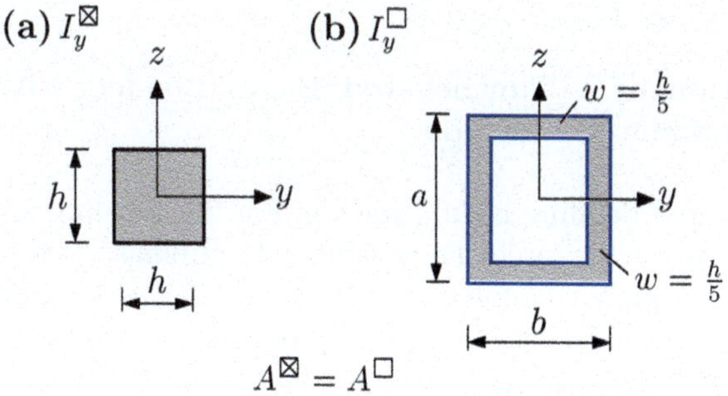

Fig. 27. Realization of the concept of lightweight design shapes based on $I_y^{\square} > I_y^{\boxtimes}$ and $A^{\boxtimes} = A^{\square}$: **a** Original cross section with $b = h$; **b** modified cross section

The optimization is to be carried out for stainless steel with the material properties according to Table 1.

2. Optimization of a Bending Beam Along Its Axis

For a cantilevered bending beam (bending stiffness $E \times I(x)$, length L) with a single force F_0 at the free end, optimize the tubular cross section along the beam axis x taking into account a stress criterion (0.2% strain limit $R_{\mathrm{p}0.2}$). The cross-sectional dimensions at the clamping point $x = 0$ (see Fig. 28a) are given by the outside diameter d_0 and the wall thickness $s_0 = \frac{d_0}{5}$. Determine:

(a) the beam contour $d = d(x)$ along the beam axis (including sketch),
(b) the ratio of the masses of the optimized cross-section with a beam of constant cross-section according to Fig. 28a,
(c) the ratio of the lightweight index of the optimized cross section with a beam of constant cross-section according to Fig. 28a,
(d) the ratio of the specific energy absorption of the optimized cross section with a beam of constant cross section according to Fig. 28a.

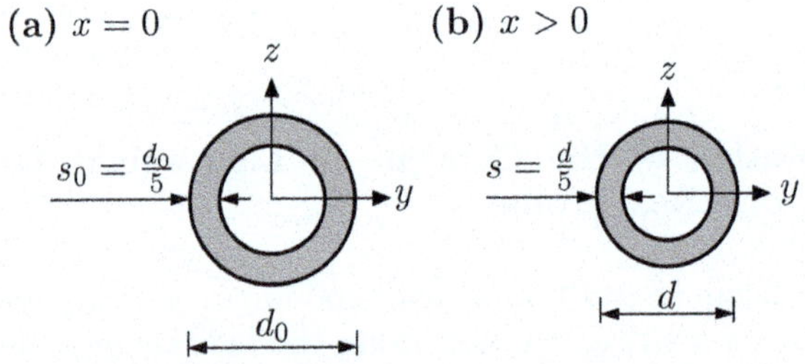

Fig. 28. Beam with changing cross section along the axis: **a** location of support; **b** arbitrary position

3. Optimization of a Cantilevered Beam Loaded with Two Single Forces of the Same Magnitude

For the cantilevered bending beam shown in Fig. 29 (bending stiffness $E \times I(x)$, length L) with two single forces F_0, optimize the tubular cross section along the beam axis x taking into account a stress criterion (0.2% strain limit $R_{\mathrm{p}0.2}$). The cross-sectional dimensions at the clamping point $x = 0$ (see Fig. 28a) are given by the outside diameter d_0 and the wall thickness $s_0 = \frac{d_0}{5}$. The cross-sectional ratios at any location x are shown in Fig. 28b. Determine:

(a) the beam contour $d = d(x)$ along the beam axis (including sketch),
(b) the specific energy absorption SEA of the optimized cross section,
(c) the problem in considering the lightweight index M in this case.

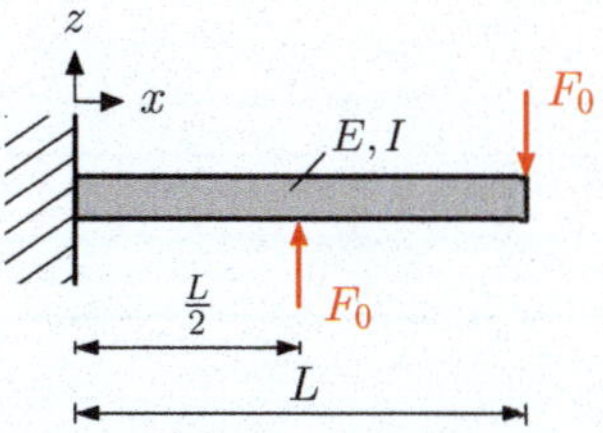

Fig. 29. Beam loaded with two single forces (same magnitude)

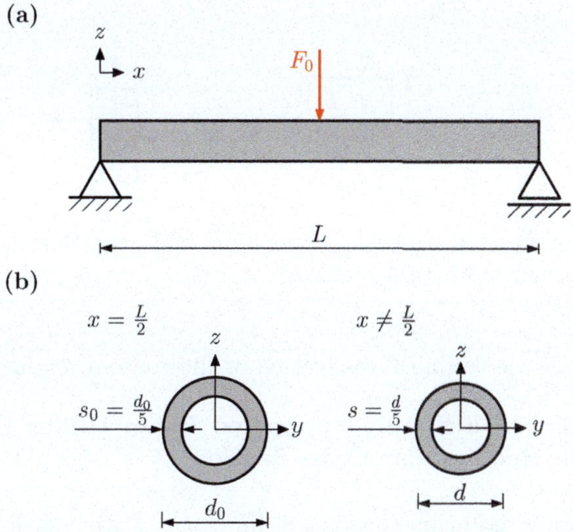

Fig. 30. Beam under 3-point bending: **a** General configuration and **b** cross sections

4. Optimization of a Beam Under 3-Point Bending: Tubular Cross Section

For a beam under 3-point bending (bending stiffness $E \times I(x)$, length L, see Fig. 30a) which is loaded by a central single force F_0, optimize the tubular cross section along the beam axis x taking into account a stress criterion (0.2% strain limit $R_{p0.2}$), see Fig. 30b.

The cross-sectional dimensions in the middle at $x = L/2$ are given by the outer diameter d_0 and the wall thickness $s_0 = d_0/5$, and the cross-sectional ratios at any point $x \neq L/2$ are modified in the same ratio, see Fig. 30b. Determine:

(a) the beam contour $d = d(x)$ along the beam axis (including sketch),
(b) the specific energy absorption SEA of the optimized cross section. Compare this value with the specific energy absorption for a beam with a constant diameter d_0,
(c) the lightweight index M of the optimized cross section. Compare this value with the lightweight index for a beam with a constant diameter d_0,

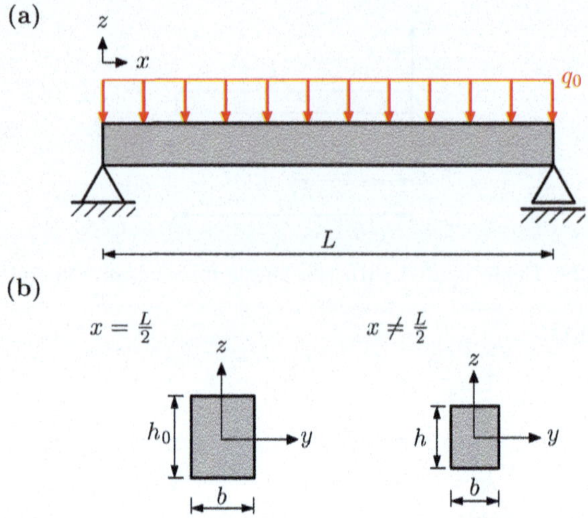

Fig. 31. Optimization of a simply supported beam under constant distributed load: **a** General configuration and **b** cross sections

(d) the problem of the technical realization of the contour $d = d(x)$.

5. Optimization of a Simply Supported Beam Under Constant Distributed Load: Rectangular Cross Section

For a simply supported beam (bending stiffness $E \times I(x)$, length L, see Fig. 31a) which is loaded by a constant distributed load q_0, optimize the rectangular cross section along the beam axis x taking into account a stress criterion (0.2% strain limit $R_{\text{p0.2}}$), see Fig. 31b.
The cross-sectional dimensions in the middle at $x = L/2$ are given by the height h_0 and the constant width b, and the cross-sectional ratios at any point $x \neq L/2$ are given by the height $h(x)$ and the constant width b, see Fig. 31b. Determine:

(a) the optimized beam contour $h = h(x)$ along the beam axis x (including sketch),
(b) the specific energy absorption SEA of the optimized cross section $(h(x) \times b)$. Compare this value with the specific energy absorption for a beam with a constant cross section $(h_0 \times b)$,
(c) the problem of the technical realization of the contour $h = h(x)$.

Consider:

$$\int_0^L \left(-\left[\tfrac{x}{L}\right]^2 + \left[\tfrac{x}{L}\right]^1\right)^{\frac{1}{2}} \mathrm{d}x = \tfrac{\pi L}{8} \text{ oder } \int_0^L \left(-x^2 + Lx^1\right)^{\frac{1}{2}} \mathrm{d}x = \tfrac{\pi L^2}{8}.$$

6. Optimization of a Beam with a Square Box Profile Along the Longitudinal Axis Under the Influence of a Constant Distributed Load

For a simply supported beam (bending stiffness $E \times I(x)$, length L, see Fig. 32a) which is loaded by a constant distributed load q_0, optimize the rectangular cross section along the beam axis x taking into account a stress criterion (0.2-% strain limit $R_{\text{p0.2}}$), see Fig. 32b. The cross-sectional dimensions in the middle at $x = L/2$ are given by the height d_0, the width d_0 and the constant thickness $s_0 = d_0/6$, and the cross-sectional ratios at any point $x \neq L/2$ are given by the height $d(x)$, the width $d(x)$ and the thickness $s(x) = d(x)/6$, see Fig. 32b. *Consider the comments on integration at the end of the problem!*

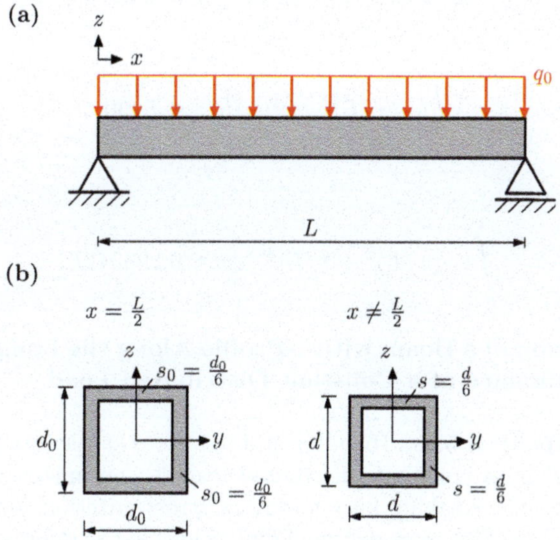

Fig. 32. Optimization of a simply supported beam under constant distributed load: **a** General configuration and **b** cross sections

Determine:

(a) The optimized beam contour $d = d(x)$ along the beam axis x (including sketch).
(b) An approximate value for the lightweight index M_0 of the cross section with constant values (d_0, d_0, s_0).
(c) An approximate value for the lightweight index M of the optimized cross section $(d(x), d(x), s(x))$.
(d) The specific energy absorption SEA_0 of the cross-section with constant values (d_0, d_0, s_0).
(e) The specific energy absorption SEA of the optimized cross section $(d(x), d(x), s(x))$.
(f) What can be said about the technical realization of the contour $d = d(x)$?

Remarks:

$$\int_0^L (-(x/L)^2 + x/L)^2 \mathrm{d}x \approx L/30$$

$$\int_0^L (-(x/L)^2 + x/L)^{2/3} \mathrm{d}x \approx 0.293342L.$$

7. Optimization of a Beam with I-Profile Along the Longitudinal Axis Under the Influence of a Constant Distributed Load

For a simply supported beam (bending stiffness $E \times I(x)$, length L, see Fig. 33a) which is loaded by a constant distributed load q_0, optimize the cross-section along the beam axis x taking into account a stress criterion (0.2% strain limit $R_{\text{p0.2}}$), see Fig. 33b. The cross-sectional dimensions in the middle at $x = L/2$ are given by the outer dimensions $\alpha a_0 \times a_0$ and the constant thickness $w_0 = a_0/6$. The Cross-sectional ratios at any point $x \neq L/2$ are given by the external dimensions $\alpha a(x) \times a(x)$ and the thickness $w(x) = a(x)/6$, see Fig. 33b. *Consider the comments on integration at the end of the problem!*

Determine for $\alpha = 2$ and alternatively for $\alpha = 3$:

(a) The optimized beam contour $a = a(x)$ along the beam axis x (including sketch).
(b) An approximate value for the lightweight index M_0 of the cross-section with constant values $(\alpha a_0, a_0, w_0)$.
(c) An approximate value for the lightweight index M of the optimized cross-section $(\alpha a(x), a(x), w(x))$.
(d) The specific energy absorption SEA_0 of the cross section with constant values $(\alpha a_0, a_0, w_0)$.

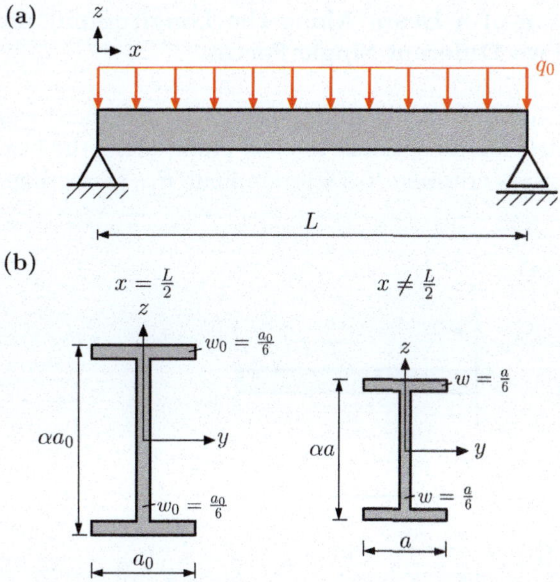

Fig. 33. Optimization of a simply supported beam under constant distributed load: **a** General configuration and **b** cross sections

(e) The specific energy absorption SEA of the optimized cross section with variable values $(\alpha a(x), a(x), w(x))$.
(f) What can be said about the technical realization of the contour $a = a(x)$?

Remarks:

$$\int_0^L (-(x/L)^2 + x/L)^2 \mathrm{d}x \approx L/30$$

$$\int_0^L (-(x/L)^2 + x/L)^{2/3} \mathrm{d}x \approx 0.293342L.$$

8. Optimization of a Beam with I-Profile Along the Longitudinal Axis and the Height-to-Width Ratio Under the Influence of a Constant Distributed Load

For Problem 7, determine the optimal α to maximize the specific energy absorption SEA of the optimized cross-section.

9. Optimization of a Beam Along the Longitudinal Axis Under the Influence of Two Different Single Forces

For a cantilever beam (bending stiffness $E \times I(x)$, length L) which is loaded by two different single forces, optimize the box profile along the beam axis x taking into account a stress criterion (0.2% strain limit $R_{p0.2}$), see Fig. 34a.

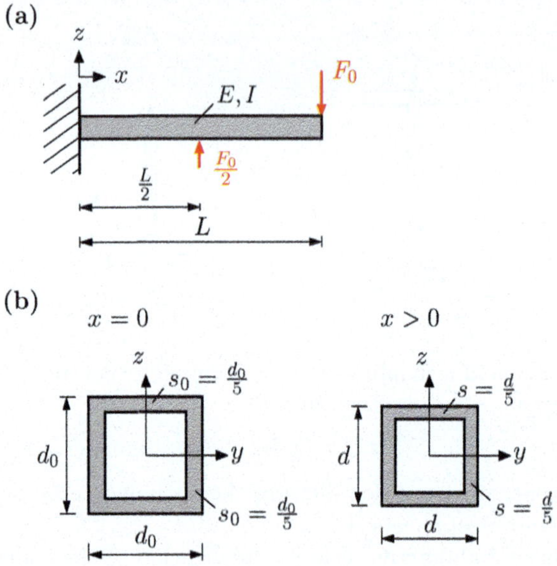

Fig. 34. Cantilever beam loaded by two different point loads: **a** General configuration and **b** cross sections

The cross-sectional dimensions at the fixed support at $x = 0$ (see Fig. 34b) are given by the outer diameter d_0 and the wall thickness $s_0 = d_0/5$. The cross-sectional ratios at any point $x > 0$ are shown in Fig. 34b. Determine:

(a) the beam contour $d = d(x)$ along the beam axis (including sketch),
(b) the specific energy absorption SEA of the optimized cross section.

10. Optimization of a Beam Along the Longitudinal Axis Under the Influence of a Single Force and a Single Moment

For a cantilever beam (bending stiffness $E \times I(x)$, length L) loaded by a single force F_0 and a single moment $M_0 = \frac{F_0 L}{6}$, optimize the I-profile along the beam axis x taking into account a stress criterion (0.2% strain limit $R_{p0.2}$), see Fig. 35a. The cross-sectional dimensions at the fixed support at $x = 0$ (see Fig. 35b) are given by the external dimensions a_0 and the wall thicknesses $w_0 = \frac{a_0}{6}$. The cross-sectional ratios at any point $x > 0$ are shown in Fig. 35b. Determine:

(a) the beam contour $a = a(x)$ along the beam axis (including sketch),
(b) the specific energy absorption SEA of the optimized cross section,
(c) compare the specific energy absorption SEA of the optimized cross-section with the specific energy absorption for a beam with a constant cross section (a_0 and w_0).

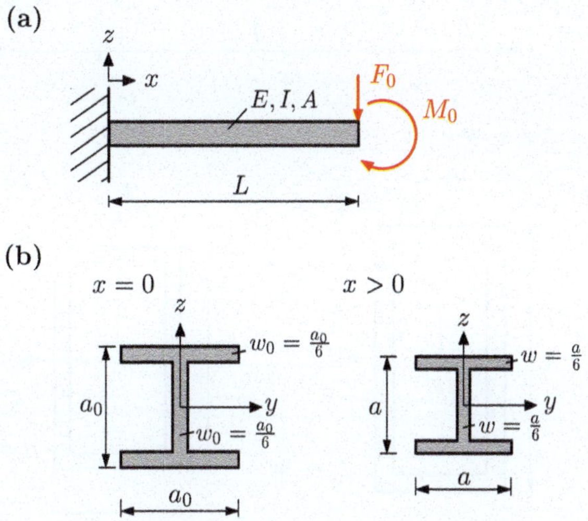

Fig. 35. Cantilever beam under the influence of a single force and a single moment: **a** General configuration and **b** cross sections

11. Optimization of a Simply Supported Beam Under Constant Distributed Load: Box Cross Section

For a simply supported beam (bending stiffness $E \times I(x)$, length L, see Fig. 36a) which is loaded by a constant distributed load q_0, optimize the rectangular cross section along the beam axis x taking into account a stress criterion (0.2% strain limit $R_{p0.2}$), see Fig. 36b. The cross-sectional dimensions in the middle at $x = L/2$ are given by the height a_0, the width $b_0 = a_0/2$ and the constant thickness $d_0 = a_0/6$, and the cross-sectional ratios at any point $x \neq L/2$ are given by the height $a(x)$, the width $b(x) = a(x)/2$ and the thickness $d(x) = a(x)/6$, see Fig. 36b.

Determine the optimized beam contour $a = a(x)$ along the beam axis x (including sketch).

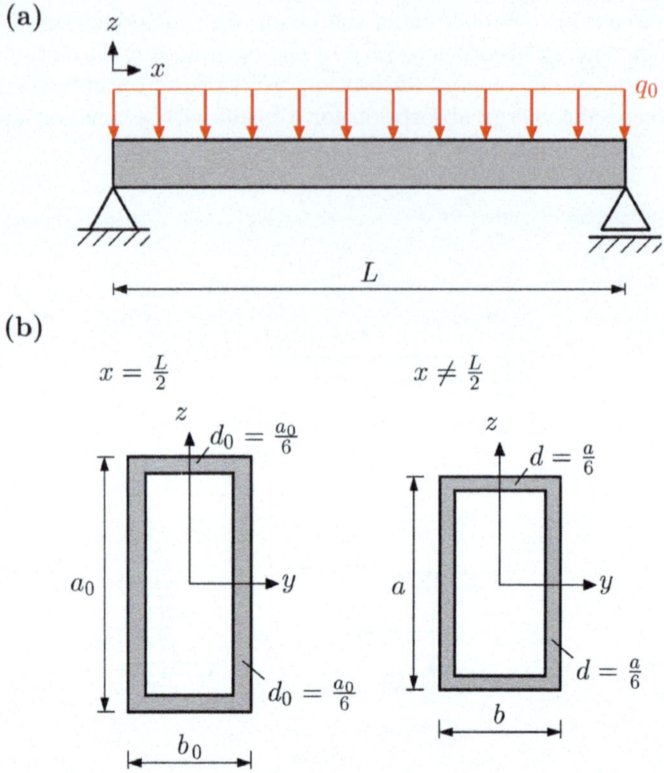

Fig. 36. Optimization of a simply supported beam under constant distributed load: **a** General configuration and **b** cross sections

12. Optimization of a Simply Supported Bending Under Constant Distributed Load: Box Cross Section. Numerical Calculation of the Specific Energy Absorption

For Problem 11 determine the specific energy absorption taking into account a stress criterion (0.2% strain limit $R_{\mathrm{p}0.2}$). The numerical calculation is to be carried out for the reference cross section and the optimized cross section.

13. Optimal Design of a Cantilever Beam (Classical Optimization Problem)

Consider a cantilever beam fixed at one end as shown in Fig. 37. The beam is loaded by a single force F_0 and has constant material (E, ϱ) and geometric (I) properties along its longitudinal axis. The material is isotropic and homogeneous and the beam theory for thin beams (Euler-Bernoulli) can be used for this example.

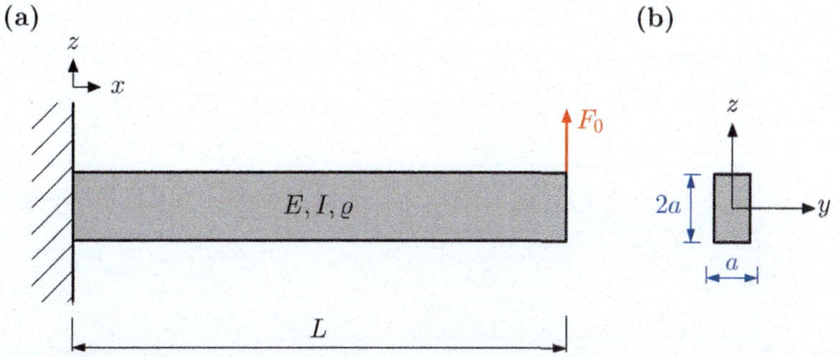

Fig. 37. **a** General configuration of the beam problem; **b** beam cross section

Given are:

- Geometric dimension: $L = 2540\,\text{mm}$.
- Material properties of the beam: Elastic modulus $E = 68{,}948\,\text{MPa}$, mass density $\varrho = 2691\,\text{kg/m}^3$, 0.2% strain limit $R_{\text{p}0.2} = 247\,\text{MPa}$.
- Load: $F_0 = 2667\,\text{N}$.

Determine the optimized cross-sectional dimension a under the condition that the effective stress does not exceed the initial yield stress. Furthermore, a maximum deflection of $u_z(L) = r_1 L$ with $r_1 = 0.03$ should not be exceeded. Based on a graphical representation of the objective function and the corresponding inequality constraints, the optimum is to be determined analytically.

14. Optimal Design of a Compression Strut (Classical Optimization Problem)

A compression strut is considered according to Fig. 38. The strut is loaded by a single force F_0 and has constant material (E, ϱ) and geometric (I, A) properties along its longitudinal axis. Furthermore, the material can be considered isotropic and homogeneous.
Given are:

- Geometric dimension: $L = 1000\,\text{mm}$.
- Material properties of the strut: Elastic modulus $E = 70{,}000\,\text{MPa}$, mass density $\varrho = 2691\,\text{kg/m}^3$, 0.2% strain limit $R_{\text{p}0.2} = 247\,\text{MPa}$.
- Load: $F_0 = 2667\,\text{N}$.

Determine the optimized cross-sectional dimension a under the condition that the effective stress does not exceed the initial yield stress. Furthermore, no buckling should occur in the elastic region. The optimum is to be determined analytically based on a graphical representation of the objective function and the corresponding inequality constraints.

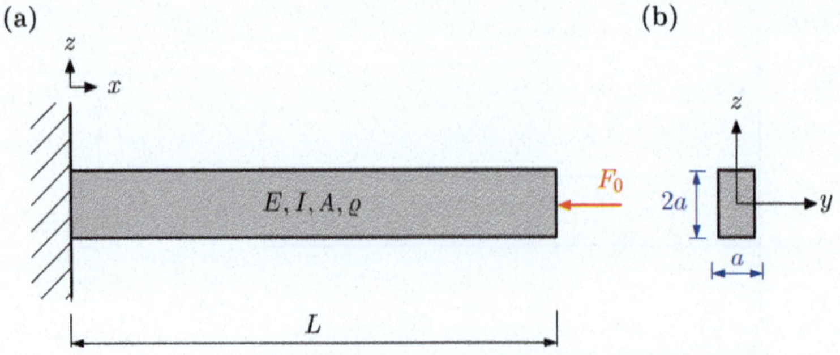

Fig. 38. **a** General configuration of the compression strut problem; **b** cross section

15. Optimal Design of a Short Cantilever Beam (Classical Optimization Problem)

Consider a short cantilever beam as shown in Fig. 39. The beam is loaded by a single force F_0 and has constant material (E, ϱ) and geometric (I) properties along its longitudinal axis. The material is isotropic and homogeneous and the beam theory for thick beams (i.e., consideration of the acting shear stress) can be used for this example.

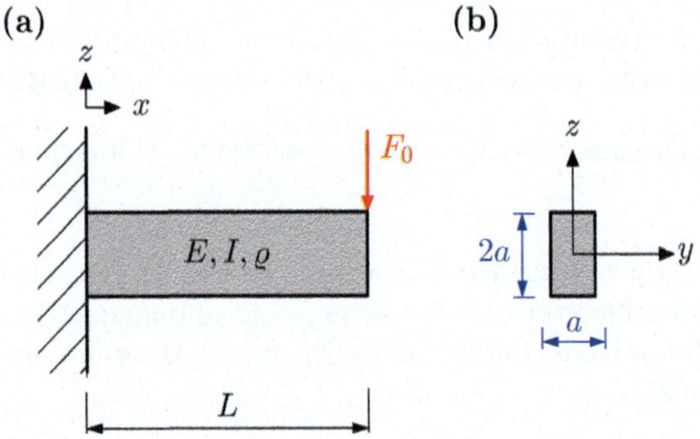

Fig. 39. **a** General configuration of the short cantilever beam problem; **b** cross section

Given are:

- Geometric dimension: $L = 846{,}33\,\text{mm}$.
- Material properties of the beam: elastic modulus $E = 68{,}948\,\text{MPa}$, mass density $\varrho = 2691\,\text{kg/m}^3$, 0.2% strain limit $R_{\text{p}0.2} = 247\,\text{MPa}$.
- Load: $F_0 = 2667\,\text{N}$.

Determine the optimized cross-sectional dimension a under the condition that the acting normal and shear stresses do not exceed the corresponding initial yield stress. The initial shear yield stress can be approximated using the Tresca hypothesis. The normal stress distribution can be assumed to be linear and the shear stress distribution to be parabolic over the beam height. Based on a graphical representation of the objective function and the corresponding inequality restrictions, the optimum is to be determined analytically.

16. Optimal Design of a Cantilever Beam: Constant Rectangular Cross Section (Classical Optimization Problem)

Consider a cantilever beam clamped at one end as shown in Fig. 40. The beam is loaded by a single force F_0 and has constant material (E, ϱ) and geometric (I) properties along its longitudinal axis. The material is isotropic and homogeneous and the beam theory for thin beams (Euler-Bernoulli) can be used for this problem.

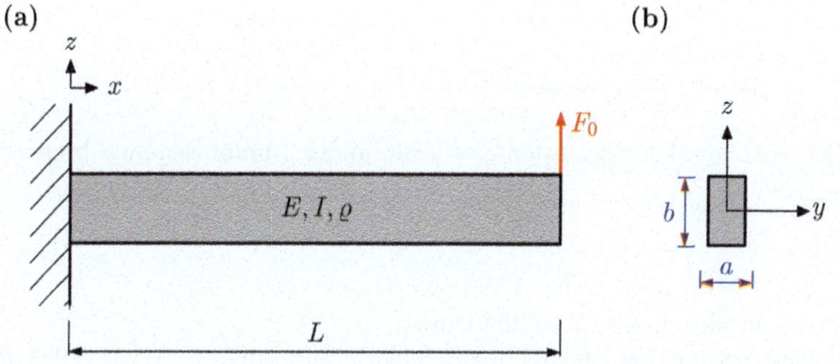

Fig. 40. a General configuration of the beam problem; **b** beam cross section

Given are:

- Geometric dimension: $L = 2540\,\text{mm}$.
- Material properties of the beam: elastic modulus $E = 68{,}948\,\text{MPa}$, mass density $\varrho = 2691\,\text{kg/m}^3$, 0.2% strain limit $R_{\text{p0.2}} = 247\,\text{MPa}$.
- Load: $F_0 = 2667\,\text{N}$.

Determine the optimized cross-sectional dimensions a and b under the condition that the acting normal stress does not exceed the initial yield stress. Furthermore, a maximum deflection of $u_z(L) = r_1 L$ with $r_1 = 0.03$ should not be exceeded. Finally, the height-to-width ratio should be limited to $b \leq 20a$ in order to avoid instabilities. Based on a graphical representation of the objective function and the corresponding inequality constraints, the optimum is to be determined analytically. Compare the optimized geometry with the results from Problem 13.

17. Optimal Design of a Simply Supported Beam: Constant Rectangular Cross Section (Classical Optimization Problem)

Consider a simply supported beam as shown in Fig. 41. The beam is loaded in the middle by a single force F_0 and has constant material (E, ϱ) and geometric (I) properties along its longitudinal axis. The material is isotropic and homogeneous and the beam theory for thin beams (Euler-Bernoulli) can be used for this problem.

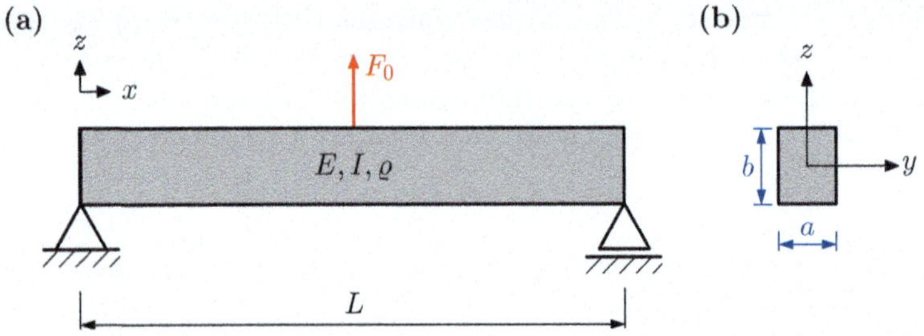

Fig. 41. **a** General configuration of a beam under 3-point bending; **b** cross section

Given are:

- Geometric dimension: $L = 2540$ mm.
- Material properties of the beam: elastic modulus $E = 68{,}948$ MPa, mass density $\varrho = 2691\,\text{kg/m}^3$, 0.2% strain limit $R_{\text{p0.2}} = 247$ MPa, shear yield strength $\tau_{\text{p}} = R_{\text{p0.2}}/2$.
- Load: $F_0 = 2667$ N.

Determine the optimized cross-sectional dimensions a and b under the condition that the acting normal and shear stresses do not exceed the corresponding initial yield stresses. Furthermore, a maximum deflection of $u_z(L) = r_1 L$ with $r_1 = 0.03$ or $r_1' = 0.003$ should not be exceeded. Finally, the height-to-width ratio should be limited to $b \leq 20a$ in order to avoid instabilities. Based on a graphical representation of the objective function and the corresponding inequality constraints, the optimum is to be determined analytically.

18. Optimal Design of a Compression Strut: Constant Rectangular Cross Section (Classical Optimization Problem)

A compression strut is considered according to Fig. 42. The strut is loaded by a single force F_0 and has constant material (E, ϱ) and geometric (I, A) properties along its longitudinal axis. Furthermore, the material can be considered isotropic and homogeneous.
Given are:

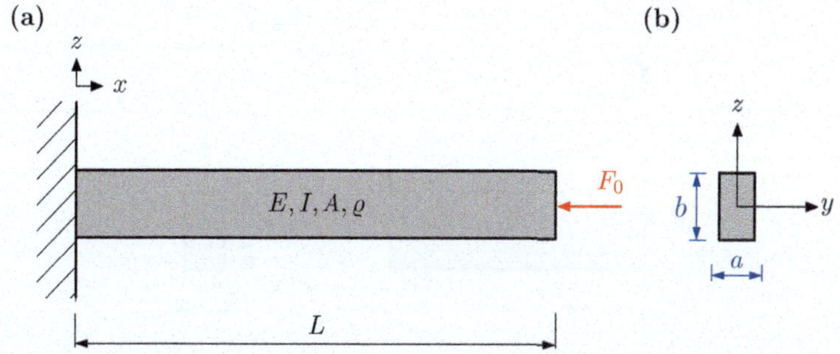

Fig. 42. **a** General configuration of the compression strut problem; **b** cross section

- Geometric dimension: $L = 1000\,\text{mm}$.
- Material properties of the rod: elastic modulus $E = 70{,}000\,\text{MPa}$, mass density $\varrho = 2691\,\text{kg/m}^3$, 0.2% strain limit $R_{\text{p}0.2} = 247\,\text{MPa}$.
- Load: $F_0 = 2667\,\text{N}$.

Determine the optimized cross-sectional dimensions a and b under the condition that the effective stress does not exceed the initial yield stress. Furthermore, no buckling should occur in the elastic range. Based on a graphical representation of the objective function and the corresponding inequality constraints, the optimum is to be determined analytically. Compare the optimized geometry with the results from Problem 14.

19. Optimal Design of a Short Cantilever Beam: Constant Rectangular Cross Section (classical optimization problem)

Consider a short, cantilevered bending beam as shown in Fig. 43. The beam is loaded by a single force F_0 and has constant material (E, ϱ) and geometric (I) properties along its longitudinal axis. The material is isotropic and homogeneous and the beam theory for thick beams (i.e., consideration of the acting shear stress) can be used for this example.
Given are:

- Geometric dimension: $L = 846.33\,\text{mm}$.
- Material properties of the beam: elastic modulus $E = 68{,}948\,\text{MPa}$, mass density $\varrho = 2691\,\text{kg/m}^3$, 0.2% strain limit $R_{\text{p}0.2} = 247\,\text{MPa}$.
- Load: $F_0 = 2667\,\text{N}$.

Determine the optimized cross-sectional dimensions a and b under the condition that the acting normal and shear stresses do not exceed the corresponding initial yield stresses. The initial shear yield stress can be approximated using the Tresca hypothesis. The normal stress distribution can be assumed to be linear and the shear stress distribution to be parabolic over the beam height. Furthermore, the height-to-width ratio should be limited to $b \leq 20a$ in order to

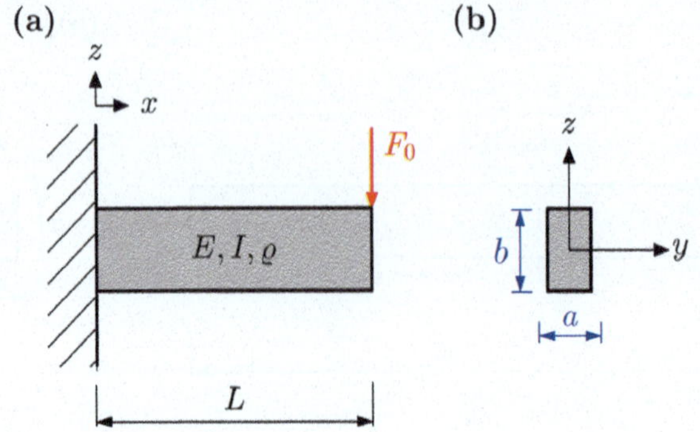

Fig. 43. **a** General configuration of the short cantilever beam problem; **b** cross section

avoid instabilities. Based on a graphical representation of the objective function and the corresponding inequality constraints, the optimum is to be determined analytically. Compare the optimized geometry with the results from Problem 15.

20. Optimal Design of a Stepped Cantilever Beam with Two Sections (Classical Optimization Problem)

Consider a cantilever beam that is fixed on one side as shown in Fig. 44, which is loaded at the right-hand end by a single force F_0 in the negative Z-direction. The beam is divided into two sections of length $L_{\mathrm{I}} = L_{\mathrm{II}} = L/2$. The material of each section is the same, i.e. $E_{\mathrm{I}} = E_{\mathrm{II}} = E$, but the cross sections are different in both sections. Optimize the two cross sections, i.e. the heights b_i ($i = \mathrm{I}, \mathrm{II}$), whereby the width of the two sections (a) remains identical.
Given are:

- Geometric dimensions: $L = 2540\,\mathrm{mm}$, $a = 45.16\,\mathrm{mm}$.
- Material properties of the beam: elastic modulus $E = 68{,}948\,\mathrm{MPa}$, mass density $\varrho = 2691\,\mathrm{kg/m^3}$, 0.2% strain limit $R_{\mathrm{p}0.2} = 247\,\mathrm{MPa}$.
- Load: $F_0 = 2667\,\mathrm{N}$.

Determine the optimized cross-sectional dimensions b_i under the condition that the acting normal and shear stresses do not exceed the corresponding initial yield stresses. The initial shear yield stress can be approximated using the Tresca hypothesis. The normal stress distribution can be assumed to be linear and the shear stress distribution to be parabolic over the beam height. Furthermore, a maximum deflection of $|u_z(L)| = r_1 L$ with $r_1 = 0.06$ should not be exceeded. Finally, the height-to-width ratio should be limited to $b \leq 20a$ to avoid instabilities. Based on a graphical representation of the objective function and the corresponding inequality constraints, the optimum is to be determined analytically. Use a finite element approach ('hand calculation') to determine the deformations and internal reactions, see [7,12,15].

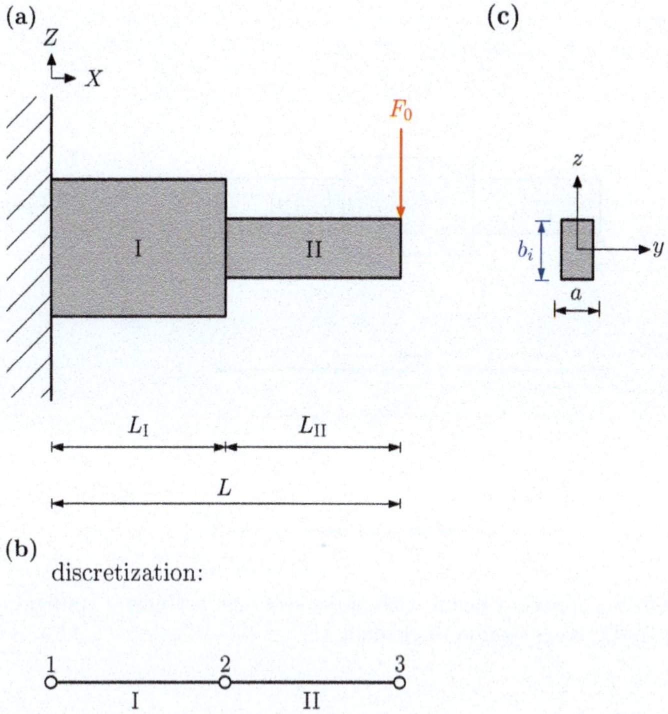

Fig. 44. Cantilever stepped beam with two sections: **a** General configuration, **b** discretization, and **c** cross section of element i

21. Optimal Design of a Simply Supported Beam with Three Sections (Classical Optimization Problem)

Consider a simply supported beam as shown in Fig. 45, which is loaded in the middle by a single force F_0 in the negative Z-direction. The beam is divided into three sections of length $L_{\text{I}} = L_{\text{II}} = L_{\text{III}} = L/3$. The material of each section is the same, i.e. $E_{\text{I}} = E_{\text{II}} = E_{\text{III}} = E$, but the middle cross section has different dimensions than the two outer sections (I = III). Optimize the cross sections, i.e. the heights b_i ($i = \text{I}, \text{II}, \text{III}$), whereby the width of the three sections (a) remains identical. For simplification, consider the symmetry of the problem.
Given are:

- Geometric dimensions: $L = 2540\,\text{mm}$, $a = 45.16\,\text{mm}$.
- Material properties of the beam: elastic modulus $E = 68{,}948\,\text{MPa}$, mass density $\varrho = 2691\,\text{kg/m}^3$, 0.2% strain limit $R_{\text{p0.2}} = 247\,\text{MPa}$.
- Load: $F_0 = 2667\,\text{N}$.

Determine the optimized cross-sectional dimensions b_i under the condition that the acting normal and shear stresses do not exceed the corresponding initial yield stresses. The initial shear yield stress can be approximated using the

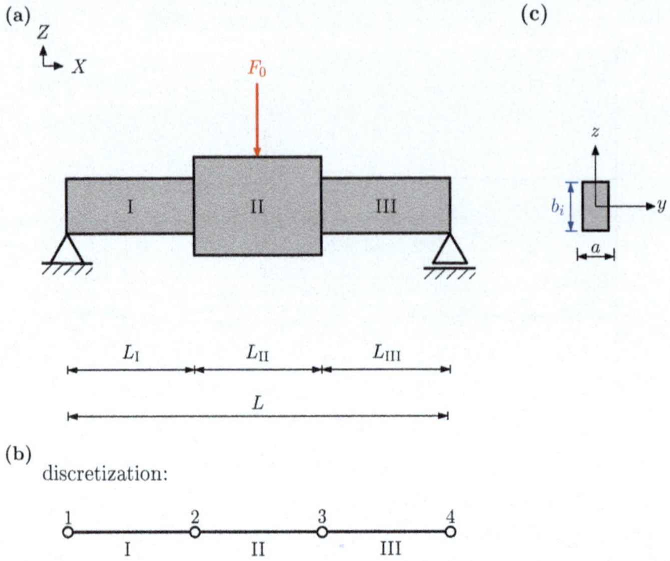

Fig. 45. Simply supported beam with three sections: **a** General configuration, **b** discretization, and **c** cross section of element i

Tresca hypothesis. The normal stress distribution can be assumed to be linear and the shear stress distribution to be parabolic over the beam height. Furthermore, a maximum deflection of $|u_z(L/2)| = r_1 L$ with $r_1 = 0.01$ should not be exceeded. Finally, the height-to-width ratio should be limited to $b \leq 20a$ to avoid instabilities. Based on a graphical representation of the objective function and the corresponding inequality constraints, the optimum is to be determined analytically. Use a finite element approach ('hand calculation') to determine the deformations and internal reactions, see [7,12,16].

3.5 Solutions to the Calculation Problems

1 Implementation of the Concept of Lightweight Design Shapes: Optimization of a Box Profile

- Geometric condition $A^{\boxtimes} = A^{\square}$:

The equality of the cross-sectional areas results in the following relationship:

$$h^2 = ab - (b-2w)(a-2w), \tag{87}$$

$$\overset{w=h/5}{=} ab - \left(b - \frac{2h}{5}\right)\left(a - \frac{2h}{5}\right), \tag{88}$$

$$= \frac{2}{5}bh + \frac{2}{5}ah - \frac{4}{25}h^2, \tag{89}$$

$$\rightarrow h = \frac{2}{5}b + \frac{2}{5}a - \frac{4}{25}h, \tag{90}$$

or

$$b = \frac{29}{10}h - a, \tag{91}$$

or also as:

$$\frac{a}{h} = \frac{29}{10}\left(\frac{b}{a} + 1\right)^{-1}, \tag{92}$$

or rearranged according to the width-to-height ratio:

$$\frac{a}{b} = \frac{10}{29}\left(\frac{h}{a} - 1\right)^{-1} = \frac{10}{29}\left(\left(\frac{a}{h}\right)^{-1} - 1\right)^{-1}. \tag{93}$$

- Axial second moment of area of the box profile:

$$I_y^{\square} = \frac{b\,a^3}{12} - \frac{1}{12}\,(b-2w)\,(a-2w)^3, \tag{94}$$

$$= -\frac{4w^4}{3} + \frac{2b\,w^3}{3} + 2a\,w^3 - ab\,w^2 - a^2\,w^2 + \frac{a^2bw}{2} + \frac{a^3w}{6}, \tag{95}$$

or by means of $w = \frac{h}{5}$

$$I_y^{\square} = -\frac{4h^4}{1875} + \frac{2b\,h^3}{375} + \frac{2a\,h^3}{125} - \frac{ab\,h^2}{25} - \frac{a^2\,h^2}{25} + \frac{a^2bh}{10} + \frac{a^3h}{30}, \tag{96}$$

or by means of $b = \frac{29h}{10} - a$ (see Eq. (91))

$$I_y^{\square} = \frac{h^4}{75} - \frac{79a\,h^3}{750} + \frac{29a^2\,h^2}{100} - \frac{a^3h}{15}, \tag{97}$$

or in normalized representation:

$$\frac{I_y^{\square}}{h^4} = \frac{1}{75} - \frac{79}{750}\left(\frac{a}{h}\right) + \frac{29}{100}\left(\frac{a}{h}\right)^2 - \frac{1}{15}\left(\frac{a}{h}\right)^3. \tag{98}$$

The course of the axial second moment of area is shown in Fig. 46.

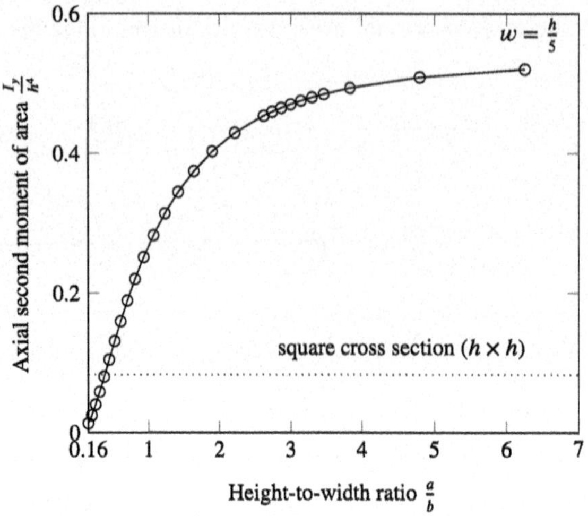

Fig. 46. Axial second moment of area for different box sections. Same cross-sectional area ($A^{\boxtimes} = A^{\square}$) and thus same mass ($m^{\boxtimes} = m^{\square}$) assumed

- Lightweight index:

The definition equation of the lightweight index (see [14]) results in:

$$M = \frac{F_0}{F_\mathrm{G}}. \tag{99}$$

The single force (F_0) is replaced by a stress criterion:

$$\sigma_\mathrm{max} = \frac{|M_{y,\mathrm{max}}|}{I_y} \times z_\mathrm{max} = \frac{F_0 L}{I_y} \times \frac{a}{2} = R_\mathrm{p0.2}, \tag{100}$$

or solved for the single force:

$$F_0 = \frac{2 I_y R_\mathrm{p0.2}}{aL}. \tag{101}$$

This ultimately results in the lightweight index:

$$M == \frac{R_\mathrm{St}}{\varrho_\mathrm{St} g L^2 h^2} \times \frac{I_y^{\square}}{\frac{a}{2}} \tag{102}$$

$$= \frac{R_\mathrm{St}}{\varrho_\mathrm{St} g L^2 h^2} \times \frac{\frac{h^4}{75} - \frac{79ah^3}{750} + \frac{29a^2h^2}{100} - \frac{a^3h}{15}}{\frac{a}{2}} \tag{103}$$

$$= \frac{R_\mathrm{St}}{\varrho_\mathrm{St} g L^2 h^2} \times \frac{\left(\frac{I_y^{\square}}{h^4}\right) \times h^4}{\frac{1}{2}\left(\frac{a}{h}\right) \times h}. \tag{104}$$

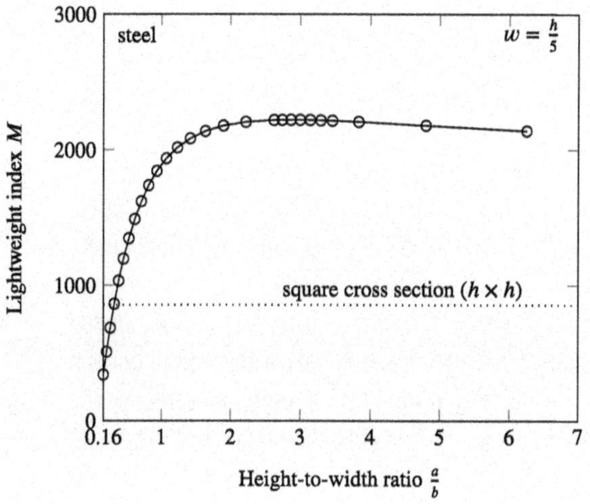

Fig. 47. Lightweight index for different box sections and stress criterion ($\sigma_{\max} = R_{\mathrm{p}0.2}$). Same cross-sectional area ($A^{\boxtimes} = A^{\square}$) and thus same mass ($m^{\boxtimes} = m^{\square}$) assumed

The course of the lightweight index is shown in Fig. 47.

The graphical representation according to Figs. 46 and 47 was created using the following procedure: The markers in both figures correspond to a calculated coordinate in the respective coordinate system, which was then approximated using a best-fit curve. By specifying numerical values of $\frac{a}{h}$, the width-to-height ratio $\frac{a}{b}$ can be calculated using Eq. (93). The standardized axial area moment is obtained for a given $\frac{a}{h}$ according to Eq. (98) and then the lightweight index can be calculated according to Eq. (104).

The pairwise mapping for creating the diagrams is summarized in Table 4.

Finally, it should be noted that the highest lightweight construction index of $M_{\max} = 2220.84$ is achieved for a width-to-height ratio of $\frac{a}{b} = 2.867$.

2 Optimization of a Bending Beam Along Its Axis

The bending moment distribution can be taken, for example, from [16]:

$$M_y(x) = F_0 L \left(1 - \frac{x}{L}\right) = F_0 \left(L - x\right). \tag{105}$$

This gives the maximum bending moment at the fixed support ($x = 0$) as:

$$M_{y,\max} = M_y(x = 0) = F_0 L. \tag{106}$$

Table 4. Pairs of values for the generation of Figs. 46–47

$\frac{a}{h}$	$\frac{a}{b}$	$\frac{I_y}{h^4}$	M
0.4	0.16	0.013333	342.403095
0.5	0.2083333	0.024833	510.180611
0.6	0.260870	0.040133	687.088877
0.7	0.318182	0.058833	863.344946
0.8	0.380952	0.080533	1034.057346
0.9	0.45	0.104833	1196.508592
1.0	0.526316	0.131333	1349.068193
1.1	0.611111	0.159633	1490.698564
1.2	0.705882	0.189333	1620.707982
1.3	0.8125	0.220033	1738.617560
1.4	0.933333	0.251333	1844.085239
1.5	1.071429	0.2828333	1936.860172
1.6	1.230769	0.314133	2016.754228
1.7	1.416667	0.344833	2083.623538
1.8	1.636364	0.374533	2137.356207
1.9	1.9	0.402833	2177.863894
2.0	2.222222	0.429333	2205.075930
2.1	2.625	0.453633	2218.935103
2.125	2.741935	0.459318	2220.307832
2.15	2.866667	0.464833	2220.842398
2.175	3.0	0.470174	2220.538185
2.2	3.142857	0.475333	2219.394605
2.225	3.296296	0.480305	2217.411098
2.25	3.461538	0.485083	2214.587127
2.3	3.833333	0.494033	2206.415768
2.4	4.8	0.509333	2179.966370
2.5	6.25	0.520833	2140.019342

- Reference configuration:

The cross-sectional area of the reference configuration is:

$$\begin{aligned} A_0^{\circledcirc} &= \frac{\pi}{4}\left(d_0^2 - d_i^2\right) = \frac{\pi}{4}\left(d_0^2 - (d_0 - 2s_0)^2\right) = \frac{\pi}{4}\left(d_0^2 - (d_0 - \tfrac{2d_0}{5})^2\right) \\ &= \frac{4\pi}{25}d_0^2 . \end{aligned} \quad (107)$$

The axial second moment of area of the reference configuration is:

$$I_{y,0}^{\circledcirc} = \frac{\pi}{64}\left(d_0^4 - d_\mathrm{i}^4\right) = \frac{\pi}{64}\left(d_0^4 - (d_0 - 2s_0)^4\right) = \frac{\pi}{64}\left(d_0^4 - (d_0 - \tfrac{2d_0}{5})^4\right)$$
$$= \frac{17}{1250}\pi d_0^4. \tag{108}$$

The mass of the reference configuration is:

$$m_0^{\circledcirc} = \varrho V_0^{\circledcirc} = \varrho A_0^{\circledcirc} L = \varrho \frac{4\pi}{25} d_0^2 L. \tag{109}$$

Using the stress criterion

$$\sigma_{x,\mathrm{max}} = \frac{M_{y,\mathrm{max}}}{I_{y,0}^{\circledcirc}} \times \frac{d_0}{2} \overset{!}{=} R_{\mathrm{p}0.2}, \tag{110}$$

the maximum value of the force is obtained as:

$$F_0 = \frac{R_{\mathrm{p}0.2} I_{y,0}^{\circledcirc}}{\frac{1}{2} d_0 L}. \tag{111}$$

The definition equation of the lightweight index (see [14]) results in:

$$M = \frac{F_0}{F_\mathrm{G}}. \tag{112}$$

From this, using Eqs. (111) and (109), the lightweight index of the reference configuration is:

$$M_0 = \frac{F_0}{F_\mathrm{G}} = \frac{17}{100} \times \frac{R_{\mathrm{p}0.2}}{\varrho g \frac{L^2}{d_0}}. \tag{113}$$

The total strain energy is calculated according to [14] for a bending load case as follows:

$$\Pi = \int_0^L \frac{M_y(x)^2}{2EI_y}\,\mathrm{d}x. \tag{114}$$

For the reference configuration in our case this results in:

$$\Pi_0 = \int_0^L \frac{M_y(x)^2}{2EI_{y,0}^{\circledcirc}}\,\mathrm{d}x = \frac{F_0^2}{2EI_{y,0}^{\circledcirc}} \int_0^L \left(L^2 - 2Lx + x^2\right)\mathrm{d}x \tag{115}$$

$$= \frac{F_0^2}{2EI_{y,0}^{\circledcirc}} \left[L^2 x - L x^2 + \frac{1}{3}x^3\right]_0^L \tag{116}$$

$$= \frac{F_0^2 L^3}{6EI_{y,0}^{\circledcirc}}. \tag{117}$$

Using Eq. (111) we get:

$$\Pi_0 = \frac{L^3}{6EI^{\circledcirc}_{y,0}} \times \frac{R^2_{\text{p}0.2} I^{\circledcirc,2}_{y,0}}{\frac{1}{4} d_0^2 L^2} = \frac{L R^2_{\text{p}0.2} I^{\circledcirc}_{y,0}}{\frac{3}{2} E d_0^2}. \tag{118}$$

The specific energy absorption is generally given as [14]:

$$SEA = \frac{\Pi}{m}. \tag{119}$$

For the reference configuration in our case this results in:

$$SEA_0 = \frac{\Pi_0}{m_0} = \frac{L R^2_{\text{p}0.2} I^{\circledcirc}_{y,0}}{\frac{3}{2} E d_0^2} \times \frac{1}{\varrho \frac{4\pi}{25} d_0^2 L} = \frac{17}{300} \times \frac{R^2_{\text{p}0.2}}{E \varrho}. \tag{120}$$

- Optimized configuration:

Using the stress criterion

$$\sigma_{x,\max} = \frac{M_y(x)}{I^{\circledcirc}_y(x)} \times \frac{d(x)}{2} = \frac{F_0 L (1 - \frac{x}{L})}{\frac{17}{1250} \pi d^4(x)} \times \frac{d(x)}{2} \overset{!}{=} R_{\text{p}0.2}, \tag{121}$$

results in

$$d^3(x) = \frac{F_0 L}{\frac{17}{1250} \pi 2 R_{\text{p}0.2}} \times \left(1 - \tfrac{x}{L}\right). \tag{122}$$

By evaluating the stress criterion at $x = 0$, i.e., for the reference configuration, we get:

$$\sigma_{x,\max} = \frac{M_y(x=0)}{I^{\circledcirc}_{y,0}(x)} \times \frac{d_0}{2} = \frac{F_0 L}{\frac{17}{1250} \pi d_0^4} \times \frac{d_0}{2} \overset{!}{=} R_{\text{p}0.2}, \tag{123}$$

or resolved according to the reference diameter:

$$d_0^3 = \frac{F_0 L}{\frac{17}{1250} \pi 2 R_{\text{p}0.2}}. \tag{124}$$

This gives the following expression from Eq. (122):

$$d^3(x) = d_0^3 \left(1 - \tfrac{x}{L}\right). \tag{125}$$

Thus, the outer circle diameter of the beam cross-section is (see Fig. 48):

$$d(x) = d_0 \times \sqrt[3]{1 - \frac{x}{L}}. \tag{126}$$

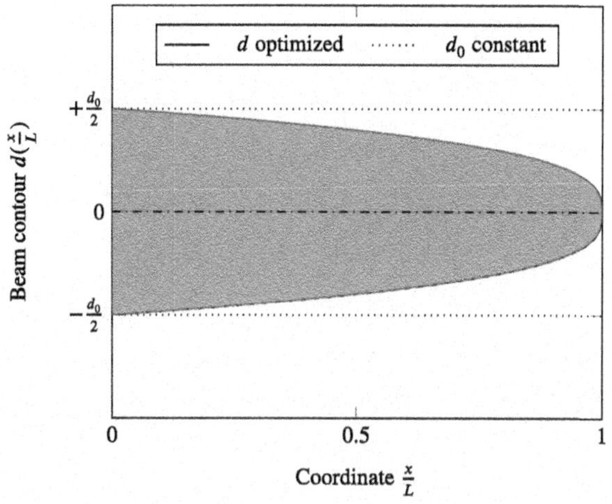

Fig. 48. Beam diameter (outside) along the beam axis for a cantilever beam (bending stiffness $E \times I(x)$, length L) loaded by a single force at the free end

The volume of the optimized configuration must be calculated by integration, since the cross-section changes continuously with the x-coordinate:

$$V = \int_0^L A(x)\mathrm{d}x = \int_0^L \frac{4\pi}{25} d^2(x)\mathrm{d}x = \frac{4\pi d_0^2}{25} \int_0^L \left(1 - \frac{x}{L}\right)^{\frac{2}{3}} \mathrm{d}x \tag{127}$$

$$= \frac{4\pi d_0^2}{25} \left[\frac{3}{5} \left(1 - \frac{x}{L}\right)^{\frac{5}{3}} (-L) \right]_0^L \tag{128}$$

$$= \frac{12\pi}{125} d_0^2 L. \tag{129}$$

With this volume, the mass of the optimized configuration is:

$$m = \varrho V = \frac{12\pi}{125} d_0^2 L \varrho. \tag{130}$$

Thus, the lightweight index of the optimized configuration is (the force F_0 can still be replaced using Eq. (111)):

$$M = \frac{F_0}{F_\mathrm{G}} = \frac{R_{\mathrm{p}0.2} I_{y,0}^{\circledcirc}}{\frac{1}{2} d_0 L} \times \frac{1}{\varrho g \frac{12\pi}{125} d_0^2 L} = \frac{17}{60} \times \frac{R_{\mathrm{p}0.2}}{\varrho g \frac{L^2}{d_0}}. \tag{131}$$

The total strain energy of the optimized configuration is calculated as:

$$\Pi = \int_0^L \frac{M_y(x)^2}{2EI_y^{\circledcirc}(x)}\mathrm{d}x = \frac{F_0^2L^2}{2EI_{y,0}^{\circledcirc}}\int_0^L \frac{(1-\frac{x}{L})^2}{(1-\frac{x}{L})^{\frac{4}{3}}}\mathrm{d}x = \frac{F_0^2L^2}{2EI_{y,0}^{\circledcirc}}\int_0^L (1-\tfrac{x}{L})^{\frac{2}{3}}\mathrm{d}x$$

$$= \frac{F_0^2L^2}{2EI_{y,0}^{\circledcirc}}\left[\frac{3}{5}\left(1-\frac{x}{L}\right)^{\frac{5}{3}}(-L)\right]_0^L \tag{132}$$

$$= \frac{3}{10}\times\frac{F_0^2L^3}{EI_{y,0}^{\circledcirc}}. \tag{133}$$

Using Eq. (111) we get:

$$\Pi = \frac{3}{10}\frac{L^3}{EI_{y,0}^{\circledcirc}}\times\frac{R_{\mathrm{p}0.2}^2 I_{y,0}^{\circledcirc,2}}{\frac{1}{4}d_0^2L^2} = \frac{6}{5}\times\frac{LR_{\mathrm{p}0.2}^2 I_{y,0}^{\circledcirc}}{Ed_0^2}, \tag{134}$$

or for the specific energy absorption of the optimized configuration:

$$SEA = \frac{\Pi}{m} = \frac{6}{5}\times\frac{LR_{\mathrm{p}0.2}^2 I_{y,0}^{\circledcirc}}{Ed_0^2}\times\frac{1}{\frac{12\pi}{125}d_0^2L\varrho} = \frac{17}{100}\times\frac{R_{\mathrm{p}0.2}^2}{E\varrho}. \tag{135}$$

- Ratios:

$$\frac{m}{m_0} = \frac{3}{5} \approx 0.6, \tag{136}$$

$$\frac{M}{M_0} = \frac{5}{3} \approx 1.667, \tag{137}$$

$$\frac{SEA}{SEA_0} = 3.0. \tag{138}$$

3 Optimization of a Cantilevered Beam Loaded with Two Single Forces of the Same Magnitude

- Internal reactions:

To determine the internal reactions, cut the beam under the influence of two single forces at an arbitrary location x to the left and then to the right of the left single force, see Fig. 49.

The moment equilibrium around the reference point at the location x in section 1 ($0 \leq x \leq \frac{L}{2}$) gives the distribution of the internal bending moment, see Fig. 49a:

$$\overset{\curvearrowright}{\sum} M_y = 0: \quad +F_0(L-x) - F_0\left(\tfrac{L}{2}-x\right) - M_y(x) = 0, \tag{139}$$

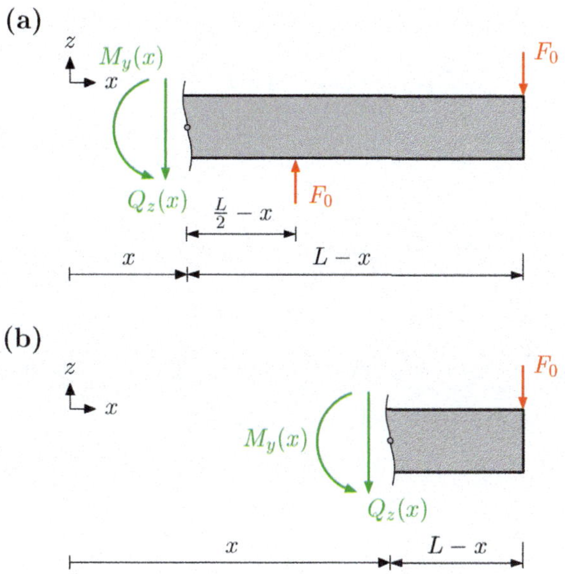

Fig. 49. Internal reactions of a beam under the influence of two single forces: **a** section 1 and **b** section 2

$$M_y(x) = \frac{F_0 L}{2} \quad \text{for} \quad 0 \le x \le \frac{L}{2}. \tag{140}$$

The moment equilibrium around the reference point at the location x in section 2 ($\frac{L}{2} \le x \le L$) results in the distribution of the internal bending moment, see Fig. 49b:

$$\overset{\curvearrowright}{\sum} M_y = 0: \quad +F_0(L-x) - M_y(x) = 0, \tag{141}$$

$$M_y(x) = F_0(L-x) \quad \text{for} \quad \frac{L}{2} \le x \le L. \tag{142}$$

- Optimized diameter:

Using the stress criterion in section 1 ($0 \le x \le \frac{L}{2}$)

$$\sigma_{x,\max} = \frac{M_y(x)}{I_y^{\circledcirc}(x)} \times \frac{d(x)}{2} = \frac{\frac{F_0 L}{2}}{\frac{17}{1250}\pi d^4(x)} \times \frac{d(x)}{2} \overset{!}{=} R_{\text{p0.2}}, \tag{143}$$

results in (the evaluation is also valid at the location $x = 0$)

$$d^3(x) = \frac{625 F_0 L}{34\pi R_{\text{p0.2}}} = d_0^3. \tag{144}$$

Using the stress criterion in section 2 ($\frac{L}{2} \leq x \leq L$)

$$\sigma_{x,\max} = \frac{M_y(x)}{I_y^{\circledcirc}(x)} \times \frac{d(x)}{2} = \frac{F_0(L-x)}{\frac{17}{1250}\pi d^4(x)} \times \frac{d(x)}{2} \overset{!}{=} R_{\mathrm{p}0.2}, \tag{145}$$

results in

$$d^3(x) = \frac{625 F_0(L-x)}{17\pi R_{\mathrm{p}0.2}} = 2 \times d_0^3 \left(1 - \frac{x}{L}\right). \tag{146}$$

In summary, the following distribution results for both sections:

$$d(x) = d_0 \quad \text{for} \quad 0 \leq x \leq \frac{L}{2}, \tag{147}$$

$$d(x) = d_0 \left(2 - \frac{2x}{L}\right)^{\frac{1}{3}} \quad \text{for} \quad \frac{L}{2} \leq x \leq L, \tag{148}$$

or as a single equation using the Föppl symbol (see Fig. 50):

$$d(x) = d_0 \left(1 + \left(1 - \frac{2x}{L}\right) \left\langle x - \frac{L}{2} \right\rangle^0 \right)^{\frac{1}{3}}. \tag{149}$$

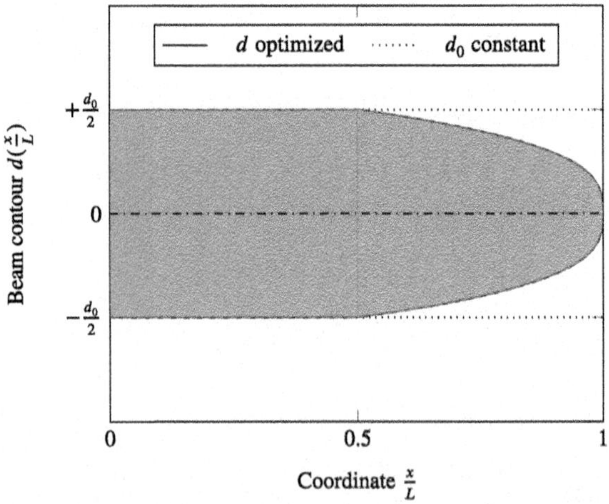

Fig. 50. Beam diameter (outside) along the beam axis for a cantilever beam (bending stiffness $E \times I(x)$, length L) loaded by two single forces of the same magnitude

- Axial second moment of area:

For a tube with outer diameter d and wall thickness $s = d/5$, the relationship for the axial second moment of area can be taken from the solution of task 2. This gives the following for both sections:

$$I_y^{\circledcirc} = \frac{17}{1250} \times \pi d_0^4 \quad \text{for} \quad 0 \leq x \leq \frac{L}{2}, \tag{150}$$

$$I_y^{\circledcirc} = \frac{17}{1250} \times \pi d_0^4 (16)^{\frac{1}{3}} \left(1 - \frac{x}{L}\right)^{\frac{4}{3}} \quad \text{for} \quad \frac{L}{2} \leq x \leq L, \tag{151}$$

or as a single equation using the Föppl symbol:

$$I_y^{\circledcirc} = \frac{17}{1250} \times \pi d_0^4 \left(1 + \left(1 - \frac{2x}{L}\right) \left\langle x - \frac{L}{2} \right\rangle^0 \right)^{\frac{4}{3}}. \tag{152}$$

- Total strain energy:

$$\Pi = \int_0^{\frac{L}{2}} \frac{M_y^2}{2EI_y^{\circledcirc}} \mathrm{d}x + \int_{\frac{L}{2}}^{L} \frac{M_y(x)^2}{2EI_y^{\circledcirc}(x)} \mathrm{d}x \tag{153}$$

$$= \frac{\frac{1}{4} F_0^2 L^2}{2EI_y^{\circledcirc}} \int_0^{\frac{L}{2}} 1 \, \mathrm{d}x + \frac{F_0^2}{2E \frac{17}{1250} \pi d_0^4 (16)^{\frac{1}{3}}} \int_{\frac{L}{2}}^{L} \frac{L^2 (1 - \frac{x}{L})^2}{(1 - \frac{x}{L})^{\frac{4}{3}}} \mathrm{d}x \tag{154}$$

$$= \frac{F_0^2 L^3}{16EI_y^{\circledcirc}} + \frac{F_0^2 L^2}{2EI_y^{\circledcirc}(16)^{\frac{1}{3}}} \int_{\frac{L}{2}}^{L} \left(1 - \frac{x}{L}\right)^{\frac{2}{3}} \mathrm{d}x \tag{155}$$

$$= \frac{F_0^2 L^3}{16EI_y^{\circledcirc}} + \frac{F_0^2 L^2}{2EI_y^{\circledcirc}(16)^{\frac{1}{3}}} \left[\frac{3}{5} \left(1 - \frac{x}{L}\right)^{\frac{5}{3}} (-L) \right]_{\frac{L}{2}}^{L} \tag{156}$$

$$= \frac{125}{17} \times \frac{F_0^2 L^3}{\pi d_0^4 E}. \tag{157}$$

- Volume:

For a tube with outer diameter d and wall thickness $s = d/5$, the relationship for the cross-sectional area can be taken from the solution of task 2. This gives the total volume:

$$V = \int_0^{\frac{L}{2}} A_0 \, dx + \int_{\frac{L}{2}}^{L} A(x) \, dx \tag{158}$$

$$= \int_0^{\frac{L}{2}} \frac{4\pi}{25} d_0^2 \, dx + \int_{\frac{L}{2}}^{L} \frac{4\pi}{25} d_0^2 2^{\frac{2}{3}} \left(1 - \frac{x}{L}\right)^{\frac{2}{3}} dx \tag{159}$$

$$= \frac{16}{125} \times \pi d_0^2 L. \tag{160}$$

- Specific energy absorption:

The specific energy absorption is generally given as [14]:

$$SEA = \frac{\Pi}{m}. \tag{161}$$

The expression for the force F_0 in the total strain energy according to Eq. (157) can be replaced by the stress criterion at $x = 0$:

$$\sigma_{x,\max} = \frac{M_y(x)}{I_y^{\circledcirc}(x)} \times \frac{d(x)}{2} = \frac{\frac{F_0 L}{2}}{\frac{17}{1250}\pi d_0^4} \times \frac{d_0}{2} \overset{!}{=} R_{\mathrm{p}0.2}, \tag{162}$$

or solved for the force:

$$F_0 = \frac{34}{625} \times \frac{\pi d_0^3 R_{\mathrm{p}0.2}}{L}. \tag{163}$$

This gives the specific energy absorption:

$$SEA = \frac{\Pi}{m} = \frac{125^2}{16 \times 17} \times \frac{F_0^2 L^2}{\pi^2 d_0^6 \varrho E} = \frac{17}{100} \times \frac{R_{\mathrm{p}0.2}^2}{\varrho E}. \tag{164}$$

4 Optimization of a Beam Under 3-Point Bending: Tubular Cross Section

- Internal reactions:

The bending moment distribution can be taken, for example, from [16]:

$$M_y(x) = -\frac{F_0}{2} \times x \text{ for } 0 \le x \le \frac{L}{2}, \tag{165}$$

Due to the symmetry of the problem, the following moment distribution results for the right-hand part of the beam:

$$M_y(x) = -\frac{F_0}{2} \times (L - x) \text{ for } \frac{L}{2} \leq x \leq L, \tag{166}$$

where the maximum value of the bending moment is in the middle of the beam:

$$M_{y,\text{max}} = M_y\left(x = \tfrac{L}{2}\right) = -\frac{F_0 L}{4}. \tag{167}$$

- Axial second moment of area and cross-sectional area:

For a tube with outer diameter d_0 and wall thickness $s_0 = d_0/5$, the relationship for the axial second moment of area and the cross-sectional area can be taken from the solution of Problem 2:

$$I_{y,0}^{\circledcirc} = \frac{17}{1250}\pi d_0^4. \tag{168}$$

$$A_0^{\circledcirc} = \frac{4\pi}{25} d_0^2. \tag{169}$$

- Optimized diameter:

Using the stress criterion in section 1 $(0 \leq x \leq \frac{L}{2})$

$$\sigma_{x,\text{max}} = \frac{M_y(x)}{I_y^{\circledcirc}(x)} \times \frac{d(x)}{2} = \frac{\frac{F_0 x}{2}}{\frac{17}{1250}\pi d^4(x)} \times \frac{d(x)}{2} \stackrel{!}{=} R_{\text{p0.2}}, \tag{170}$$

results in

$$d^3(x) = \frac{1250 F_0 L}{4 \times 17\pi R_{\text{p0.2}}} \times x. \tag{171}$$

The evaluation of the stress criterion in the reference cross section in the middle of the beam at $x = \frac{L}{2}$, i.e.

$$\sigma_{x,\text{max}} = \frac{M_{y,\text{max}}}{I_y^{\circledcirc}} \times \frac{d_0}{2} = \frac{\frac{F_0 L}{4}}{\frac{17}{1250}\pi d_0^4} \times \frac{d_0}{2} \stackrel{!}{=} R_{\text{p0.2}}, \tag{172}$$

gives the reference diameter:

$$d_0 = \left(\frac{F_0 L}{2 \times \frac{68}{1250}\pi R_{\text{p0.2}}}\right)^{\frac{1}{3}}. \tag{173}$$

Using the stress criterion in section 2 $(\frac{L}{2} \leq x \leq L)$

$$\sigma_{x,\max} = \frac{M_y(x)}{I_y^{\circledcirc}(x)} \times \frac{d(x)}{2} = \frac{\frac{F_0}{2}(L-x)}{\frac{17}{1250}\pi d^4(x)} \times \frac{d(x)}{2} \stackrel{!}{=} R_{\mathrm{p}0.2}, \tag{174}$$

results in

$$d^3(x) = 2 \times \frac{1250 F_0 (L-x)}{2 \times 4 \times 17 \pi R_{\mathrm{p}0.2}} = 2 \times d_0^3 \left(1 - \frac{x}{L}\right). \tag{175}$$

In summary, the following distribution results for both sections (see also Fig. 51):

$$d(x) = d_0 \left(\frac{2x}{L}\right)^{\frac{1}{3}} \quad \text{for} \quad 0 \le x \le \frac{L}{2}, \tag{176}$$

$$d(x) = d_0 \left(2\left(1 - \frac{x}{L}\right)\right)^{\frac{1}{3}} \quad \text{for} \quad \frac{L}{2} \le x \le L. \tag{177}$$

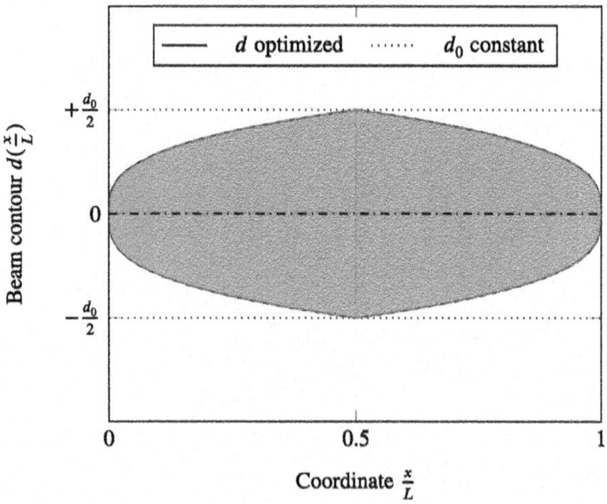

Fig. 51. Beam diameter (outside) along the axis for a beam under symmetrical 3-point bending

- Total strain energy:

The total strain energy is calculated according to [14] for a bending load case as follows:

$$\Pi = \int_0^L \frac{M_y(x)^2}{2EI_y} \mathrm{d}x. \tag{178}$$

Due to the symmetry of the problem, we get:

$$\Pi = 2 \times \int_0^{\frac{L}{2}} \frac{M_y(x)^2}{2EI_y^{\circledcirc}} \mathrm{d}x \tag{179}$$

$$= \frac{F_0^2}{4 \times \frac{17}{1250} E\pi d_0^4 \left(\frac{2}{L}\right)^{4/3}} \int_0^{\frac{L}{2}} x^{\frac{2}{3}} \, \mathrm{d}x \tag{180}$$

$$= \frac{375}{272} \times \frac{F_0^2 L^3}{\pi d_0^4 E} \approx 1.379 \times \frac{F_0^2 L^3}{\pi d_0^4 E}. \tag{181}$$

- Volume:

$$V = 2 \times \int_0^{\frac{L}{2}} A(x) \, \mathrm{d}x \tag{182}$$

$$= \frac{8 \times 2^{2/3} \pi d_0^2}{25 \, L^{2/3}} \int_0^{\frac{L}{2}} x^{\frac{2}{3}} \, \mathrm{d}x \tag{183}$$

$$= \frac{12}{125} \times \pi d_0^2 L. \tag{184}$$

- Specific energy absorption:

The specific energy absorption is generally given as [14]:

$$SEA = \frac{\Pi}{m}. \tag{185}$$

Using the total strain energy according to Eq. (181) we get:

$$SEA = \frac{125 \times 375}{12 \times 272} \times \frac{F_0^2 L^2}{\pi^2 d_0^6 \varrho E} \approx 14.361 \frac{F_0^2 L^2}{\pi^2 d_0^6 \varrho E}. \tag{186}$$

- Lightweight index:

The definition equation of the lightweight index (see [14]) results in:

$$M = \frac{F_0}{F_{\mathrm{G}}}. \tag{187}$$

This results in:

$$M = \frac{F_0}{V \varrho g} = \frac{125}{12} \times \frac{F_0}{\pi d_0^2 L \varrho g} \approx 10.417 \times \frac{F_0}{\pi d_0^2 L \varrho g}. \tag{188}$$

- Values for the beam with constant diameter:

$$\Pi_0 = 2 \times \int_0^{\frac{L}{2}} \frac{M_y(x)^2}{2EI_{y,0}^{\circledcirc}} \mathrm{d}x = \frac{1250}{1632} \times \frac{F_0^2 L^3}{\pi d_0^4 E}, \tag{189}$$

$$V_0 = A_0 L = \frac{4}{25} \times \pi d_0^2 L, \tag{190}$$

$$SEA_0 = \frac{\Pi_0}{m_0} = \frac{31250}{6528} \times \frac{F_0^2 L^2}{\pi^2 d_0^6 \varrho E} \approx 4.787 \times \frac{F_0^2 L^2}{\pi^2 d_0^6 \varrho E}, \tag{191}$$

$$M_0 = \frac{F_0}{F_{\mathrm{G},0}} = \frac{25 F_0}{4\pi d_0^2 L \varrho g}. \tag{192}$$

- Comparison:

$$\frac{SEA}{SEA_0} = 3.00, \tag{193}$$

$$\frac{M}{M_0} = 1.67. \tag{194}$$

5 Optimization of a Simply Supported Beam Under Constant Distributed Load: Rectangular Cross Section

- Moment distribution:

The bending moment distribution is obtained as:

$$M_y(x) = \frac{q_0}{2}(x^2 - Lx) = \frac{q_0 L^2}{2} \left(\left[\frac{x}{L}\right]^2 - \left[\frac{x}{L}\right]^1 \right) < 0. \tag{195}$$

The maximum bending moment in the middle of the beam is:

$$M_{y,\max} = M_y\left(\tfrac{L}{2}\right) = -\frac{q_0 L^2}{8}. \tag{196}$$

- Geometric properties of the cross section:

For a rectangular cross-section of width b and height h we get:

$$I_y = \frac{bh^3}{12}, \tag{197}$$

$$A = bh. \tag{198}$$

- Optimized diameter:

Using the stress criterion

$$\sigma_{x,\max} = \frac{M_y(x)}{I_y(x)} \times \frac{h(x)}{2} = \frac{\frac{q_0 L^2}{2}\left(\left[\frac{x}{L}\right] - \left[\frac{x}{L}\right]^2\right)}{\frac{bh(x)^3}{12}} \times \frac{h(x)}{2} \stackrel{!}{=} R_{\mathrm{p}0.2}, \tag{199}$$

results in

$$h^2(x) = \frac{q_0 L^2\left(\left[\frac{x}{L}\right] - \left[\frac{x}{L}\right]^2\right)}{4 \times \frac{1}{12} b R_{\mathrm{p}0.2}}. \tag{200}$$

The evaluation of the stress criterion in the reference cross-section at location $x = L/2$, i.e.

$$\sigma_{x,\max} = \frac{M_{y,\max}}{I_y} \times \frac{h_0}{2} = \frac{\frac{q_0 L^2}{8}}{\frac{bh_0^3}{12}} \times \frac{h_0}{2} \stackrel{!}{=} R_{\mathrm{p}0.2}, \tag{201}$$

provides the reference height:

$$h_0^2 = \frac{q_0 L^2}{16 \times \frac{b}{12} R_{\mathrm{p}0.2}}. \tag{202}$$

With this result, the optimized cross-sectional height is finally calculated as (see also Fig. 52):

$$h(x) = 2h_0 \left(\left[\frac{x}{L}\right] - \left[\frac{x}{L}\right]^2\right)^{\frac{1}{2}}. \tag{203}$$

- Optimized cross section:

The volume is determined by evaluating the integral:

$$V = \int_0^L A(x)\,\mathrm{d}x = 2bh_0 \underbrace{\int_0^L \left(\left[\frac{x}{L}\right] - \left[\frac{x}{L}\right]^2\right)^{\frac{1}{2}} \mathrm{d}x}_{\frac{\pi L}{8}} = \frac{1}{4} \times \pi b h_0 L. \tag{204}$$

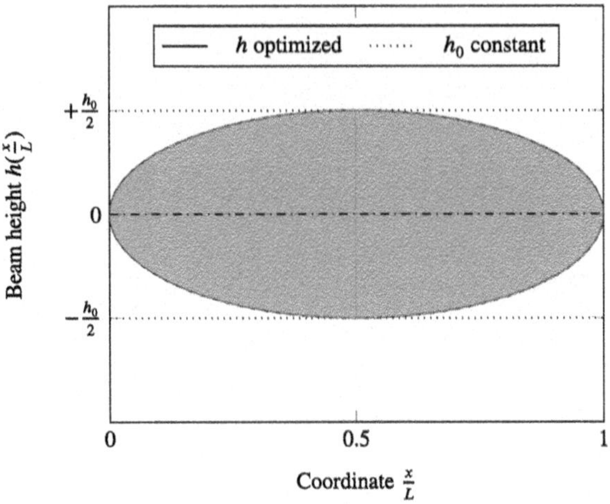

Fig. 52. Beam height along the longitudinal axis for a simply supported beam under constant distributed load with a rectangular cross section

The total strain energy is calculated according to [14] for a bending load case as follows:

$$\Pi = \int_0^L \frac{M_y(x)^2}{2EI_y} \mathrm{d}x. \tag{205}$$

For our example we get:

$$\Pi = \frac{1}{2E} \int_0^L \frac{\frac{q_0^2 L^4}{4} \left(\left[\frac{x}{L}\right] - \left[\frac{x}{L}\right]^2 \right)^2}{\frac{1}{12} b 2^3 h_0^3 \left(\left[\frac{x}{L}\right] - \left[\frac{x}{L}\right]^2 \right)^{\frac{3}{2}}} \mathrm{d}x \tag{206}$$

$$= \frac{3 q_0^2 L^4}{16 E b h_0^3} \underbrace{\int_0^L \left(\left[\frac{x}{L}\right] - \left[\frac{x}{L}\right]^2 \right)^{\frac{1}{2}} \mathrm{d}x}_{\frac{\pi L}{8}} \tag{207}$$

$$= \frac{3\pi}{128} \times \frac{L^5 q_0^2}{E b h_0^3}. \tag{208}$$

From the relationship for the stress criterion at the location $x = L/2$ (see Eq. (202)) the expression for the distributed load results in:

$$q_0 = \frac{16 b h_0^2 R_{\mathrm{p}0.2}}{12\, L^2}. \tag{209}$$

In summary, we obtain for the optimized cross section:

$$V = \frac{b\pi L h_0}{4}, \tag{210}$$

$$\Pi = \frac{3\pi L^5 q_0^2}{128 E b h_0^3}, \tag{211}$$

$$SEA = \frac{3}{32} \times \frac{L^4 q_0^2}{E b^2 h_0^4 \varrho} = \frac{1}{6} \times \frac{R_{\text{p0.2}}^2}{E\varrho}. \tag{212}$$

- Reference cross section:

The total strain energy for the reference cross section is obtained by integration as:

$$\Pi_0 = \frac{1}{2E} \int\limits_0^L \frac{\frac{q_0^2 L^4}{4} \left(\left[\frac{x}{L}\right] - \left[\frac{x}{L}\right]^2 \right)^2}{\frac{1}{12} b h_0^3} \mathrm{d}x, \tag{213}$$

$$= \frac{3 q_0^2 L^4}{2 E b h_0^3} \underbrace{\int\limits_0^L \left(\left[\frac{x}{L}\right]^2 - 2\left[\frac{x}{L}\right]^3 + \left[\frac{x}{L}\right]^4 \right) \mathrm{d}x}_{\frac{L}{30}}, \tag{214}$$

$$= \frac{q_0^2 L^5}{20 E b h_0^3}. \tag{215}$$

In summary, this gives for the reference beam:

$$V_0 = b h_0 L, \tag{216}$$

$$\Pi_0 = \frac{L^5 q_0^2}{20 E b h_0^3}, \tag{217}$$

$$SEA_0 = \frac{1}{20} \times \frac{L^4 q_0^2}{E b^2 h_0^4 \varrho} = \frac{4}{45} \times \frac{R_{\text{p0.2}}^2}{E\varrho}. \tag{218}$$

- Comparison:

$$\frac{SEA}{SEA_0} = \frac{15}{8} \approx 1.875. \tag{219}$$

6 Optimization of a Beam with a Square Box Profile Along the Longitudinal Axis Under the Influence of a Constant Distributed Load

- Moment distribution:

The bending moment distribution is obtained as:

$$M_y(x) = \frac{q_0}{2}\left(x^2 - Lx\right) = \frac{q_0 L^2}{2}\left(\left[\frac{x}{L}\right]^2 - \left[\frac{x}{L}\right]^1\right) < 0. \tag{220}$$

The maximum bending moment acts in the middle of the beam:

$$M_{y,\mathrm{max}} = M_y\left(\tfrac{L}{2}\right) = -\frac{q_0 L^2}{8}. \tag{221}$$

- Geometric dimensions of the cross section (I_y, A):

$$I_y = \frac{65}{972} d_0^4, \tag{222}$$

$$A = \frac{5}{9} d_0^2. \tag{223}$$

- Reference beam with constant cross section:

The definition equation of the lightweight index (see [14]) results in:

$$M = \frac{F_0}{F_\mathrm{G}}. \tag{224}$$

This gives us under consideration of the approximation $F_0 \approx q_0 L$:

$$M_0 \approx \frac{q_0 L}{\varrho g A_0 L}. \tag{225}$$

The evaluation of the stress criterion in the reference cross section at the location $x = L/2$, i.e.

$$\sigma_{x,\mathrm{max}} = \frac{M_{y,\mathrm{max}}}{I_y} \times \frac{d_0}{2} = \frac{\frac{q_0 L^2}{8}}{\frac{65 d_0^4}{972}} \times \frac{d_0}{2} \overset{!}{=} R_{\mathrm{p}0.2}, \tag{226}$$

gives the expression for the distributed load as:

$$q_0 = \frac{260}{243} \times \frac{d_0^3 R_{\mathrm{p}0.2}}{L^2}. \tag{227}$$

This gives the lightweight index as:

$$M_0 \approx \frac{52}{27} \times \frac{d_0 R_{\mathrm{p}0.2}}{\varrho g L^2}. \tag{228}$$

The total strain energy for the reference cross section is obtained by integration as:

$$\Pi_0 = \frac{1}{2E}\int_0^L \frac{\frac{q_0^2 L^4}{4}\left(\left[\frac{x}{L}\right]-\left[\frac{x}{L}\right]^2\right)^2}{\frac{65}{972}d_0^4}\mathrm{d}x, \tag{229}$$

$$= \frac{972 q_0^2 L^4}{8\times 65 E d_0^4}\underbrace{\int_0^L \left(\left[\frac{x}{L}\right]^2 - 2\left[\frac{x}{L}\right]^3 + \left[\frac{x}{L}\right]^4\right)\mathrm{d}x}_{\frac{L}{30}}, \tag{230}$$

$$= \frac{972 q_0^2 L^5}{15600 E d_0^4}. \tag{231}$$

In summary, this gives the reference beam:

$$A_0 = \frac{5d_0^2}{9}, \tag{232}$$

$$I_0 = \frac{65 d_0^4}{972}, \tag{233}$$

$$q_0 = \frac{260 d_0^3 R_{\mathrm{p}0.2}}{243 L^2}, \tag{234}$$

$$M_0 \approx \frac{52}{27}\times\frac{d_0 R_{\mathrm{p}0.2}}{\varrho g L^2}, \tag{235}$$

$$\approx 1.92593\times\frac{d_0 R_{\mathrm{p}0.2}}{\varrho g L^2}, \tag{236}$$

$$\Pi_0 = \frac{52}{729}\times\frac{L d_0^2 R_{\mathrm{p}0.2}^2}{E}, \tag{237}$$

$$SEA_0 = \frac{52}{405}\times\frac{R_{\mathrm{p}0.2}^2}{\varrho E}, \tag{238}$$

$$\approx 0.12840\times\frac{R_{\mathrm{p}0.2}^2}{\varrho E}. \tag{239}$$

- Beams with optimized cross section:

Using the stress criterion

$$\sigma_{x,\max} = \frac{M_y(x)}{I_y(x)}\times\frac{d(x)}{2} = \frac{\frac{q_0 L^2}{2}\left(\left[\frac{x}{L}\right]-\left[\frac{x}{L}\right]^2\right)}{\frac{65 d(x)^4}{972}}\times\frac{h(x)}{2} \overset{!}{=} R_{\mathrm{p}0.2}, \tag{240}$$

results in

$$d^3(x) = \frac{972 q_0 L^2}{1260 R_{\mathrm{p0.2}}} \times \left(\left[\frac{x}{L}\right] - \left[\frac{x}{L}\right]^2 \right). \tag{241}$$

The evaluation of the stress criterion in the reference cross section at the location $x = L/2$ yields the reference dimension (see Eq. (227)):

$$d_0^3 = \frac{243 q_0 L^2}{260 R_{\mathrm{p0.2}}}. \tag{242}$$

With this result follows:

$$d^3(x) = 4 \times \underbrace{\frac{243 q_0 L^2}{260 R_{\mathrm{p0.2}}}}_{d_0^3} \times \left(\left[\frac{x}{L}\right] - \left[\frac{x}{L}\right]^2 \right). \tag{243}$$

The volume is determined by evaluating the integral:

$$V = \int_0^L A(x)\,\mathrm{d}x = \frac{5 \times 4^{\frac{2}{3}} d_0^2}{9} \underbrace{\int_0^L \left(\left[\frac{x}{L}\right] - \left[\frac{x}{L}\right]^2 \right)^{\frac{2}{3}} \mathrm{d}x}_{0.2933414631064508\,L}. \tag{244}$$

The lightweight index is again approximately:

$$M \approx \frac{q_0 L}{\varrho g V} = \frac{q_0 L}{\varrho g \times \frac{5}{9} \times 4^{2/3} d_0^2 \times 0.293341\,L}. \tag{245}$$

The total strain energy is:

$$\Pi = \frac{1}{2E} \int_0^L \frac{\frac{q_0^2 L^4}{4} \left(\left[\frac{x}{L}\right] - \left[\frac{x}{L}\right]^2 \right)^2}{\frac{65}{972} 4^{4/3} d_0^4 \left(\left[\frac{x}{L}\right] - \left[\frac{x}{L}\right]^2 \right)^{\frac{4}{3}}} \mathrm{d}x \tag{246}$$

$$= \frac{972 q_0^2 L^4}{8 \times 65 \times 4^{4/3} E d_0^4} \underbrace{\int_0^L \left(\left[\frac{x}{L}\right] - \left[\frac{x}{L}\right]^2 \right)^{\frac{2}{3}} \mathrm{d}x}_{0.2933414631064508\,L}. \tag{247}$$

In summary, one obtains for the optimized cross section:

$$d(x) = 4^{\frac{1}{3}} d_0 \left(\left[\frac{x}{L}\right]^1 - \left[\frac{x}{L}\right]^2 \right)^{\frac{1}{3}}, \tag{248}$$

$$d_0 = \frac{1}{4^{\frac{1}{3}}} \left(\frac{243}{65} \times \frac{q_0 L^2}{R_{\mathrm{p0.2}}} \right)^{\frac{1}{3}}, \tag{249}$$

$$V = \frac{5}{9} \times 4^{\frac{2}{3}} \times 0.293342 d_0^2 L, \tag{250}$$

$$\approx 0.41 \times d_0^2 L, \tag{251}$$

$$M \approx \frac{52}{27} \times \frac{1}{4^{\frac{2}{3}} \times 0.293342} \times \frac{d_0 R_{\mathrm{p0.2}}}{\varrho g L^2}, \tag{252}$$

$$\approx 2.60551 \times \frac{d_0 R_{\mathrm{p0.2}}}{\varrho g L^2}, \tag{253}$$

$$\Pi = \frac{2 \times 52 \times 260 \times 0.293342}{4^{\frac{7}{3}} \times 13 \times 243} \times \frac{L d_0^2 R_{\mathrm{p0.2}}^2}{E}, \tag{254}$$

$$SEA = \frac{13}{54} \times \frac{R_{\mathrm{p0.2}}^2}{\varrho E} \approx 0.24074 \times \frac{R_{\mathrm{p0.2}}^2}{\varrho E}. \tag{255}$$

The optimized beam is shown in Fig. 53.

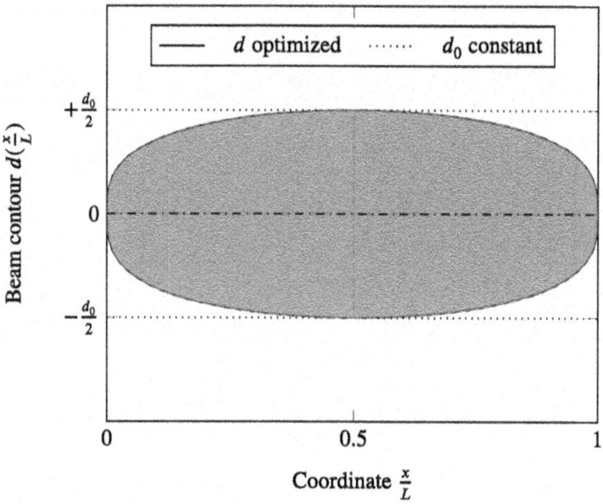

Fig. 53. Beam contour along the longitudinal axis of a beam with squared box profile under constant distributed load

7 Optimization of a Beam with I-Profile Along the Longitudinal Axis Under the Influence of a Constant Distributed Load

- Moment distribution:

The bending moment distribution is obtained as:

$$M_y(x) = \frac{q_0}{2}(x^2 - Lx) = \frac{q_0 L^2}{2}\left(\left[\frac{x}{L}\right]^2 - \left[\frac{x}{L}\right]^1\right) < 0. \tag{256}$$

The maximum bending moment is obtained in the middle of the beam:

$$M_{y,\mathrm{max}} = M_y\left(\tfrac{L}{2}\right) = -\frac{q_0 L^2}{8}. \tag{257}$$

- Geometric dimensions of the cross section (I_y, A):

For a box profile with external dimensions $a \times \alpha a$ and constant wall thickness $w = a/6$ the following results:

$$I_y = \frac{1}{12}a(\alpha a)^3 - 2 \times \frac{1}{12}\left(\frac{a}{2} - \frac{w}{2}\right)(\alpha a - 2w)^3 \tag{258}$$

$$= \frac{1}{12}\alpha^3 a^4 - \frac{5}{72}\left(\frac{3\alpha - 1}{3}\right)^3 a^4 \tag{259}$$

$$A = \alpha a^2 - 2\left(\frac{a}{2} - \frac{w}{2}\right)(\alpha a - 2w) \tag{260}$$

$$= \alpha a^2 - \frac{5}{6}\left(\frac{3\alpha - 1}{3}\right)a^2. \tag{261}$$

▶ Case $\alpha = 2$:

- Reference beam with constant cross section:

The evaluation of the stress criterion in the reference cross section at the location $x = L/2$, i.e.,

$$\sigma_{x,\mathrm{max}} = \frac{M_{y,\mathrm{max}}}{I_y} \times a_0 = \frac{\frac{q_0 L^2}{8}}{\frac{671 a_0^4}{1944}} \times a_0 \overset{!}{=} R_{\mathrm{p0.2}}, \tag{262}$$

gives the expression for the distributed load as:

$$q_0 = \frac{671}{243} \times \frac{a_0^3 R_{\mathrm{p0.2}}}{L^2}. \tag{263}$$

The definition equation of the lightweight index (see [14]) results in:

$$M = \frac{F_0}{F_{\mathrm{G}}}. \tag{264}$$

This gives us under consideration of the approximation $F_0 \approx q_0 L$:

$$M_0 \approx \frac{q_0 L}{\varrho g A_0 L} = \frac{q_0 L}{\varrho g L \frac{11}{18} a_0^2} = \frac{122}{27} \times \frac{a_0 R_{\mathrm{p0.2}}}{\varrho g L^2}. \tag{265}$$

The total strain energy is calculated according to [14] for a bending load case as follows:

$$\Pi = \int_0^L \frac{M_y(x)^2}{2EI_y} \mathrm{d}x. \tag{266}$$

For our example, the total strain energy for the reference cross section is obtained by integration as:

$$\Pi_0 = \frac{1}{2E} \int_0^L \frac{\frac{q_0^2 L^4}{4} \left(\left[\frac{x}{L}\right] - \left[\frac{x}{L}\right]^2 \right)^2}{\frac{671}{1944} a_0^4} \mathrm{d}x, \tag{267}$$

$$= \frac{1944 q_0^2 L^4}{8 \times 671 \times 4 E a_0^4} \underbrace{\int_0^L \left(\left[\frac{x}{L}\right]^2 - 2\left[\frac{x}{L}\right]^3 + \left[\frac{x}{L}\right]^4 \right) \mathrm{d}x}_{\frac{L}{30}}, \tag{268}$$

$$= \frac{1944 q_0^2 L^5}{8 \times 30 \times 671 E a_0^4} = \frac{671}{7290} \times \frac{L a_0^2 R_{\mathrm{p0.2}}^2}{E}. \tag{269}$$

In summary, this gives for the reference beam:

$$A_0 = \frac{11 a_0^2}{18}, \tag{270}$$

$$I_0 = \frac{671 a_0^4}{1944}, \tag{271}$$

$$q_0 = \frac{671 a_0^3 R_{\mathrm{p0.2}}}{243 L^2}, \tag{272}$$

$$M_0 = \frac{122}{27} \times \frac{a_0 R_{\mathrm{p0.2}}}{\varrho g L^2}, \tag{273}$$

$$\approx 4.518519 \times \frac{a_0 R_{\mathrm{p}0.2}}{\varrho g L^2}, \tag{274}$$

$$\Pi_0 = \frac{671}{7290} \times \frac{L a_0^2 R_{\mathrm{p}0.2}^2}{E}, \tag{275}$$

$$\approx 0.0920439 \times \frac{L a_0^2 R_{\mathrm{p}0.2}^2}{E}, \tag{276}$$

$$SEA_0 = \frac{61}{405} \times \frac{R_{\mathrm{p}0.2}^2}{\varrho E}, \tag{277}$$

$$\approx 0.150617 \times \frac{R_{\mathrm{p}0.2}^2}{\varrho E}. \tag{278}$$

- Beam with optimized cross section:

Using the stress criterion

$$\sigma_{x,\max} = \frac{M_y(x)}{I_y(x)} \times a(x) = \frac{\frac{q_0 L^2}{2}\left(\left[\frac{x}{L}\right] - \left[\frac{x}{L}\right]^2\right)}{\frac{671 a(x)^4}{1944}} \times a(x) \overset{!}{=} R_{\mathrm{p}0.2}, \tag{279}$$

results in

$$a^3(x) = \frac{1944 q_0 L^2}{2 \times 671 R_{\mathrm{p}0.2}} \times \left(\left[\frac{x}{L}\right] - \left[\frac{x}{L}\right]^2\right). \tag{280}$$

The evaluation of the stress criterion in the reference cross-section at the location $x = L/2$ yields the reference dimension (see Eq. (263)):

$$a_0^3 = \frac{1944 q_0 L^2}{8 \times 671 R_{\mathrm{p}0.2}}. \tag{281}$$

With this result follows:

$$a^3(x) = 4 \times \underbrace{\frac{1944 q_0 L^2}{8 \times 671 R_{\mathrm{p}0.2}}}_{a_0^3} \times \left(\left[\frac{x}{L}\right] - \left[\frac{x}{L}\right]^2\right). \tag{282}$$

The volume is determined by evaluating the integral:

$$V = \int_0^L A(x)\,\mathrm{d}x = \frac{11 \times 4^{\frac{2}{3}} a_0^2}{18} \underbrace{\int_0^L \left(\left[\frac{x}{L}\right] - \left[\frac{x}{L}\right]^2\right)^{\frac{2}{3}} \mathrm{d}x}_{0.2933414631064508\,L}. \tag{283}$$

The lightweight index is again approximately:

$$M \approx \frac{q_0 L}{\varrho g V} = \frac{q_0 L}{\varrho g \times \frac{11}{18} \times 4^{2/3} a_0^2 \times 0.293341\, L}. \tag{284}$$

The total strain energy is:

$$\Pi = \frac{1}{2E} \int\limits_0^L \frac{\frac{q_0^2 L^4}{4}\left(\left[\frac{x}{L}\right] - \left[\frac{x}{L}\right]^2\right)^2}{\frac{671}{1944} 4^{4/3} a_0^4 \left(\left[\frac{x}{L}\right] - \left[\frac{x}{L}\right]^2\right)^{\frac{4}{3}}} \mathrm{d}x \tag{285}$$

$$= \frac{243 q_0^2 L^4}{671 \times 4^{4/3} E a_0^4} \underbrace{\int\limits_0^L \left(\left[\frac{x}{L}\right] - \left[\frac{x}{L}\right]^2\right)^{\frac{2}{3}} \mathrm{d}x}_{0.2933414631064508\, L}. \tag{286}$$

In summary, this gives for the optimized cross section:

$$a(x) = 4^{\frac{1}{3}} a_0 \left(-\left[\frac{x}{L}\right]^2 + \left[\frac{x}{L}\right]^1\right)^{\frac{1}{3}}, \tag{287}$$

$$a_0 = \left(\frac{1}{4} \times \frac{972}{671} \times \frac{q_0 L^2}{R_{\mathrm{p}0.2}}\right)^{\frac{1}{3}}, \tag{288}$$

$$V \approx \frac{11}{18} \times 4^{\frac{2}{3}} \times 0.293342 a_0^2 L, \tag{289}$$

$$\approx 0.452 \times a_0^2 L, \tag{290}$$

$$M \approx \frac{122}{27} \times \frac{1}{4^{\frac{2}{3}} \times 0.293342} \times \frac{a_0 R_{\mathrm{p}0.2}}{\varrho g L^2}, \tag{291}$$

$$\approx 6.112917 \times \frac{a_0 R_{\mathrm{p}0.2}}{\varrho g L^2}, \tag{292}$$

$$\Pi \approx \frac{671 \times 0.293342}{243 \times 4^{\frac{4}{3}}} \times \frac{L a_0^2 R_{\mathrm{p}0.2}^2}{E}, \tag{293}$$

$$\approx 0.127569 \times \frac{L a_0^2 R_{\mathrm{p}0.2}^2}{E}, \tag{294}$$

$$SEA \approx 0.282407 \times \frac{R_{\mathrm{p}0.2}^2}{\varrho E}. \tag{295}$$

The optimized beam is shown in Fig. 54.

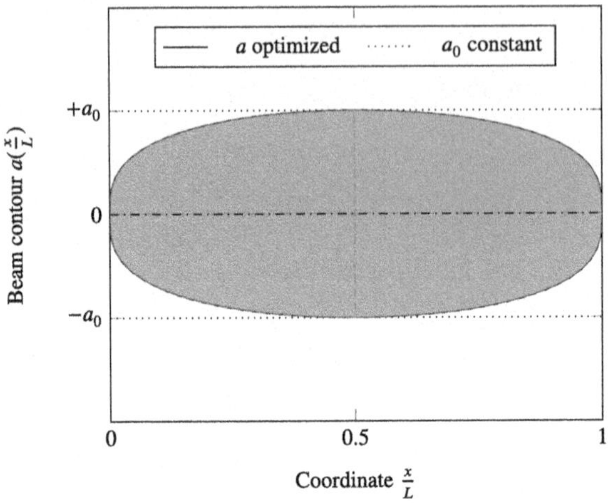

Fig. 54. Beam contour along the longitudinal axis

▶ Case $\alpha = 3$:

- Reference beam with constant cross section:

The evaluation of the stress criterion in the reference cross section at the location $x = L/2$, i.e.,

$$\sigma_{x,\max} = \frac{M_{y,\max}}{I_y} \times a_0 = \frac{\frac{q_0 L^2}{8}}{\frac{907 a_0^4}{972}} \times \frac{3a_0}{2} \stackrel{!}{=} R_{\text{p0.2}}, \tag{296}$$

gives the expression for the distributed load as:

$$q_0 = \frac{3628}{729} \times \frac{a_0^3 R_{\text{p0.2}}}{L^2}. \tag{297}$$

The lightweight index is calculated with the approximation $F_0 \approx q_0 L$ as:

$$M_0 \approx \frac{q_0 L}{\varrho g A_0 L} = \frac{q_0 L}{\varrho g L \frac{7}{9} a_0^2} = \frac{3628}{567} \times \frac{a_0 R_{\text{p0.2}}}{\varrho g L^2}. \tag{298}$$

For our example, the total strain energy for the reference cross section is obtained by integration as:

$$\Pi_0 = \frac{1}{2E}\int_0^L \frac{\frac{q_0^2 L^4}{4}\left(\left[\frac{x}{L}\right]-\left[\frac{x}{L}\right]^2\right)^2}{\frac{907}{972}a_0^4}\mathrm{d}x, \tag{299}$$

$$= \frac{972 q_0^2 L^4}{8\times 907 E a_0^4}\underbrace{\int_0^L \left(\left[\frac{x}{L}\right]^2 - 2\left[\frac{x}{L}\right]^3 + \left[\frac{x}{L}\right]^4\right)\mathrm{d}x}_{\frac{L}{30}}, \tag{300}$$

$$= \frac{972 q_0^2 L^5}{8\times 30\times 907 E a_0^4} = \frac{3628}{32805}\times\frac{L a_0^2 R_{\mathrm{p}0.2}^2}{E}. \tag{301}$$

In summary, this gives for the reference beam:

$$A_0 = \frac{7a_0^2}{9}, \tag{302}$$

$$I_0 = \frac{907 a_0^4}{972}, \tag{303}$$

$$q_0 = \frac{3628 a_0^3 R_{\mathrm{p}0.2}}{729 L^2}, \tag{304}$$

$$M_0 = \frac{3628}{567}\times\frac{a_0 R_{\mathrm{p}0.2}}{\varrho g L^2}, \tag{305}$$

$$\approx 6.398589\times\frac{a_0 R_{\mathrm{p}0.2}}{\varrho g L^2}, \tag{306}$$

$$\Pi_0 = \frac{3628}{32805}\times\frac{L a_0^2 R_{\mathrm{p}0.2}^2}{E}, \tag{307}$$

$$\approx 0.110593\times\frac{L a_0^2 R_{\mathrm{p}0.2}^2}{E}, \tag{308}$$

$$SEA_0 = \frac{3628}{25515}\times\frac{R_{\mathrm{p}0.2}^2}{\varrho E}, \tag{309}$$

$$\approx 0.142191\times\frac{R_{\mathrm{p}0.2}^2}{\varrho E}. \tag{310}$$

- Beam with optimized cross section:

Using the stress criterion

$$\sigma_{x,\max} = \frac{M_y(x)}{I_y(x)}\times a(x) = \frac{\frac{q_0 L^2}{2}\left(\left[\frac{x}{L}\right]-\left[\frac{x}{L}\right]^2\right)}{\frac{907 a(x)^4}{972}}\times\frac{3a(x)}{2} \overset{!}{=} R_{\mathrm{p}0.2}, \tag{311}$$

results in

$$a^3(x) = \frac{729 q_0 L^2}{907 R_{\text{p0.2}}} \times \left(\left[\frac{x}{L}\right] - \left[\frac{x}{L}\right]^2 \right). \tag{312}$$

The evaluation of the stress criterion in the reference cross section at the location $x = L/2$ yields the reference dimension (see Eq. (297)):

$$a_0^3 = \frac{729 q_0 L^2}{3628 R_{\text{p0.2}}}. \tag{313}$$

With this result follows:

$$a^3(x) = 4 \times \underbrace{\frac{729 q_0 L^2}{4 \times 907 R_{\text{p0.2}}}}_{a_0^3} \times \left(\left[\frac{x}{L}\right] - \left[\frac{x}{L}\right]^2 \right). \tag{314}$$

The volume is determined by evaluating the integral:

$$V = \int_0^L A(x)\,\mathrm{d}x = \frac{7 \times 4^{\frac{2}{3}} a_0^2}{9} \underbrace{\int_0^L \left(\left[\frac{x}{L}\right] - \left[\frac{x}{L}\right]^2 \right)^{\frac{2}{3}} \mathrm{d}x}_{0.2933414631064508\,L}. \tag{315}$$

The lightweight index is again approximately:

$$M \approx \frac{q_0 L}{\varrho g V} = \frac{q_0 L}{\varrho g \times \frac{9}{9} \times 4^{2/3} a_0^2 \times 0.293341\,L}. \tag{316}$$

The total strain energy is:

$$\Pi = \frac{1}{2E} \int_0^L \frac{\frac{q_0^2 L^4}{4} \left(\left[\frac{x}{L}\right] - \left[\frac{x}{L}\right]^2 \right)^2}{\frac{907}{972} 4^{4/3} a_0^4 \left(\left[\frac{x}{L}\right] - \left[\frac{x}{L}\right]^2 \right)^{\frac{4}{3}}} \mathrm{d}x \tag{317}$$

$$= \frac{972 q_0^2 L^4}{907 \times 8 \times 4^{4/3} E a_0^4} \underbrace{\int_0^L \left(\left[\frac{x}{L}\right] - \left[\frac{x}{L}\right]^2 \right)^{\frac{2}{3}} \mathrm{d}x}_{0.2933414631064508\,L}. \tag{318}$$

In summary, this gives for the optimized cross section:

$$a(x) = 4^{\frac{1}{3}} a_0 \left(- \left[\frac{x}{L}\right]^2 + \left[\frac{x}{L}\right]^1 \right)^{\frac{1}{3}}, \tag{319}$$

$$a_0 = \left(\frac{1}{4} \times \frac{729}{907} \times \frac{q_0 L^2}{R_{\mathrm{p0.2}}} \right)^{\frac{1}{3}}, \tag{320}$$

$$V \approx \frac{7}{9} \times 4^{\frac{2}{3}} \times 0.293342 a_0^2 L, \tag{321}$$

$$\approx 0.575 \times a_0^2 L, \tag{322}$$

$$M \approx \frac{3628}{567} \times \frac{1}{4^{\frac{2}{3}} \times 0.293342} \times \frac{a_0 R_{\mathrm{p0.2}}}{\varrho g L^2}, \tag{323}$$

$$\approx 8.656387 \times \frac{a_0 R_{\mathrm{p0.2}}}{\varrho g L^2}, \tag{324}$$

$$\Pi \approx \frac{7256 \times 0.293342}{2187 \times 4^{\frac{4}{3}}} \times \frac{L a_0^2 R_{\mathrm{p0.2}}^2}{E}, \tag{325}$$

$$\approx 0.153277 \times \frac{L a_0^2 R_{\mathrm{p0.2}}^2}{E}, \tag{326}$$

$$SEA \approx 0.266608 \times \frac{R_{\mathrm{p0.2}}^2}{\varrho E}. \tag{327}$$

The optimized beam is shown in Fig. 54.

8 Optimization of a Beam with I-Profile Along the Longitudinal Axis and the Height-to-Width Ratio Under the Influence of a Constant Distributed Load

The basic procedure can be found in the solution to Problem 7. Depending on the height-to-width ratio, the following relationships result:

$$I_y(x) = \left(\frac{\alpha^3}{12} - \frac{5}{72} \left(\frac{3\alpha - 1}{3} \right)^3 \right) a^4(x), \tag{328}$$

$$A(x) = \left(\alpha - \frac{5}{6} \left(\frac{3\alpha - 1}{3} \right) \right) a^2(x) \tag{329}$$

$$a_0^3 = \frac{\alpha q_0 L^2}{16 \left(\frac{\alpha^3}{12} - \frac{5}{72} \left(\frac{3\alpha - 1}{3} \right)^3 \right) R_{\mathrm{p0.2}}}, \tag{330}$$

$$a(x) = 4^{\frac{1}{3}} a_0 \left(\left[\frac{x}{L}\right]^1 - \left[\frac{x}{L}\right]^2 \right)^{\frac{1}{3}}, \tag{331}$$

$$V = \left(\alpha - \frac{5}{6}\left(\frac{3\alpha - 1}{3}\right)\right) 4^{\frac{2}{3}} a_0^2 \times 0.293342 L, \tag{332}$$

$$\Pi = \frac{0.293342 \times L^5 q_0^2}{8 \times 4^{4/3} E a_0^4 \left(\frac{\alpha^3}{12} - \frac{5}{72}\left(\frac{3\alpha-1}{3}\right)^3\right)} \tag{333}$$

$$= \frac{32 \times 0.293342}{4^{4/3}} \times \frac{\left(\frac{\alpha^3}{12} - \frac{5}{72}\left(\frac{3\alpha-1}{3}\right)^3\right)}{\alpha^2} \times \frac{R_{\text{p0.2}}^2 a_0^2 L}{E}, \tag{334}$$

$$SEA = 2 \times \frac{R_{\text{p0.2}}^2}{\varrho E} \times \frac{\frac{\alpha^3}{12} - \frac{5}{72}\left(\frac{3\alpha-1}{3}\right)^3}{\alpha^2\left(\alpha - \frac{5}{6}\left(\frac{3\alpha-1}{3}\right)\right)}. \tag{335}$$

The evaluation of the specific energy absorption SEA of the optimized cross-section for different values of α is shown in Fig. 55.

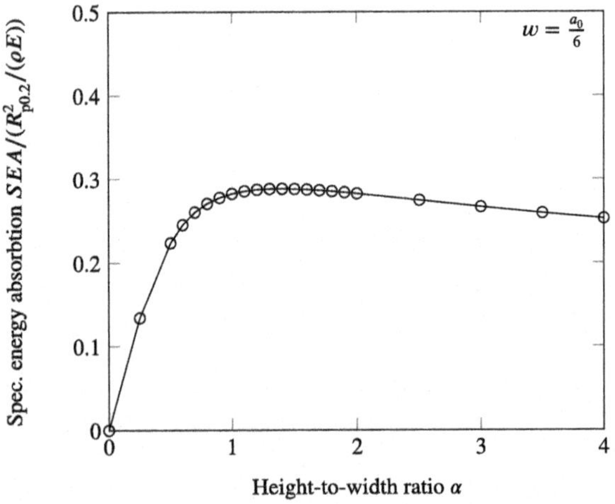

Fig. 55. Specific energy absorbtion for different values of α

9 Optimization of a Beam Along the Longitudinal Axis Under the Influence of Two Different Single Forces

- Internal reactions:

To determine the internal reactions, cut the beam under the influence of two different single forces at an arbitrary location x to the left and then to the right of the left single force, see Fig. 56.

The moment equilibrium around the reference point at the location x in section 1 $(0 \leq x \leq \frac{L}{2})$ gives the distribution of the internal bending moment, see Fig. 56a:

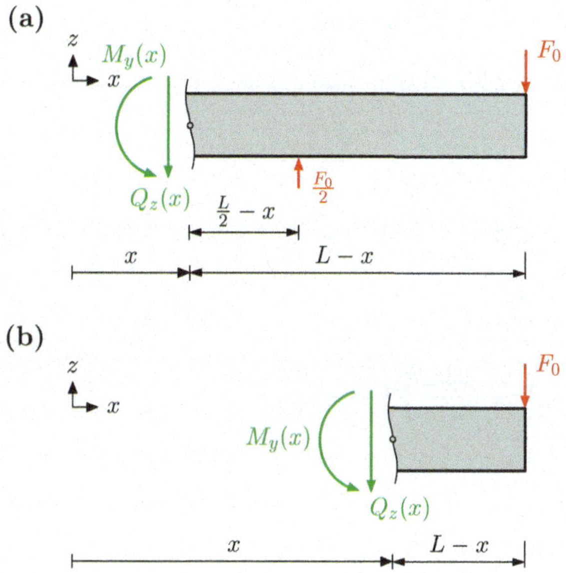

Fig. 56. Internal reactions of a beam under the influence of two different single forces: **a** section 1 and **b** section 2

$$\overset{\frown}{\sum} M_y = 0: \quad +F_0(L-x) - \frac{F_0}{2}\left(\frac{L}{2} - x\right) - M_y(x) = 0, \tag{336}$$

$$M_y(x) = \frac{3F_0L}{4}\left(1 - \frac{2x}{3L}\right) \quad \text{for} \quad 0 \leq x \leq \frac{L}{2}. \tag{337}$$

The moment equilibrium around the reference point at the location x in Sect. 2 ($\frac{L}{2} \leq x \leq L$) results in the distribution of the internal bending moment, see Fig. 56b:

$$\overset{\frown}{\sum} M_y = 0: \quad +F_0(L-x) - M_y(x) = 0, \tag{338}$$

$$M_y(x) = F_0(L-x) \quad \text{for} \quad \frac{L}{2} \leq x \leq L. \tag{339}$$

In summary, the two moment distributions can be given as follows:

$$M_{y,\mathrm{I}}(x) = \frac{3F_0L}{4}\left(1 - \frac{2x}{3L}\right) \quad \text{for} \quad 0 \leq x \leq \frac{L}{2}, \tag{340}$$

$$M_{y,\mathrm{II}}(x) = F_0L\left(1 - \frac{x}{L}\right) \quad \text{for} \quad \frac{L}{2} \leq x \leq L. \tag{341}$$

- Geometric dimensions of the cross section (I_y, A):

For a box profile with external dimensions $d \times d$ and constant wall thickness $s = d/5$ we get:

$$I_y = \frac{1}{12}\left(d_0^2 + \left(d_0 - \frac{2d_0}{5}\right)^2\right)\left(d_0 - \frac{3d_0}{5}\right)\left(d_0 + \frac{3d_0}{5}\right) \tag{342}$$

$$= \frac{136}{1875} d_0^5, \tag{343}$$

$$A = \left(d_0 - \frac{3d_0}{5}\right)\left(d_0 + \frac{3d_0}{5}\right) \tag{344}$$

$$= \frac{16}{25} d_0^2. \tag{345}$$

- Optimized diameter:

Using the stress criterion in Sect. 1 ($0 \leq x \leq \frac{L}{2}$)

$$\sigma_{x,\text{max}} = \frac{M_y(x)}{I_y(x)} \times \frac{d(x)}{2} = \frac{\frac{3F_0L}{4}\left(1 - \frac{2x}{3L}\right)}{\frac{136}{1875} d^4(x)} \times \frac{d(x)}{2} \stackrel{!}{=} R_{\text{p0.2}}, \tag{346}$$

results in

$$d^3(x) = \frac{3 \times 1875 F_0 L}{8 \times 136 R_{\text{p0.2}}} \times \left(1 - \frac{2x}{3L}\right). \tag{347}$$

The evaluation of the stress criterion in the reference cross-section, i.e. for $x = 0$, results in:

$$\sigma_{x,\text{max}} = \frac{M_y(x=0)}{I_y} \times \frac{d_0}{2} = \frac{\frac{3F_0L}{4}}{\frac{136}{1875} d_0^4} \times \frac{d_0}{2} \stackrel{!}{=} R_{\text{p0.2}}, \tag{348}$$

or

$$d_0^3 = \frac{3 \times 1875 F_0 L}{8 \times 136 R_{\text{p0.2}}}. \tag{349}$$

With this result, the optimized cross section in the first section is calculated as:

$$d^3(x) = d_0^3 \times \left(1 - \frac{2x}{3L}\right). \tag{350}$$

Using the stress criterion in Sect. 2 ($\frac{L}{2} \leq x \leq L$)

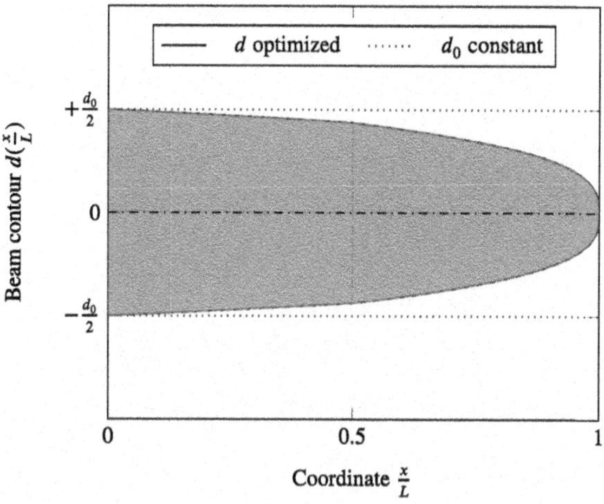

Fig. 57. Beam contour (outside) along the beam axis

$$\sigma_{x,\max} = \frac{M_y(x)}{I_y(x)} \times \frac{d(x)}{2} = \frac{F_0 L \left(1 - \frac{x}{L}\right)}{\frac{136}{1875} d^4(x)} \times \frac{d(x)}{2} \stackrel{!}{=} R_{\text{p0.2}}, \tag{351}$$

results in

$$d^3(x) = \frac{4}{3} \times \frac{3 \times 1875 F_0 L}{8 \times 136 R_{\text{p0.2}}} \left(1 - \frac{x}{L}\right) = \frac{4}{3} \times d_0^3 \left(1 - \frac{x}{L}\right). \tag{352}$$

In summary, the following distribution results for both sections (see also Fig. 57):

$$d_{\text{I}}(x) = d_0 \times \sqrt[3]{1 - \frac{2x}{3L}} \quad \text{for} \quad 0 \leq x \leq \frac{L}{2}, \tag{353}$$

$$d_{\text{II}}(x) = \sqrt[3]{\frac{4}{3}} \times d_0 \times \sqrt[3]{1 - \frac{x}{L}} \quad \text{for} \quad \frac{L}{2} \leq x \leq L, \tag{354}$$

where the reference diameter at the fixed support is as follows:

$$d_0 = \sqrt[3]{\frac{\frac{3 F_0 L}{4}}{\frac{272}{1875} R_{\text{p0.2}}}}. \tag{355}$$

- Total strain energy:

$$\Pi = \int_0^L \frac{(M_y(x))^2}{2EI} dx = \int_0^{\frac{L}{2}} \frac{M_{y,\mathrm{I}}(x)^2}{2EI_{\mathrm{I}}} dx + \int_{\frac{L}{2}}^L \frac{(M_{y,\mathrm{II}}(x))^2}{2EI_{\mathrm{II}}} dx \tag{356}$$

$$= \frac{1}{2E} \int_0^{\frac{L}{2}} \frac{\frac{9F_0^2L^2}{16}\left(1-\frac{2x}{3L}\right)^2}{\frac{136}{1875} d_0^4 \left(1-\frac{2x}{3L}\right)^{4/3}} dx + \frac{1}{2E} \int_{\frac{L}{2}}^L \frac{F_0^2 L^2 \left(1-\frac{x}{L}\right)^2}{\frac{136}{1875} d_0^4 \left(\frac{4}{3}\right)^{4/3} \left(1-\frac{x}{L}\right)^{4/3}} dx \tag{357}$$

$$= \frac{16875 \times F_0^2 L^2 \left(\frac{9L}{10} - \frac{2^{\frac{2}{3}} 3^{\frac{1}{3}} L}{5}\right)}{4352 \times E d_0^4} + \frac{125 \times 3^{\frac{10}{3}} F_0^2 L^3}{17 \times 2^{\frac{5}{3}} 4^{\frac{10}{3}} E d_0^4} \tag{358}$$

$$\approx 2.602043 \times \frac{F_0^2 L^3}{E d_0^4}. \tag{359}$$

- Total mass:

$$m = \int_0^L \varrho A(x) dx = \int_0^{\frac{L}{2}} \varrho A_{\mathrm{I}}(x) dx + \int_{\frac{L}{2}}^L \varrho A_{\mathrm{II}}(x) dx \tag{360}$$

$$= \varrho \frac{16}{25} d_0^2 \int_0^{\frac{L}{2}} \left(1 - \frac{2x}{3L}\right)^{\frac{2}{3}} dx + \varrho \frac{16}{25} \left(\frac{4}{3}\right)^{\frac{2}{3}} d_0^2 \int_{\frac{L}{2}}^L \left(1 - \frac{x}{L}\right)^{\frac{2}{3}} dx \tag{361}$$

$$= \frac{16}{25} \times \left(\frac{9L}{10} - \frac{2^{\frac{2}{3}} 3^{\frac{1}{3}} L}{5}\right) d_0^2 \varrho + \frac{3^{\frac{1}{3}} 4^{\frac{8}{3}} L d_0^2 \varrho}{125 \times 2^{\frac{5}{3}}} \tag{362}$$

$$\approx 0.429477 \times L\, {d_0}^2 \varrho. \tag{363}$$

- Specific energy absorption:

$$SEA = \frac{\Pi}{m} = 6.058637 \times \frac{F_0^2 L^2}{E d_0^6 \varrho}. \tag{364}$$

10 Optimization of a Beam Along the Longitudinal Axis Under the Influence of a Single Force and a Single Moment

- Internal reactions:

To determine the internal reactions, the beam is cut free at an arbitrary location x inside the beam under the influence of a single force and a single moment, see Fig. 58.

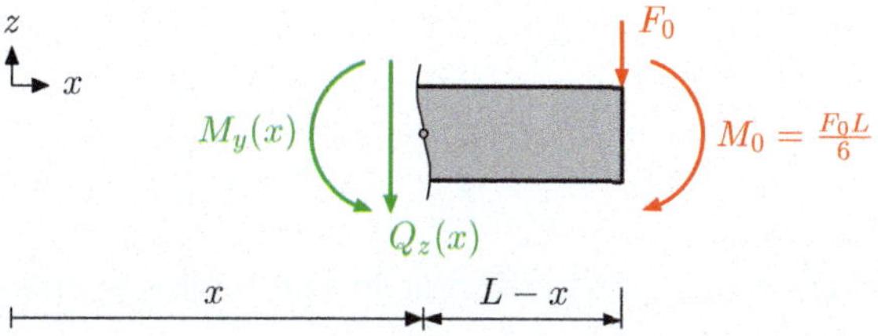

Fig. 58. Internal reactions of a beam under the influence of a single force and a single moment

The moment equilibrium around the reference point at the location x results in the distribution of the internal bending moment:

$$\overset{\curvearrowright}{\sum} M_y = 0 : \quad +F_0(L-x) + M_0 - M_y(x) = 0, \tag{365}$$

$$M_y(x) = F_0 L \left(1 - \frac{x}{L}\right) + M_0 = F_0 L \left(1 - \frac{x}{L}\right) + \frac{F_0 L}{6} \tag{366}$$

$$= F_0 L \left(\frac{7}{6} - \frac{x}{L}\right). \tag{367}$$

- Geometric dimensions of the cross section (I_y, A):

The geometric dimensions can be taken from the solution to Problem 8:

$$I_y(x) = \left(\frac{\alpha^3}{12} - \frac{5}{72}\left(\frac{3\alpha - 1}{3}\right)^3\right) a^4(x), \tag{368}$$

$$A(x) = \left(\alpha - \frac{5}{6}\left(\frac{3\alpha - 1}{3}\right)\right) a^2(x). \tag{369}$$

Assigning $\alpha = 1$, one gets:

$$I(x) = \frac{61 a(x)^4}{972}, \tag{370}$$

$$A(x) = \frac{4 a(x)^2}{9}. \tag{371}$$

- Beam contour $a = a(x)$ along the beam axis:

Using the stress criterion

$$\sigma_{x,\max} = \frac{M_y(x)}{I_y(x)} \times \frac{a(x)}{2} = \frac{F_0 L\left(\frac{7}{6} - \frac{x}{L}\right)}{\frac{61a(x)^4}{972}} \times \frac{a(x)}{2} \overset{!}{=} R_{\text{p}0.2}, \tag{372}$$

results in

$$a^3(x) = \frac{972F_0L}{122R_{\text{p}0.2}} \times \left(\frac{7}{3} - \frac{x}{L}\right). \tag{373}$$

The evaluation of the stress criterion in the reference cross section at the location $x = 0$ yields the reference dimension:

$$a_0^3 = \frac{7 \times 81F_0L}{61R_{\text{p}0.2}}. \tag{374}$$

With this result it follows:

$$a^3(x) = \frac{6}{7} \times \underbrace{\frac{7 \times 972F_0L}{6 \times 122R_{\text{p}0.2}}}_{a_0^3} \times \left(\frac{7}{6} - \frac{x}{L}\right), \tag{375}$$

or (see Fig. 59)

$$a(x) = \left(\frac{6}{7}\right)^{\frac{1}{3}} a_0 \left(\frac{7}{6} - \frac{x}{L}\right)^{\frac{1}{3}} \approx 0.9499 \times a_0 \left(\frac{7}{6} - \frac{x}{L}\right)^{\frac{1}{3}}, \tag{376}$$

where the reference cross-section has the following representation:

$$a_0 = \left(\frac{567F_0L}{61R_{\text{p}0.2}}\right)^{\frac{1}{3}}. \tag{377}$$

- Constant cross section:

The total strain energy is calculated according to [14] for a bending load case as follows:

$$\Pi = \int_0^L \frac{M_y(x)^2}{2EI_y} \mathrm{d}x. \tag{378}$$

For our example, the distortion energy for the reference cross section is obtained by integration as:

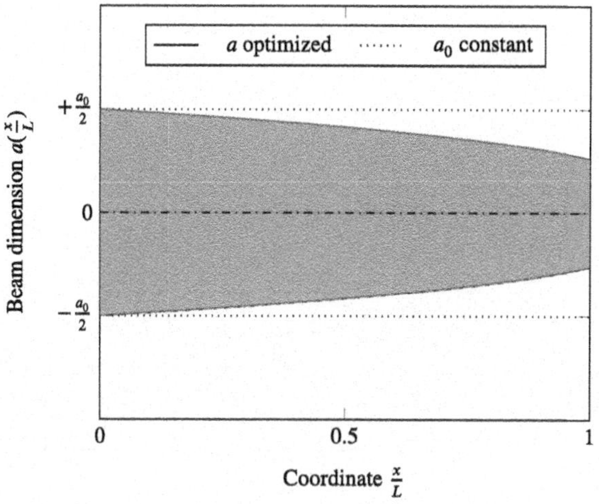

Fig. 59. Beam contour along the beam axis

$$\Pi_0 = \frac{1}{2E}\int_0^L \frac{F_0^2 L^2 \left(\frac{7}{6} - \frac{x}{L}\right)^2}{\frac{61}{972}a_0^4}\mathrm{d}x, \tag{379}$$

$$= \frac{972 F_0^2 L^2}{122 E a_0^4} \underbrace{\int_0^L \left(\frac{7}{6} - \frac{x}{L}\right)^2 \mathrm{d}x}_{\frac{19L}{36}}, \tag{380}$$

$$= \frac{513 F_0^2 L^3}{122 E a_0^4} \approx 4.2049180 \times \frac{F_0^2 L^3}{E a_0^4} \tag{381}$$

$$= \frac{1159}{23814} \times \frac{L a_0^2 R_{\mathrm{p}0,2}^2}{E} \approx 0.0486689 \times \frac{L a_0^2 R_{\mathrm{p}0.2}^2}{E}. \tag{382}$$

The specific energy absorption is generally given as [14]:

$$SEA = \frac{\Pi}{m}, \tag{383}$$

or taking into account the expressions for the total strain energy (see Eq. (382)) and mass ($m = \varrho L \frac{4}{9} a_0^2$):

$$SEA_0 = \frac{1159}{10584} \times \frac{R_{\mathrm{p}0.2}^2}{E\varrho} \approx 0.109505 \times \frac{R_{\mathrm{p}0.2}^2}{E\varrho}. \tag{384}$$

- Optimized cross section:

The total strain energy for the reference cross section is obtained by integration as:

$$\Pi = \frac{1}{2E}\int_0^L \frac{F_0^2 L^2 \left(\frac{7}{6}-\frac{x}{L}\right)^2}{\frac{61}{972}\left(\frac{6}{7}\right)^{\frac{4}{3}} a_0^4 \left(\frac{7}{6}-\frac{x}{L}\right)^{\frac{4}{3}}}\,\mathrm{d}x \tag{385}$$

$$= \frac{972}{122\left(\frac{6}{7}\right)^{4/3}} \times \frac{F_0^2 L^2}{a_0^4} \underbrace{\int_0^L \left(\frac{7}{6}-\frac{x}{L}\right)^{\frac{2}{3}} \mathrm{d}x}_{0.852346\,L} \tag{386}$$

$$\approx 7.294636 \times \frac{F_0^2\,L^3}{E a_0^4} \approx 0.084430 \times \frac{R_{\mathrm{p}0.2}^2 L a_0^2}{E}. \tag{387}$$

The volume is determined by evaluating the integral:

$$V = \int_0^L A(x)\,\mathrm{d}x = \int_0^L \frac{4}{9} a^2(x)\,\mathrm{d}x \approx 0.298966 a_0^2 L. \tag{388}$$

With these values, the specific energy absorption is:

$$SEA \approx 0.282407 \times \frac{R_{\mathrm{p}0.2}^2}{E\varrho}. \tag{389}$$

- Ratio:

$$\frac{SEA}{SEA_0} = \frac{0.282407}{0.109505} \approx 2.5789. \tag{390}$$

11 Optimization of a Simply Supported Beam Under Constant Distributed Load: Box Cross Section

- Moment distribution:

The bending moment distribution is obtained as:

$$M_y(x) = \frac{q_0 L^2}{2}\left(\left[\frac{x}{L}\right]^2 - \left[\frac{x}{L}\right]^1\right) < 0. \tag{391}$$

The maximum bending moment is obtained in the middle of the beam:

$$M_{y,\max} = M_y\left(x = \frac{L}{2}\right) = -\frac{q_0 L^2}{8}. \tag{392}$$

- Geometric dimensions of the cross section (I_y, A):

For a box profile with external dimensions $a/2 \times a$ and constant wall thickness $d = a/6$ the following results:

$$I_y = \frac{1}{12}\frac{a}{2}a^3 - \frac{1}{12}\left(\frac{a}{2} - 2\frac{a}{6}\right)\left(a - 2\frac{a}{6}\right)^3 \tag{393}$$

$$= \frac{a^3}{24} - \frac{1}{12}\frac{a}{6}\left(\frac{2a}{3}\right)^3 = \frac{73}{1944}a^4, \tag{394}$$

$$A = a\frac{a}{2} - \left(\frac{a}{2} - 2\frac{a}{6}\right)\left(a - 2\frac{a}{6}\right) \tag{395}$$

$$= \frac{7}{18}a^2. \tag{396}$$

- Beam contour $a = a(x)$ along the beam axis:

Using the stress criterion

$$\sigma_{x,\max} = \frac{M_y(x)}{I_y(x)} \times \frac{a(x)}{2} = \frac{\frac{q_0 L^2}{2}\left(\left[\frac{x}{L}\right] - \left[\frac{x}{L}\right]^2\right)}{\frac{73a(x)^4}{1944}} \times \frac{a(x)}{2} \overset{!}{=} R_{\text{p0.2}}, \tag{397}$$

results in

$$a^3(x) = \frac{1944 q_0 L^2}{4 \times 73 R_{\text{p0.2}}} \times \left(\left[\frac{x}{L}\right] - \left[\frac{x}{L}\right]^2\right). \tag{398}$$

The evaluation of the stress criterion in the reference cross section at the location $x = L/2$ yields the reference dimension:

$$a_0^3 = \frac{1944 q_0 L^2}{4 \times 4 \times 73 R_{\text{p0.2}}}. \tag{399}$$

With this result follows (see Fig. 60):

$$a(x) = (4)^{\frac{1}{3}}\, a_0 \left(-\left(\frac{x}{L}\right)^2 + \frac{x}{L}\right)^{\frac{1}{3}} \approx 1.587 \times a_0 \left(-\left(\frac{x}{L}\right)^2 + \frac{x}{L}\right)^{\frac{1}{3}} > 0, \tag{400}$$

where the reference cross section has the following dimensions:

$$a_0 = \left(\frac{486 q_0 L^2}{4 \times 73 R_{\text{p0.2}}}\right)^{\frac{1}{3}}. \tag{401}$$

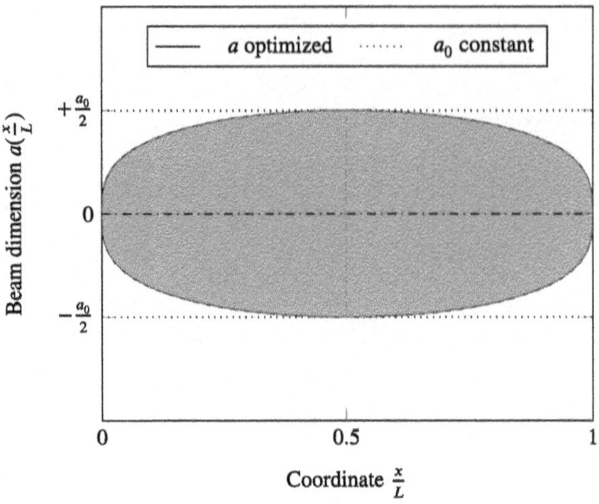

Fig. 60. Beam dimension a along the beam axis

12 Optimization of a Simply Supported Bending Under Constant Distributed Load: Box Cross Section. Numerical Calculation of the Specific Energy Absorption

- Moment distribution:

The bending moment distribution can be taken from the solution of Problem 11:

$$M_y(x) = \frac{q_0 L^2}{2}\left(\left[\frac{x}{L}\right]^2 - \left[\frac{x}{L}\right]^1\right) < 0. \tag{402}$$

The maximum bending moment is obtained in the middle of the beam:

$$M_{y,\max} = M_y\left(x = \frac{L}{2}\right) = -\frac{q_0 L^2}{8}. \tag{403}$$

- Geometric dimensions of the cross section (I_y, A):

For a box profile with external dimensions $a/2 \times a$ and constant wall thickness $d = a/6$ we get (see Problem 11):

$$I_y = \frac{73}{1944}a^4, \tag{404}$$

$$A = \frac{7}{18}a^2. \tag{405}$$

- Constant cross section:

The total strain energy is calculated according to [14] for a bending load case as follows:

$$\Pi = \int_0^L \frac{M_y(x)^2}{2EI_y} \mathrm{d}x. \tag{406}$$

For our example, the total strain energy for the reference cross section is obtained by integration as:

$$\Pi_0 = \frac{1}{2E} \int_0^L \frac{\frac{q_0^2 L^4}{4}\left(\left[\frac{x}{L}\right] - \left[\frac{x}{L}\right]^2\right)^2}{\frac{73}{1944} a_0^4} \mathrm{d}x, \tag{407}$$

$$= \frac{1944 q_0^2 L^4}{8 \times 73 E a_0^4} \underbrace{\int_0^L \left(\left[\tfrac{x}{L}\right] - \left[\tfrac{x}{L}\right]^2\right)^2 \mathrm{d}x}_{\frac{L}{30}}, \tag{408}$$

$$= \frac{81 q_0^2 L^5}{730 E a_0^4}. \tag{409}$$

The specific energy absorption is generally given as [14]:

$$SEA = \frac{\Pi}{m}, \tag{410}$$

or taking into account the expressions for the distortion energy (see Eq. (409)) and mass ($m = \varrho L \frac{7}{18} a_0^2$):

$$SEA_0 = \frac{729 q_0^2 L^4}{2555 E a_0 \varrho} \approx 0.285323 \times \frac{q_0^2 L^4}{E a_0 \varrho}. \tag{411}$$

In summary, this gives for the reference beam:

$$A_0 = \frac{7{a_0}^2}{18}, \tag{412}$$

$$I_0 = \frac{73{a_0}^4}{1944}, \tag{413}$$

$$m_0 = \frac{8\, L\, {a_0}^2 \varrho}{9}, \tag{414}$$

$$\Pi_0 = \frac{81 L^5\, {q_0}^2}{730 E\, {a_0}^4}, \tag{415}$$

$$SEA_0 = \frac{729 L^4\, {q_0}^2}{2555 E\, {a_0}^6 \varrho}. \tag{416}$$

- Optimized cross section:

The total strain energy for the optimized beam is obtained by integration as:

$$\Pi = \frac{1}{2E} \int_0^L \frac{\frac{q_0^2 L^4}{4} \left(\left[\frac{x}{L}\right] - \left[\frac{x}{L}\right]^2 \right)^2}{\frac{73}{1944} 4^{4/3} a_0^4 \left(\left[\frac{x}{L}\right] - \left[\frac{x}{L}\right]^2 \right)^{\frac{4}{3}}} \mathrm{d}x \tag{417}$$

$$= \frac{1944}{8 \times 73 \times 4^{4/3}} \times \frac{q_0^2 L^4}{E a_0^4} \underbrace{\int_0^L \left(\left[\frac{x}{L}\right] - \left[\frac{x}{L}\right]^2 \right)^{\frac{2}{3}} \mathrm{d}x}_{0.2933415685302212 \times L \ (n=4112)} \tag{418}$$

$$= 0.1537837221422918 \times \frac{q_0^2 L^5}{E a_0^4}. \tag{419}$$

The volume is determined by evaluating the integral:

$$V = \int_0^L A(x)\,\mathrm{d}x = \int_0^L \frac{7}{18} a^2(x)\,\mathrm{d}x = \frac{7}{18} 4^{\frac{2}{3}} a_0^2 \underbrace{\int_0^L \left(\left[\frac{x}{L}\right] - \left[\frac{x}{L}\right]^2 \right)^{\frac{2}{3}}}_{0.2933415685302212 \times L \ (n=4112)} \tag{420}$$

$$\approx 0.2874567243336484 \times a_0^2 L. \tag{421}$$

In summary, this gives for the optimized cross-section (see Fig. 60):

$$a(x) = (4)^{\frac{1}{3}} a_0 \left(-\left(\frac{x}{L}\right)^2 + \frac{x}{L} \right)^{\frac{1}{3}}, \tag{422}$$

$$A(x) = \frac{7}{18} a(x)^2, \tag{423}$$

$$I(x) = \frac{73}{1944} a(x)^4, \tag{424}$$

$$V = \frac{7}{18} \times 4^{\frac{2}{3}} \times a_0^2 \times 0.2933415685302212\,L, \tag{425}$$

$$m = \frac{7}{18} \times 4^{\frac{2}{3}} \times a_0^2 \times \varrho \times 0.2933415685302212\,L, \tag{426}$$

$$\Pi = \frac{0.1537837221422918 L^5 \, {q_0}^2}{E\, {a_0}^4}, \tag{427}$$

$$SEA = \frac{0.5349804305283756 L^4 \, {q_0}^2}{E\, {a_0}^6 \varrho}. \tag{428}$$

- Ratios:

$$\frac{m}{m_0} = 0.739, \tag{429}$$

$$\frac{SEA}{SEA_0} = 1.875. \tag{430}$$

13 Optimal Design of a Cantilever Beam (Classical Optimization Problem)

- Geometric dimensions of the cross section (I_y, A):

For a rectangular cross section with external dimensions $a \times 2a$ we get:

$$I_y = \frac{1}{12}a(2a)^3 = \frac{2}{3}a^4, \tag{431}$$

$$A = 2a^2. \tag{432}$$

- Objective function:

The objective function of the optimization problem, i.e. the mass of the beam, can be represented as a function of the design variable $X = a$ as follows:

$$F(X) = m(X) = \varrho V(X) = \varrho L A(X) = 2\varrho L X^2. \tag{433}$$

- Inequality constraints:

The minimization is to be carried out taking into account the following inequality constraints, see [13]:

The maximum deflection can be taken, for example, from [11]. It should be noted that the current force acts in the positive z-direction:

$$u_{z,\max} = \frac{F_0 L^3}{3EI_y}, \tag{434}$$

or formulated as an inequality constraint:

$$g_1(X) = \frac{F_0 L^3}{2EX^4} - \underbrace{u_{z,\max}}_{r_1 L} \leq 0. \tag{435}$$

The bending moment distribution is obtained as:

$$M_y(x) = -F_0 L\left(1 - \frac{x}{L}\right). \tag{436}$$

This gives the maximum stress at the fixed support as:

$$\sigma_{x,\max} = \frac{M_y(0)}{I_y} \times (-a) = \frac{-F_0 L}{\frac{2a^4}{3}} \times (-a) \stackrel{!}{=} R_{\mathrm{p}0.2}, \tag{437}$$

or formulated as an inequality constraint (with $a = X$):

$$g_2(X) = \frac{3F_0 L}{2X^3} - R_{\mathrm{p}0.2} \leq 0. \tag{438}$$

In summary, the optimization problem is:

Minimize the objective function

$$F(X) = 2\varrho L X^2, \tag{439}$$

taking into account the inequality constraints (see Fig. 61):

$$g_1(X) = \frac{F_0 L^3}{2EX^4} - r_1 L \leq 0 \quad \text{(displacement)}, \tag{440}$$

$$g_2(X) = \frac{3F_0 L}{2X^3} - R_{\mathrm{p}0.2} \leq 0 \quad \text{(stress)}. \tag{441}$$

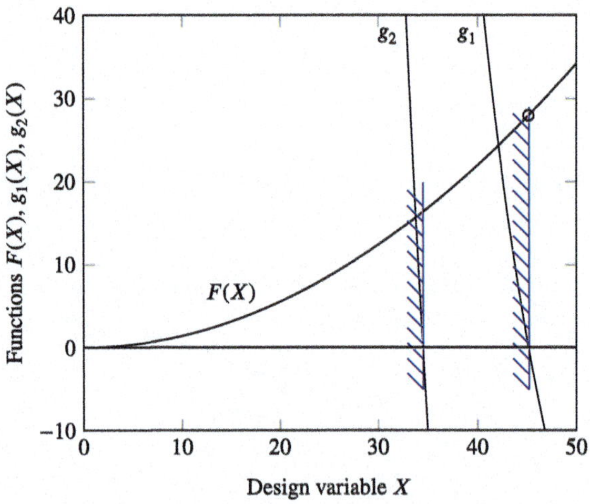

Fig. 61. Optimal design of a cantilever beam with: $L = 2540\,\mathrm{mm}$, $F_0 = 2667\,\mathrm{N}$, $E = 68{,}948\,\mathrm{MPa}$, $R_{\mathrm{p}0.2} = 247\,\mathrm{MPa}$, $\varrho = 2.691 \times 10^{-6}\,\mathrm{kg/mm^3}$, $r_1 = 0.03$ (original set of equations (439)–(441)). Exact solution for the minimum: $X_{\mathrm{extr}} = 45.160$ mm

From Fig. 61 one can see that the inequality constraint g_1 is the stricter condition. Thus, only points that lie to the right of the blue right-hand limit curve are allowed on the curve $F(X)$ (this results from the condition that $g_1 \leq 0$ must be). The optimized design variable results from the condition

$$g_1(X) \stackrel{!}{=} 0, \tag{442}$$

as analytical solution to:

$$X_{\text{extr}} = \left(\frac{F_0 L^2}{2Er_1}\right)^{\frac{1}{4}} = 45.16\,\text{mm}. \tag{443}$$

As an alternative to Eq. (439)–(441), one can also specify a normalized representation, i.e. without a specific unit. By introducing the normalized geometric dimension $a^{\text{n}} = a/L = X^{\text{n}}$ one arrives at the corresponding equations of the optimization problem:

$$F^{\text{n}}(X^{\text{n}}) = F^{\text{n}}(a^{\text{n}}) = \frac{m}{\varrho \times L^3} = 2 \times (a^{\text{n}})^2 = 2 \times (X^{\text{n}})^2, \tag{444}$$

$$g_1^{\text{n}}(X^{\text{n}}) = g_1^{\text{n}}(a^{\text{n}}) = \frac{F_0}{2EL^2 r_1 (a^{\text{n}})^4} - 1 = \frac{F_0}{2EL^2 r_1 (X^{\text{n}})^4} - 1 \leq 0, \tag{445}$$

$$g_2^{\text{n}}(X^{\text{n}}) = g_2^{\text{n}}(a^{\text{n}}) = \frac{3F_0}{2L^2 R_{\text{p}0.2} (a^{\text{n}})^3} - 1 = \frac{3F_0}{2L^2 R_{\text{p}0.2} (X^{\text{n}})^3} - 1 \leq 0. \tag{446}$$

The graphical representation of these normalized equations is shown in Fig. 62a. It can be seen that both inequality constraints again have a single root in the considered region of the abscissa. However, both slopes are extremely steep. Thus, a scaling factor β can be introduced to modify the inequality restrictions as follows:

$$g_1^{\text{n}}(X^{\text{n}})/\beta = \frac{1}{\beta}\left(\frac{F_0}{2EL^2 r_1 (X^{\text{n}})^4} - 1 \leq 0\right), \tag{447}$$

$$g_2^{\text{n}}(X^{\text{n}})/\beta = \frac{1}{\beta}\left(\frac{3F_0}{2L^2 R_{\text{p}0.2} (X^{\text{n}})^3} - 1 \leq 0\right). \tag{448}$$

It is obvious that the scaling does not change the position of the roots and a similar course of all curves is maintained, see Fig. 62b.

14 **Optimal Design of a Compression Strut (Classical Optimization Problem**

- Geometric dimensions of the cross section (I, A):

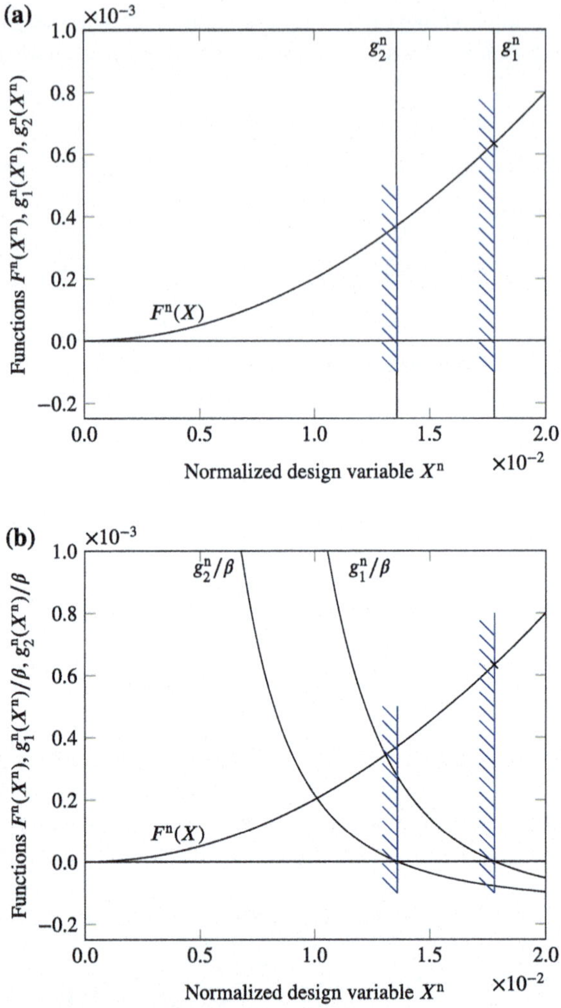

Fig. 62. Optimal design of a cantilever beam: **a** normalized equations and **b** normalized equations with scaled inequality constraints ($\beta = 7000$). Exact solution for the minimum: $X^n_{\text{extr}} = 0.01778$

For a rectangular cross section with external dimensions $a \times 2a$ we get:

$$I = \frac{1}{12}a(2a)^3 = \frac{2}{3}a^4 \quad \text{or} \quad \frac{1}{12}2a(a)^3 = \frac{1}{6}a^4, \tag{449}$$

$$A = 2a^2. \tag{450}$$

- Objective function:

The objective function of the optimization problem, i.e. the mass of the beam, can be represented as a function of the design variable $X = a$ as follows:

$$F(X) = m(X) = \varrho V(X) = \varrho LA(X) = 2\varrho LX^2. \tag{451}$$

- Inequality constraints:

The magnitude of the maximum stress is:

$$\sigma_{x,\max} = \frac{F_0}{A} \stackrel{!}{=} R_{\mathrm{p}0.2}, \tag{452}$$

or formulated as an inequality constraint (with $a = X$):

$$g_1(X) = \frac{F_0}{2X^2} - R_{\mathrm{p}0.2} \leq 0. \tag{453}$$

The critical buckling stress for Euler case 1 is:

$$F_{\mathrm{cr}} = \frac{\pi^2 EI_{\min}}{4L^2}, \tag{454}$$

or formulated as an inequality constraint (with $a = X$):

$$g_2(X) = F_0 - \frac{\pi^2 EX^4}{24L^2} \leq 0. \tag{455}$$

In summary, the optimization problem is:

Minimize the objective function

$$F(X) = m(X) = 2\varrho LX^2, \tag{456}$$

taking into account the inequality constraints (see Fig. 63):

$$g_1(X) = \frac{F_0}{2X^2} - R_{\mathrm{p}0.2} \leq 0 \quad \text{(stress)}, \tag{457}$$

$$g_2(X) = F_0 - \frac{\pi^2 EX^4}{24L^2} \leq 0 \quad \text{(buckling)}. \tag{458}$$

From Fig. 63 it can be seen that the inequality restriction g_2 is the stricter condition. Thus, only points that lie to the right of the blue right-hand limit curve are allowed on the curve $F(X)$ (this results from the condition that $g_2 \leq 0$ must be). The optimized design variable results from the condition

$$g_2(X) \stackrel{!}{=} 0, \tag{459}$$

as analytical solution to:

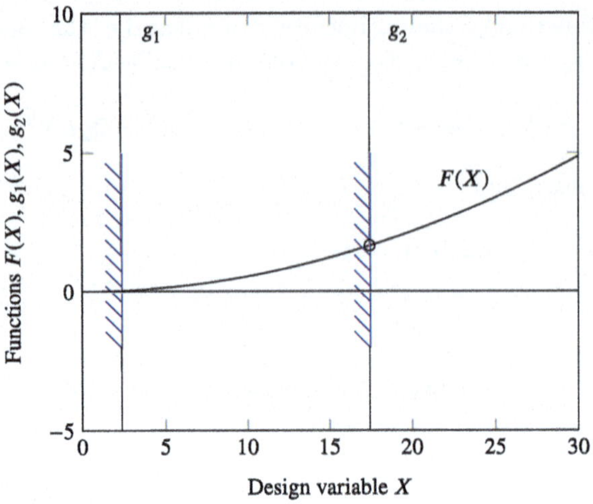

Fig. 63. Optimal design of a compression strut: $L = 1000\,\text{mm}$, $F_0 = 2667\,\text{N}$, $E = 70{,}000\,\text{MPa}$, $R_{\text{p0.2}} = 247\,\text{MPa}$, $\varrho = 2.691 \times 10^{-6}\,\text{kg/mm}^3$ (original set of equations (456)–(458)). Exact solution for the minimum: $X_{\text{extr}} = 17.447$ mm

$$X_{\text{extr}} = \left(\frac{24F_0L^2}{\pi^2 E}\right)^{\frac{1}{4}} = 17.45\,\text{mm}. \tag{460}$$

15 Optimal Design of a Short Cantilever Beam (Classical Optimization Problem)

- Internal reactions (Q_z, M_y):

The functions of the internal reactions are:

$$Q_z(x) = -F_0, \tag{461}$$

$$M_y(x) = F_0 L \left(1 - \frac{x}{L}\right), \tag{462}$$

where the maximum bending moment is obtained at the fixed support ($x = 0$):

$$M_{y,\text{max}} = M_y\,(x = 0) = F_0 L. \tag{463}$$

- Geometric dimensions of the cross section (I, A):

For a rectangular cross section with external dimensions $a \times 2a$ we get:

$$I_y = \frac{1}{12} a(2a)^3 = \frac{2}{3} a^4, \tag{464}$$

$$A = 2a^2. \tag{465}$$

- Objective function:

The objective function of the optimization problem, i.e. the mass of the beam, can be represented as a function of the design variable $X = a$ as follows:

$$F(X) = m(X) = \varrho V(X) = \varrho L A(X) = 2\varrho L X^2. \tag{466}$$

- Inequality constraints:

The minimization is to be carried out taking into account the following inequality constraints, see [13]:

The maximum normal stress is obtained at the fixed support:

$$\sigma_{x,\max} = \frac{M_y(0)}{I_y} \times (-a) = \frac{F_0 L}{\frac{2a^4}{3}} \times a \overset{!}{=} R_{\text{p0.2}}, \tag{467}$$

or formulated as an inequality constraint (with $a = X$):

$$g_1(X) = \frac{3F_0 L}{2X^3} - R_{\text{p0.2}} \leq 0. \tag{468}$$

The maximum shear stress is obtained in the middle of the cross section:

$$\tau_{xz,\max}(x,z) = \tau_{xz}(x, z=0) = \frac{Q_z(x)}{2I_y} \times (a)^2 = \frac{R_{\text{p0.2}}}{2}, \tag{469}$$

or formulated as an inequality constraint (with $a = X$):

$$g_2(X) = \frac{3F_0}{4X^2} - \frac{R_{\text{p0.2}}}{2} \leq 0. \tag{470}$$

In summary, the optimization problem is:

Minimize the objective function

$$F(X) = 2\varrho L X^2, \tag{471}$$

taking into account the inequality constraints (see Fig. 64):

$$g_1(X) = \frac{3F_0 L}{2X^3} - R_{\text{p0.2}} \leq 0 \quad \text{(normal stress)}, \tag{472}$$

$$g_2(X) = \frac{3F_0}{4X^2} - \frac{R_{\text{p0.2}}}{2} \leq 0 \quad \text{(shear stress)}. \tag{473}$$

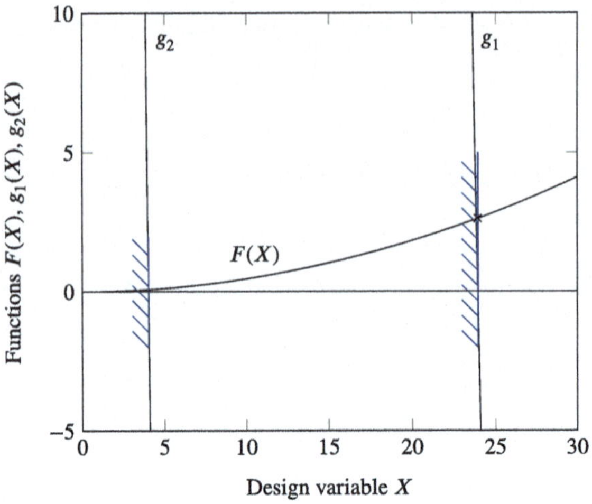

Fig. 64. Optimal design of a short cantilever beam: $L = 846.67\,\text{mm}$, $F_0 = 2667\,\text{N}$, $E = 68{,}948\,\text{MPa}$, $R_{\text{p}0.2} = 247\,\text{MPa}$, $\varrho = 2.691 \times 10^{-6}\,\text{kg/mm}^3$ (original set of equations (471)–(473)). Exact solution for the minimum: $X_{\text{extr}} = 23.936$ mm

From Fig. 64 one can see that the inequality constraint g_1 is the stricter condition. Thus, only points that lie to the right of the blue right-hand limit curve are allowed on the curve $F(X)$ (this results from the condition that $g_1 \leq 0$ must be). The optimized design variable results from the condition

$$g_1(X) \stackrel{!}{=} 0, \tag{474}$$

as analytical solution to:

$$X_{\text{extr}} = \left(\frac{3F_0L}{2R_{\text{p}0.2}} \right)^{\frac{1}{3}} = 23.94\,\text{mm}. \tag{475}$$

16 Optimal Design of a Cantilever Beam: Constant Rectangular Cross Section (Classical Optimization Problem

- Geometric dimensions of the cross section (I_y, A):

For a rectangular cross section with external dimensions $a \times b$ we get:

$$I_y = \frac{1}{12} a(b)^3 = \frac{1}{12} ab^3, \tag{476}$$

$$A = ab. \tag{477}$$

- Objective function:

The objective function of the optimization problem, i.e. the mass of the beam, can be represented as a function of the two design variables $X_1 = a$ and $X_2 = b$ as follows:

$$F(X_1, X_2) = m(X_1, X_2) = \varrho V(X_1, X_2) = \varrho L X_1 X_2. \tag{478}$$

- Inequality constraints:

The minimization is to be carried out taking into account the following inequality constraints, see [13]:

The maximum deflection can be taken, for example, from [11]. It should be noted that the current force acts in the positive z-direction:

$$u_{z,\max} = \frac{F_0 L^3}{3EI_y}, \tag{479}$$

or formulated as an inequality constraint (with $a = X_1$ and $b = X_2$):

$$g_1(X_1, X_2) = \frac{4F_0 L^3}{E X_1 X_2^3} - \underbrace{u_{z,\max}}_{r_1 L} \leq 0. \tag{480}$$

The bending moment distribution is obtained as:

$$M_y(x) = -F_0 L \left(1 - \frac{x}{L}\right). \tag{481}$$

This gives the maximum stress at the fixed support as:

$$\sigma_{x,\max} = \frac{M_y(0)}{I_y} \times \left(-\frac{b}{2}\right) = \frac{-F_0 L}{\frac{ab^3}{12}} \times \left(-\frac{b}{2}\right) \stackrel{!}{=} R_{\text{p0.2}}, \tag{482}$$

or formulated as an inequality constraint (with $a = X_1$ and $b = X_2$):

$$g_2(X_1, X_2) = \frac{6F_0 L}{X_1 X_2^2} - R_{\text{p0.2}} \leq 0. \tag{483}$$

The height-to-width ratio $b \leq 20a$ can be formulated as the following inequality constraint (with $a = X_1$ and $b = X_2$):

$$g_3(X_1, X_2) = X_2 - 20X_1 \leq 0. \tag{484}$$

In summary, the optimization problem is:
Minimize the objective function

$$F(X_1, X_2) = \varrho L X_1 X_2, \tag{485}$$

taking into account the inequality constraints (see Fig. 65):

$$g_1(X_1, X_2) = \frac{4F_0L^3}{EX_1X_2^3} - r_1L \leq 0 \quad \text{(displacement)}, \tag{486}$$

$$g_2(X_1, X_2) = \frac{6F_0L}{X_1X_2^2} - R_{\text{p0,2}} \leq 0 \quad \text{(normal stress)}, \tag{487}$$

$$g_3(X_1, X_2) = X_2 - 20X_1 \leq 0 \quad \text{(height-to-width ratio)}. \tag{488}$$

From Fig. 65 one can see that the minimum—taking into account the given inequality constraints—is the intersection point of the curves g_1 and g_3. A simple calculation gives the analytical solution to:

$$X_{1,\text{extr}} = (20)^{-\frac{3}{4}} \times \left(\frac{4F_0L^2}{Er_1}\right)^{\frac{1}{4}} = 8.031 \text{ mm}, \tag{489}$$

$$X_{2,\text{extr}} = 20 \times X_{1,\text{extr}} = 160.614 \text{ mm}. \tag{490}$$

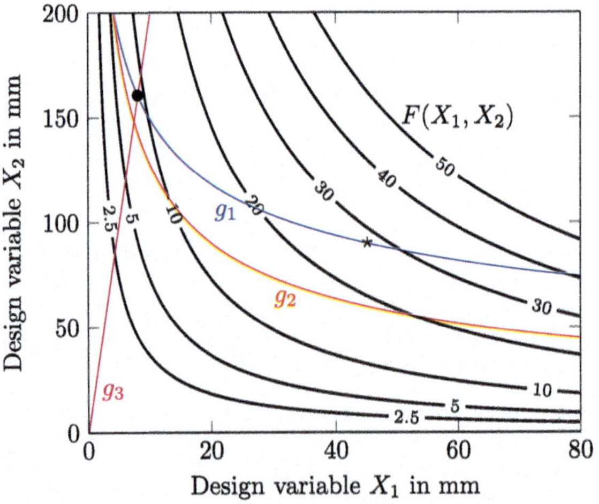

Fig. 65. Optimal design of a cantilever beam with: $L = 2540$ mm, $F_0 = 2667$ N, $E = 68{,}948$ MPa, $R_{\text{p0.2}} = 247$ MPa, $\varrho = 2.691 \times 10^{-6}$ kg/mm^3, $r_1 = 0.03$ (original set of equations (485)–(488)). Exact solution for the minimum: $X_{1,\text{extr}} = 8.031$ mm, $X_{2,\text{extr}} = 160.614$ mm (indicated by the • marker)

The result of problem 13, i.e. $a = 45.160$ and $2a = 90.3$, is marked by the ⋆ marker in Fig. 65. For the given geometric dimensions this results in a total mass of $F(a, 2a) = m(a, 2a) = 27.9$ kg, while the optimization for a variable width and height results in $F(X_{1,\text{extr}}, X_{2,\text{extr}}) = m(X_{1,\text{extr}}, X_{2,\text{extr}}) = 8.8$ kg.

17 Optimal Design of a Simply Supported Beam: Constant Rectangular Cross Section (Classical Optimization Problem)

- Geometric dimensions of the cross section (I_y, A):

For a rectangular cross section with external dimensions $a \times b$ we get:

$$I_y = \frac{1}{12}a(b)^3 = \frac{1}{12}ab^3, \tag{491}$$

$$A = ab. \tag{492}$$

- Internal reactions (M_y, Q_z):

The distributions of the internal reactions are obtained as:

$$M_y(x) = +\frac{F_0}{2} \times x \text{ for } 0 \leq x \leq \frac{L}{2}, \tag{493}$$

$$Q_z(x) = +\frac{F_0}{2} \text{ for } 0 \leq x \leq \frac{L}{2}. \tag{494}$$

- Objective function:

The objective function of the optimization problem, i.e. the mass of the beam, can be represented as a function of the two design variables $X_1 = a$ and $X_2 = b$ as follows:

$$F(X_1, X_2) = m(X_1, X_2) = \varrho V(X_1, X_2) = \varrho L X_1 X_2. \tag{495}$$

- Inequality constraints:

The minimization is to be carried out taking into account the following inequality constraints, see [13]:

The maximum normal stress in the middle of the beam is:

$$\sigma_{x,\text{max}} = \frac{M_y(L/2)}{I_y} \times \frac{b}{2} = \frac{\frac{F_0 L}{4}}{\frac{ab^3}{12}} \times \frac{b}{2} \overset{!}{=} R_{\text{p0.2}}, \tag{496}$$

or formulated as an inequality constraint (with $a = X_1$ and $b = X_2$):

$$g_1(X_1, X_2) = \frac{3F_0 L}{2X_1 X_2^2} - R_{\text{p0.2}} \leq 0. \tag{497}$$

The maximum shear stress is given in the middle of the cross section as:

$$\tau_{xz,\max}(x,z) = \tau_{xz}(x,z=0) = \frac{Q_z(x)}{2I_y} \times \left(\frac{b}{2}\right)^2 = \frac{R_{\text{p}0.2}}{2}, \tag{498}$$

or formulated as an inequality constraint (with $a = X_1$ and $b = X_2$):

$$g_2(X_1, X_2) = \frac{3F_0}{4X_1X_2} - \frac{R_{\text{p}0.2}}{2} \leq 0. \tag{499}$$

The maximum deflection in the middle of the beam is obtained by determining the bending line. Using the general solution (e.g. [11]) for constant material and geometry properties

$$Q_z(x) = -c_1, \tag{500}$$

$$M_y(x) = -c_1x - c_2, \tag{501}$$

$$\varphi_y(x) = -\frac{1}{EI_y}\left(\frac{c_1x^2}{2} + c_2x + c_3\right), \tag{502}$$

$$u_z(x) = \frac{1}{EI_y}\left(\frac{c_1x^3}{6} + \frac{c_2x^2}{2} + c_3x + c_4\right), \tag{503}$$

the integration constants, taking into account the boundary conditions $u_z(0) = 0$, $M_y(0) = 0$, $Q_z(0) = F_0/2$ and $\varphi_y(L/2) = 0$, are:

$$c_4 = 0, \tag{504}$$

$$c_2 = 0, \tag{505}$$

$$c_1 = -\frac{F_0}{2}, \tag{506}$$

$$c_3 = \frac{F_0L^2}{16}. \tag{507}$$

Thus the bending line is:

$$u_z(x) = \frac{F_0L^3}{EI_y}\left(-\frac{1}{12}\left[\frac{x}{L}\right]^3 + \frac{1}{16}\left[\frac{x}{L}\right]^1\right), \tag{508}$$

or the maximum deflection:

$$u_{z,\max} = u_z(x = L/2) = \frac{F_0L^3}{48EI_y}. \tag{509}$$

The inequality constraint is obtained from this relationship as (with $a = X_1$ and $b = X_2$):

$$g_3(X_1, X_2) = \frac{F_0L^3}{4EX_1X_2^3} - r_1L \leq 0. \tag{510}$$

The height-to-width ratio $b \leq 20a$ can be formulated as the following inequality constraint (with $a = X_1$ and $b = X_2$):

$$g_3(X_1, X_2) = X_2 - 20X_1 \leq 0. \tag{511}$$

In summary, the optimization problem is:

Minimize the objective function

$$F(X_1, X_2) = \varrho L X_1 X_2, \tag{512}$$

taking into account the inequality constraints (see Fig. 66):

$$g_1(X_1, X_2) = \frac{3F_0 L}{2X_1 X_2^2} - R_{\text{p0.2}} \leq 0 \quad \text{(normal stress)}, \tag{513}$$

$$g_2(X_1, X_2) = \frac{3F_0}{4X_1 X_2} - \frac{R_{\text{p0.2}}}{2} \leq 0 \quad \text{(shear stress)}, \tag{514}$$

$$g_3(X_1, X_2) = \frac{F_0 L^3}{4EX_1 X_2^3} - r_1 L \leq 0 \quad \text{(displacement)}, \tag{515}$$

$$g_4(X_1, X_2) = X_2 - 20X_1 \leq 0 \quad \text{(height-to-width ratio)}. \tag{516}$$

From Fig. 66 one can see that the minimum—taking into account the given inequality constraints—is the intersection of the curves g_1 and g_4 for the case $r_1 = 0.03$ and the intersection of the curves g_3 and g_4 for the case $r_1 = 0.003$. A simple calculation gives the analytical solution to:

$$X_{1,\text{extr}}\big|_{r_1=0.03} = (20)^{-\frac{2}{3}} \times \left(\frac{3F_0 L}{2R_{\text{p0.2}}}\right)^{\frac{1}{3}} = 4.685 \text{ mm}, \tag{517}$$

$$X_{2,\text{extr}}\big|_{r_1=0.03} = 20 \times X_{1,\text{extr}} = 93.704 \text{ mm}, \tag{518}$$

or

$$X_{1,\text{extr}}\big|_{r_1=0.003} = (20)^{-\frac{3}{4}} \times \left(\frac{F_0 L^2}{4Er_1}\right)^{\frac{1}{4}} = 7.140 \text{ mm}, \tag{519}$$

$$X_{2,\text{extr}}\big|_{r_1=0.003} = 20 \times X_{1,\text{extr}} = 142.809 \text{ mm}. \tag{520}$$

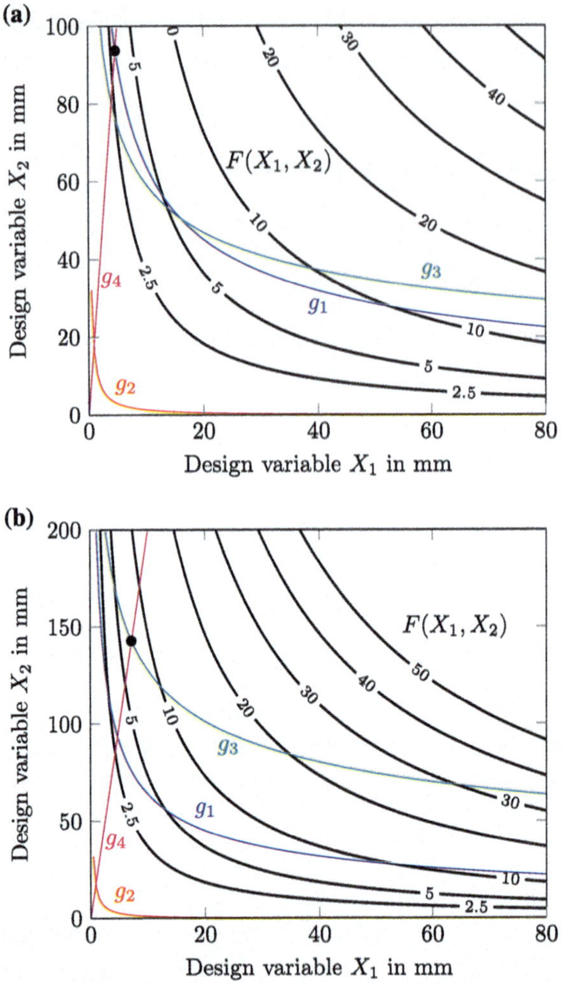

Fig. 66. Optimal design of a simply supported beam with $L = 2540\,\text{mm}$, $F_0 = 2667\,\text{N}$, $E = 68{,}948\,\text{MPa}$, $R_{\text{p}0.2} = 247\,\text{MPa}$, $\varrho = 2.691 \times 10^{-6}\,\text{kg/mm}^3$ (original set of equations (512)–(516)): **a** $r_1 = 0.03$, exact solution for the minimum: $X_{1,\text{extr}} = 4.685$ mm, $X_{2,\text{extr}} = 93.704$ mm (indicated by the • marker), **b** $r_1 = 0.003$, exact solution for the minimum: $X_{1,\text{extr}} = 7.140$ mm, $X_{2,\text{extr}} = 142.809$ mm (indicated by the • marker)

18 Optimal Design of a Compression Strut: Constant Rectangular Cross Section (Classical Optimization Problem)

- Geometric dimensions of the cross section (I, A):

For a rectangular cross section with external dimensions $a \times b$ we get:

$$I_y = \frac{1}{12} a(b)^3, \tag{521}$$

$$I_z = \frac{1}{12} b(a)^3, \tag{522}$$

$$A = ab. \tag{523}$$

- Objective function:

The objective function of the optimization problem, i.e. the mass of the compression strut, can be represented as a function of the two design variables $X_1 = a$ and $X_2 = b$ as follows:

$$F(X_1, X_2) = m(X_1, X_2) = \varrho V(X_1, X_2) = \varrho L X_1 X_2. \tag{524}$$

- Inequality constraints:

The minimization is to be carried out taking into account the following inequality constraints, see [13]:

The magnitude of the maximum stress is:

$$\sigma_{x,\text{max}} = \frac{F_0}{A} \stackrel{!}{=} R_{\text{p0.2}}, \tag{525}$$

or formulated as an inequality constraint (with $a = X_1$ and $b = X_2$):

$$g_1(X_1, X_2) = \frac{F_0}{X_1 X_2} - R_{\text{p0.2}} \leq 0. \tag{526}$$

The critical buckling force for Euler case 1 for buckling around the y-axis is (see [14]):

$$F_{\text{cr}} = \frac{\pi^2 E I_y}{4L^2}, \tag{527}$$

or formulated as an inequality constraint (with $a = X_1$ and $b = X_2$):

$$g_2(X_1, X_2) = F_0 - \frac{\pi^2 E X_1 X_2^3}{48 L^2} \leq 0. \tag{528}$$

The critical buckling force for Euler case 1 for buckling around the z-axis is (see [14]):

$$F_{\text{cr}} = \frac{\pi^2 E I_z}{4L^2}, \tag{529}$$

or formulated as an inequality constraint (with $a = X_1$ and $b = X_2$):

$$g_3(X_1, X_2) = F_0 - \frac{\pi^2 E X_2 X_1^3}{48 L^2} \leq 0. \tag{530}$$

In summary, the optimization problem is:

Minimize the objective function

$$F(X_1, X_2) = m(X_1, X_2) = \varrho V(X_1, X_2) = \varrho L X_1 X_2, \tag{531}$$

taking into account the inequality constraints (see Fig. 67):

$$g_1(X_1, X_2) = \frac{F_0}{X_1 X_2} - R_{\text{p0.2}} \leq 0 \quad \text{(normal stress)}, \tag{532}$$

$$g_2(X_1, X_2) = F_0 - \frac{\pi^2 E X_1 X_2^3}{48 L^2} \leq 0 \quad \text{(buckling around the } y\text{-axis)}, \tag{533}$$

$$g_3(X_1, X_2) = F_0 - \frac{\pi^2 E X_2 X_1^3}{48 L^2} \leq 0 \quad \text{(buckling around the } z\text{-axis)}. \tag{534}$$

From Fig. 67 one can see that the minimum—taking into account the given inequality constraints—is the intersection point of the curves g_2 and g_3. A simple calculation gives the analytical solution:

$$X_{1,\text{extr}} = \left(\frac{48 F_0 L^2}{\pi^2 E} \right)^{\frac{1}{4}} = 20.748 \text{ mm}, \tag{535}$$

$$X_{2,\text{extr}} = \frac{48 F_0 L^2}{\pi^2 E X_{1,\text{extr}}^3} = 20.748 \text{ mm}. \tag{536}$$

The result of problem 14, i.e. $a = 17.447$ and $2a = 34.894$, is marked by the $\star$ marker in Fig. 67. For the given geometric dimensions this results in a total mass of $F(a, 2a) = m(a, 2a) = 1.638$ kg, while the optimization for a variable width and height results in $F(X_{1,\text{extr}}, X_{2,\text{extr}}) = m(X_{1,\text{extr}}, X_{2,\text{extr}}) = 1.158$ kg.

19 **Optimal Design of a Short Cantilever Beam: Constant Rectangular Cross Section (Classical Optimization Problem)**

- Internal reactions (Q_z, M_y):

The distributions of the internal reactions are obtained as:

$$Q_z(x) = -F_0, \tag{537}$$

$$M_y(x) = F_0 L \left(1 - \frac{x}{L} \right), \tag{538}$$

where the maximum bending moment is obtained at the fixed support ($x = 0$):

$$M_{y,\text{max}} = M_y\,(x = 0) = F_0 L. \tag{539}$$

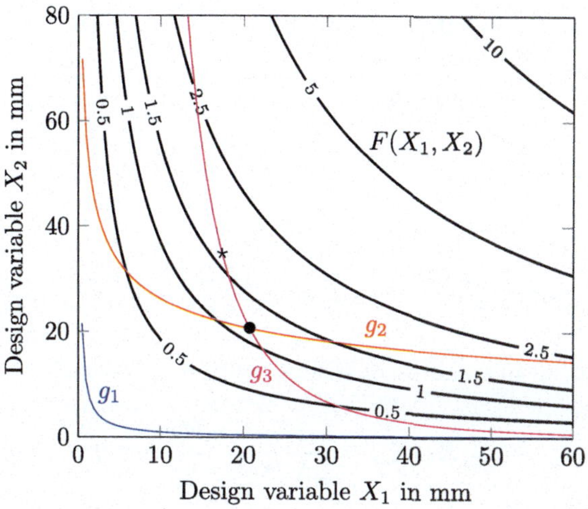

Fig. 67. Optimal design of a compression strut: $L = 1000\,\text{mm}$, $F_0 = 2667\,\text{N}$, $E = 70{,}000\,\text{MPa}$, $R_{\text{p}0.2} = 247\,\text{MPa}$, $\varrho = 2.691 \times 10^{-6}\,\text{kg/mm}^3$, (original set of equations (531)–(534)). Exact solution for the minimum: $X_{1,\text{extr}} = 20.748$ mm, $X_{2,\text{extr}} = 20.748$ mm (indicated by the • marker)

- Geometric dimensions of the cross section (I, A):

For a rectangular cross section with external dimensions $a \times b$ we get:

$$I_y = \frac{1}{12} a(b)^3, \tag{540}$$

$$A = ab. \tag{541}$$

- Objective function:

The objective function of the optimization problem, i.e. the mass of the beam, can be represented as a function of the two design variables $X_1 = a$ and $X_2 = b$ as follows:

$$F(X_1, X_2) = m(X_1, X_2) = \varrho V(X_1, X_2) = \varrho L X_1 X_2. \tag{542}$$

- Inequality constraints:

The minimization is to be carried out taking into account the following inequality constraints, see [13]:

The maximum normal stress at the fixed support is:

$$\sigma_{x,\max} = \frac{M_y(0)}{I_y} \times \frac{b}{2} = \frac{F_0 L}{\frac{ab^3}{12}} \times \frac{b}{2} \overset{!}{=} R_{\text{p}0.2}, \tag{543}$$

or formulated as an inequality constraint (with $a = X_1$ and $b = X_2$):

$$g_1(X_1, X_2) = \frac{6F_0L}{X_1X_2^2} - R_{p0.2} \leq 0. \tag{544}$$

The maximum shear stress (magnitude) is obtained in the middle of the cross section as:

$$\tau_{xz,\max}(x, z) = \tau_{xz}(x, z = 0) = \frac{Q_z(x)}{2I_y} \times \left(\frac{b}{2}\right)^2 = \frac{R_{p0.2}}{2}, \tag{545}$$

or formulated as an inequality constraint (with $a = X_1$ and $b = X_2$):

$$g_2(X_1, X_2) = \frac{3F_0}{2X_1X_2} - \frac{R_{p0.2}}{2} \leq 0. \tag{546}$$

The height-to-width ratio $b \leq 20a$ can be formulated as the following inequality constraint (with $a = X_1$ and $b = X_2$):

$$g_3(X_1, X_2) = X_2 - 20X_1 \leq 0. \tag{547}$$

In summary, the optimization problem is:
Minimize the objective function

$$F(X_1, X_2) = \varrho LX_1X_2, \tag{548}$$

taking into account the inequality constraints (see Fig. 68):

$$g_1(X_1, X_2) = \frac{6F_0L}{X_1X_2^2} - R_{p0.2} \leq 0 \quad \text{(normal stress)}, \tag{549}$$

$$g_2(X_1, X_2) = \frac{3F_0}{2X_1X_2} - \frac{R_{p0.2}}{2} \leq 0 \quad \text{(shear stress)}, \tag{550}$$

$$g_3(X_1, X_2) = X_2 - 20X_1 \leq 0 \quad \text{(height-to-width ratio)}. \tag{551}$$

From Fig. 68 one can see that the minimum—taking into account the given inequality constraints—is the intersection point of the curves g_1 and g_3. A simple calculation gives the analytical solution:

$$X_{1,\text{extr}} = 20^{-\frac{2}{3}} \times \left(\frac{6F_0L}{R_{p0.2}}\right)^{\frac{1}{3}} = 5.157 \text{ mm}, \tag{552}$$

$$X_{2,\text{extr}} = 20 \times X_{1,\text{extr}} = 103.135 \text{ mm}. \tag{553}$$

The result of problem 15, i.e. $a = 23.936$ and $2a = 47.872$, is marked by the $\star$ marker in Fig. 68. For the given geometric dimensions this results in a total mass of $F(a, 2a) = m(a, 2a) = 2.611$ kg, while the optimization for a variable

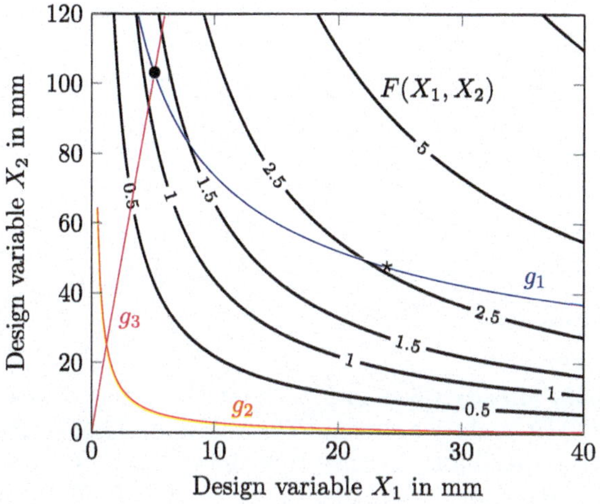

Fig. 68. Optimal design of a short cantilever beam: $L = 846.67\,\text{mm}$, $F_0 = 2667\,\text{N}$, $E = 68{,}948\,\text{MPa}$, $R_{\text{p}0.2} = 247\,\text{MPa}$, $\varrho = 2.691 \times 10^{-6}\,\text{kg/mm}^3$, (original set of equations (548)–(551)). Exact solution for the minimum: $X_{1,\text{extr}} = 5.157$ mm, $X_{2,\text{extr}} = 103.135$ mm (indicated by the • marker)

width and height results in $F(X_{1,\text{extr}}, X_{2,\text{extr}}) = m(X_{1,\text{extr}}, X_{2,\text{extr}}) = 1.212$ kg.

20 Optimal Design of a Stepped Cantilever Beam with Two Sections (Classical Optimization Problem)

The problem requires the calculation of both the deformations and the stress distributions. This task can be approached by a 'hand calculation' based on the finite element method. A single stiffness matrix of an Euler-Bernoulli beam element can be given as follows (see [12,15]):

$$\boldsymbol{K}_i^{\text{e}} = \frac{E_i I_i}{L_i^3} \begin{bmatrix} 12 & -6L_i & -12 & -6L_i \\ -6L_i & 4L_i^2 & 6L_i & 2L_i^2 \\ -12 & 6L_i & 12 & 6L_i \\ -6L_i & 2L_i^2 & 6L_i & 4L_i^2 \end{bmatrix}. \tag{554}$$

The assembly of the two element stiffness matrices $\boldsymbol{K}_i^{\text{e}}$, taking into account $E_{\text{I}} = E_{\text{II}} = E$, $L_{\text{I}} = L_{\text{II}} = \frac{L}{2}$, $I_{\text{I}} = \frac{1}{12}ab_{\text{I}}^3$ and $I_{\text{II}} = \frac{1}{12}ab_{\text{II}}^3$, results in the following global (reduced) system of equations:

$$E\begin{bmatrix} \left(\frac{8a\,b_{\text{II}}{}^3}{L^3}+\frac{8a\,b_{\text{I}}{}^3}{L^3}\right) & \left(\frac{2a\,b_{\text{I}}{}^3}{L^2}-\frac{2a\,b_{\text{II}}{}^3}{L^2}\right) & -\frac{8a\,b_{\text{II}}{}^3}{L^3} & -\frac{2a\,b_{\text{II}}{}^3}{L^2} \\ \left(\frac{2a\,b_{\text{I}}{}^3}{L^2}-\frac{2a\,b_{\text{II}}{}^3}{L^2}\right) & \left(\frac{2a\,b_{\text{II}}{}^3}{3\,L}+\frac{2a\,b_{\text{I}}{}^3}{3\,L}\right) & \frac{2a\,b_{\text{II}}{}^3}{L^2} & \frac{a\,b_{\text{II}}{}^3}{3\,L} \\ -\frac{8a\,b_{\text{II}}{}^3}{L^3} & \frac{2a\,b_{\text{II}}{}^3}{L^2} & \frac{8a\,b_{\text{II}}{}^3}{L^3} & \frac{2a\,b_{\text{II}}{}^3}{L^2} \\ -\frac{2a\,b_{\text{II}}{}^3}{L^2} & \frac{a\,b_{\text{II}}{}^3}{3\,L} & \frac{2a\,b_{\text{II}}{}^3}{L^2} & \frac{2a\,b_{\text{II}}{}^3}{3\,L} \end{bmatrix}\begin{bmatrix} u_{2z} \\ \varphi_{2y} \\ u_{3z} \\ \varphi_{3y} \end{bmatrix} = \begin{bmatrix} 0 \\ 0 \\ -F_0 \\ 0 \end{bmatrix}, \tag{555}$$

or in matrix notation as:

$$\boldsymbol{K}\boldsymbol{u} = \boldsymbol{f}. \tag{556}$$

Equation (555) already takes into account that the two degrees of freedom (i.e. u_{1z} and φ_{1y}) at node 1 are zero. The solution of the system of equations can be obtained, for example, by inverting the global stiffness matrix and then multiplying with the right-hand side, i.e. $\boldsymbol{u} = \boldsymbol{K}^{-1}\boldsymbol{f}$. This gives the column matrix of the unknown node values as:

$$\begin{bmatrix} u_{2z} \\ \varphi_{2y} \\ u_{3z} \\ \varphi_{3y} \end{bmatrix} = \begin{bmatrix} -\frac{5F_0\,L^3}{4Ea\,b_{\text{I}}{}^3} \\ \frac{9F_0\,L^2}{2Ea\,b_{\text{I}}{}^3} \\ -\frac{F_0\,L^3\left(7b_{\text{II}}{}^3+b_{\text{I}}{}^3\right)}{2Ea\,b_{\text{I}}{}^3 b_{\text{II}}{}^3} \\ \frac{3F_0\,L^2\left(3b_{\text{II}}{}^3+b_{\text{I}}{}^3\right)}{2Ea\,b_{\text{I}}{}^3 b_{\text{II}}{}^3} \end{bmatrix}. \tag{557}$$

The bending moment and shear force distributions within a single element of length L_i can generally be expressed using (the index '1' refers to the start node while the index '2' refers to the end node of an element) the following relations:

$$\begin{aligned} M_y^{\text{e}}(x) =& EI_y\left(\left[+\frac{6}{L_i^2}-\frac{12x}{L_i^3}\right]u_{1z}+\left[-\frac{4}{L_i}+\frac{6x}{L_i^2}\right]\varphi_{1y}\right. \\ &\left.+\left[-\frac{6}{L_i^2}+\frac{12x}{L_i^3}\right]u_{2z}+\left[-\frac{2}{L_i}+\frac{6x}{L_i^2}\right]\varphi_{2y}\right), \end{aligned} \tag{558}$$

$$Q_z^{\text{e}}(x) = EI_y\left(\left[-\frac{12}{L_i^3}\right]u_{1z}+\left[+\frac{6}{L_i^2}\right]\varphi_{1y}+\left[+\frac{12}{L_i^3}\right]u_{2z}+\left[+\frac{6}{L_i^2}\right]\varphi_{2y}\right). \tag{559}$$

Thus, the normal and shear stress distributions can generally be calculated as follows:

$$\sigma_x^{\text{e}}(x,z) = \frac{M_y^{\text{e}}(x)}{I_y}\times z, \tag{560}$$

$$\tau_{xz}^{\text{e}}(x,z) = \frac{Q_z^{\text{e}}(x)}{2I_y}\left[\left(\frac{b}{2}\right)^2 - z^2\right]. \tag{561}$$

The nodal values of the internal reactions for calculating the normal and shear stresses are summarized in Table 5. It is noted here that in this case the nodal values based on the finite element calculation are identical to the analytical solution.

Table 5. Nodal values of the internal reactions bending moment (M_y) and shear force (Q_z) at each node

	Element I		Element II	
	Left node	Right node	Left node	Right node
M_y	$F_0 L$	$\frac{F_0 L}{2}$	$\frac{F_0 L}{2}$	0
Q_z	$-F_0$	$-F_0$	$-F_0$	$-F_0$

From Table 5 it can be concluded that the maximum normal stress in each element is reached at the left node and that the shear stress is constant in each element. Thus, the critical stresses in each element can be calculated as follows:

$$\sigma_{x,\mathrm{I}} = \frac{6F_0 L}{a\, b_{\mathrm{I}}^{\;2}}, \tag{562}$$

$$\sigma_{x,\mathrm{II}} = \frac{3F_0 L}{a\, b_{\mathrm{II}}^{\;2}}, \tag{563}$$

or for the shear stresses:

$$\tau_{xz,\mathrm{I}} = -\frac{3F_0}{2a\, b_{\mathrm{I}}}, \tag{564}$$

$$\tau_{xz,\mathrm{II}} = -\frac{3F_0}{2a\, b_{\mathrm{II}}}. \tag{565}$$

The objective function of the optimization problem, i.e. the mass of the stepped beam, can be represented as a function of the two design variables $X_1 = b_{\mathrm{I}}$ and $X_2 = b_{\mathrm{II}}$ as follows:

$$F(X_1, X_2) = \varrho L_{\mathrm{I}} a X_1 + \varrho L_{\mathrm{II}} a X_2 = \frac{\varrho L a}{2}(X_1 + X_2), \tag{566}$$

where the minimization is to be carried out under consideration of the following seven inequality constraints:

$$g_1(X_1, X_2) = +\frac{F_0 L^3 (X_1^3 + 7X_2^3)}{2EaX_1^3 X_2^3 r_1 L} - 1 \leq 0 \quad \text{(max. deflection)}, \tag{567}$$

$$g_2(X_1, X_2) = \frac{6F_0 L}{aX_1^2} - R_{\text{p0.2}} \leq 0 \quad \text{(normal stress in I)}, \tag{568}$$

$$g_3(X_1, X_2) = \frac{3F_0 L}{aX_2^2} - R_{\text{p0.2}} \leq 0 \quad \text{(normal stress in II)}, \tag{569}$$

$$g_4(X_1, X_2) = \frac{3F_0}{2aX_1} - \frac{R_{\text{p0.2}}}{2} \leq 0 \quad \text{(shear stress in I)}, \tag{570}$$

$$g_5(X_1, X_2) = \frac{3F_0}{2aX_2} - \frac{R_{\text{p0.2}}}{2} \leq 0 \quad \text{(shear stress in II)}, \tag{571}$$

$$g_6(X_1, X_2) = X_1 - 20a \leq 0 \quad \text{(height-to-width ratio in I)}, \tag{572}$$

$$g_7(X_1, X_2) = X_2 - 20a \leq 0 \quad \text{(height-to-width ratio in II)}. \tag{573}$$

The graphical representation of the objective function and the inequality constraints in the X_1–X_2 design space is shown in Figs. 69 and 70.

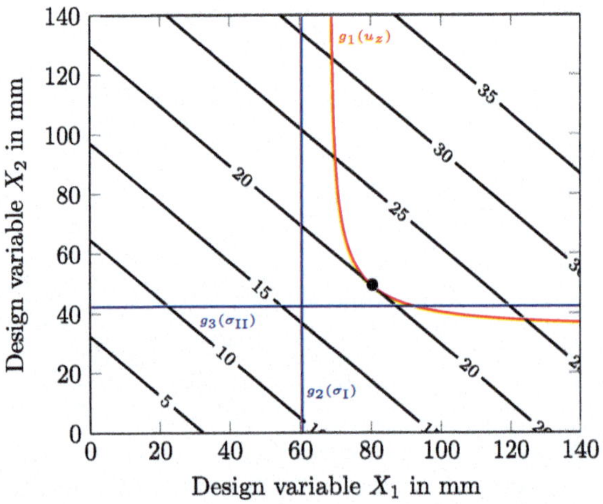

Fig. 69. Optimal design of a stepped cantilever beam with two sections: $L = 2540\,\text{mm}$, $a = 45.16\,\text{mm}$, $r_1 = 0.06$, $F_0 = 2667\,\text{N}$, $E = 68{,}948\,\text{MPa}$, $R_{\text{p0.2}} = 247\,\text{MPa}$, $\varrho = 2.691 \times 10^{-6}\,\text{kg/mm}^3$, (original set of equations (566)–(573)). Exact solution for the minimum: $X_{1,\text{extr}} = 80.442$ mm, $X_{2,\text{extr}} = 49.455$ mm (indicated by the • marker)

From Figs. 69 and 70 it can be concluded that considering g_1, g_2, and g_3 can be sufficient if a reasonable range of the design variables is chosen. Furthermore, from Fig. 69 it can be concluded that the minimum is achieved if g_1 is tangent to the objective function F. In order to derive an analytical solution for the

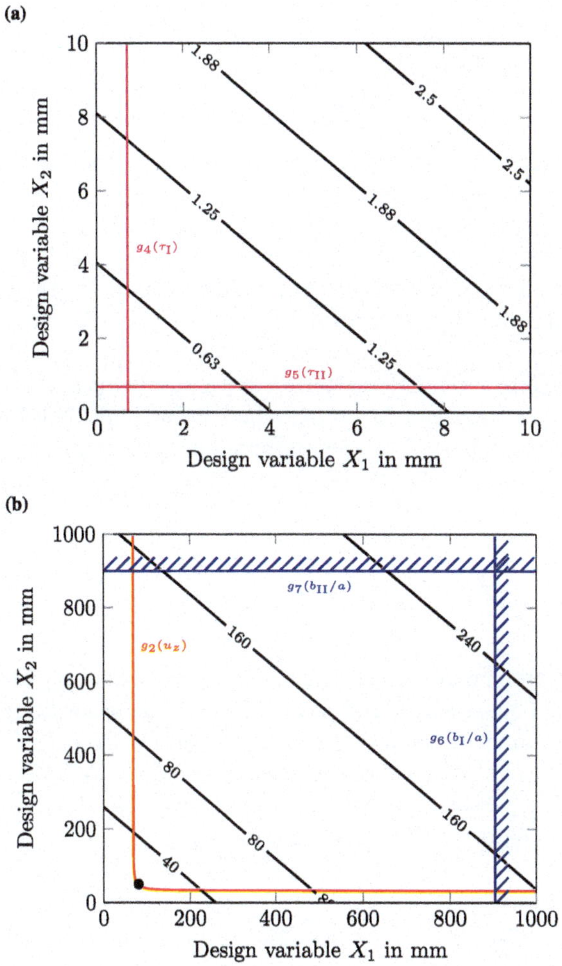

Fig. 70. Optimal design of a stepped cantilever beam with two sections: **a** magnification for smaller values of the design variables and **b** global view

minimum, one must take into account that the objective function F, in the representation $X_2(X_1) = \frac{2c}{\varrho La} - X_1$, represents straight lines with the slope of -1. Thus, the condition for the minimum is that the curve g_1 assumes a slope of -1. Based on Eq. (568), the expression for g_1 can be rewritten:

$$X_2 = \left(-\frac{F_0 L^3 X_1^3}{7F_0L^3 - 2Ear_1LX_1^3} \right)^{\frac{1}{3}}, \tag{574}$$

which, after a short calculation, gives the expression of the first-order derivative as follows:

$$\left.\frac{\mathrm{d}X_2}{\mathrm{d}X_1}\right|_{g_1} = \frac{-7F_0^{\frac{4}{3}}L^{\frac{8}{3}}}{(7F_0L^2 - 2Ear_1X_1^3)^{\frac{4}{3}}} \stackrel{!}{=} -1. \tag{575}$$

The last equation can be rearranged so that one side has a root, and Newton's method can be used to determine the root. It is important that the root is sought in a reasonable interval, e.g. $70 \leq X_{1,\,\text{extr}} \leq 85$.

21 **Optimal Design of a Simply Supported Beam with Three Sections (Classical Optimization Problem)**

The problem requires the calculation of both the deformations and the stress distributions. This task can be approached by a 'hand calculation' based on the finite element method. A single stiffness matrix of an Euler-Bernoulli beam element can be given as follows (see [12,15]):

$$\boldsymbol{K}_i^{\mathrm{e}} = \frac{E_iI_i}{L_i^3}\begin{bmatrix} 12 & -6L_i & -12 & -6L_i \\ -6L_i & 4L_i^2 & 6L_i & 2L_i^2 \\ -12 & 6L_i & 12 & 6L_i \\ -6L_i & 2L_i^2 & 6L_i & 4L_i^2 \end{bmatrix}. \tag{576}$$

This special problem allows the consideration of symmetry in terms of geometry and loading. Thus, the reduced mechanical system according to Fig. 71 allows a faster calculation than when considering the entire structure.

The assembly of the two element stiffness matrices $\boldsymbol{K}_i^{\mathrm{e}}$, taking into account $E_{\mathrm{I}} = E_{\mathrm{II}} = E$, $L_{\mathrm{I}} = L_{\mathrm{II}} = \frac{L}{3}$, $I_{\mathrm{I}} = \frac{1}{12}ab_{\mathrm{I}}^3$, and $I_{\mathrm{II}} = \frac{1}{12}ab_{\mathrm{II}}^3$, results in the following global (reduced) system of equations:

$$E\begin{bmatrix} \frac{a\,b_{\mathrm{I}}^3}{L} & \frac{9a\,b_{\mathrm{I}}^3}{2L^2} & \frac{a\,b_{\mathrm{I}}^3}{2\,L} & 0 \\ \frac{9a\,b_{\mathrm{I}}^3}{2L^2} & \left(\frac{216a\,b_{\mathrm{II}}^3}{L^3} + \frac{27a\,b_{\mathrm{I}}^3}{L^3}\right) & \left(\frac{9a\,b_{\mathrm{I}}^3}{2L^2} - \frac{18a\,b_{\mathrm{II}}^3}{L^2}\right) & -\frac{216a\,b_{\mathrm{II}}^3}{L^3} \\ \frac{a\,b_{\mathrm{I}}^3}{2\,L} & \left(\frac{9a\,b_{\mathrm{I}}^3}{2L^2} - \frac{18a\,b_{\mathrm{II}}^3}{L^2}\right) & \left(\frac{2a\,b_{\mathrm{II}}^3}{L} + \frac{a\,b_{\mathrm{I}}^3}{L}\right) & \frac{18a\,b_{\mathrm{II}}^3}{L^2} \\ 0 & -\frac{216a\,b_{\mathrm{II}}^3}{L^3} & \frac{18a\,b_{\mathrm{II}}^3}{L^2} & \frac{216a\,b_{\mathrm{II}}^3}{L^3} \end{bmatrix}\begin{bmatrix} \varphi_{1y} \\ u_{2z} \\ \varphi_{2y} \\ u_{3z} \end{bmatrix} = \begin{bmatrix} 0 \\ 0 \\ 0 \\ \frac{-F_0}{2} \end{bmatrix}. \tag{577}$$

The last equation already takes into account that the translational degree of freedom at node 1 and the rotational degree of freedom at node 3 are zero. The solution of the system of equations can be obtained, for example, by inverting the global stiffness matrix and then multiplying with the right-hand side, i.e. $\boldsymbol{u} = \boldsymbol{K}^{-1}\boldsymbol{f}$. This gives the column matrix of the unknown node values as:

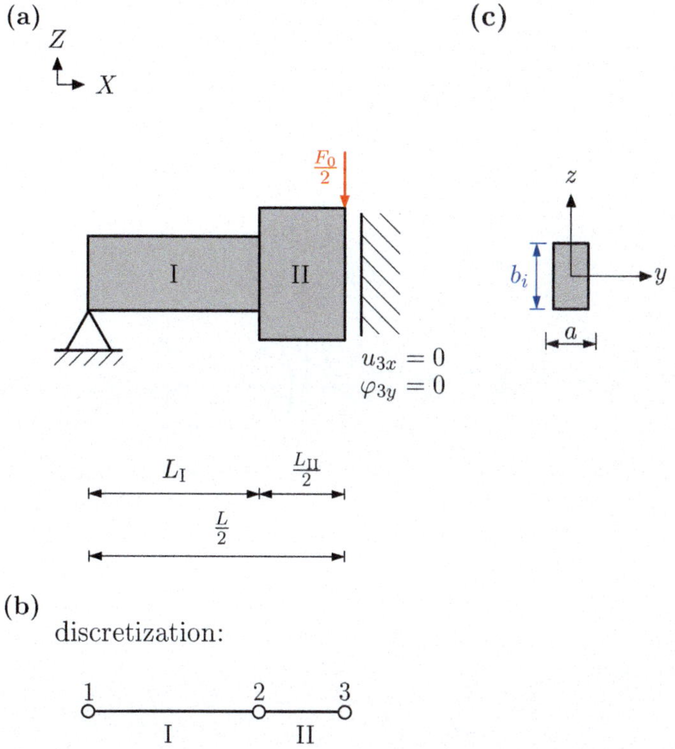

Fig. 71. Simply supported beam with three sections under consideration of symmetry: **a** general configuration, **b** discretization, and **c** cross-section of element i

$$\begin{bmatrix} \varphi_{1y} \\ u_{2z} \\ \varphi_{2y} \\ u_{3z} \end{bmatrix} = \begin{bmatrix} \frac{F_0 L^2 \left(4{b_{\mathrm{II}}}^3 + 5{b_{\mathrm{I}}}^3\right)}{12 E a\, {b_{\mathrm{I}}}^3 {b_{\mathrm{II}}}^3} \\ -\frac{F_0 L^3 \left(8{b_{\mathrm{II}}}^3 + 15{b_{\mathrm{I}}}^3\right)}{108 E a\, {b_{\mathrm{I}}}^3 {b_{\mathrm{II}}}^3} \\ \frac{5 F_0 L^2}{12 E a\, {b_{\mathrm{II}}}^3} \\ -\frac{F_0 L^3 \left(8{b_{\mathrm{II}}}^3 + 19{b_{\mathrm{I}}}^3\right)}{108 E a\, {b_{\mathrm{I}}}^3 {b_{\mathrm{II}}}^3} \end{bmatrix}. \tag{578}$$

The bending moment and shear force distributions within a single element of length L_i can be generally expressed using (the index '1' refers to the start node, while the index '2' refers to the end node of an element) the following relations:

$$\begin{aligned} M_y^{\mathrm{e}}(x) = & EI_y \left(\left[+\frac{6}{L_i^2} - \frac{12x}{L_i^3} \right] u_{1z} + \left[-\frac{4}{L_i} + \frac{6x}{L_i^2} \right] \varphi_{1y} \right. \\ & \left. + \left[-\frac{6}{L_i^2} + \frac{12x}{L_i^3} \right] u_{2z} + \left[-\frac{2}{L_i} + \frac{6x}{L_i^2} \right] \varphi_{2y} \right), \end{aligned} \tag{579}$$

$$Q_z^{\mathrm{e}}(x) = EI_y \left(\left[-\frac{12}{L_i^3} \right] u_{1z} + \left[+\frac{6}{L_i^2} \right] \varphi_{1y} + \left[+\frac{12}{L_i^3} \right] u_{2z} + \left[+\frac{6}{L_i^2} \right] \varphi_{2y} \right). \quad (580)$$

Thus, the normal and shear stress distributions can generally be calculated as follows:

$$\sigma_x^{\mathrm{e}}(x,z) = \frac{M_y^{\mathrm{e}}(x)}{I_y} \times z, \quad (581)$$

$$\tau_{xz}^{\mathrm{e}}(x,z) = \frac{Q_z^{\mathrm{e}}(x)}{2I_y} \left[\left(\frac{b}{2} \right)^2 - z^2 \right]. \quad (582)$$

The nodal values of the internal reactions for calculating the normal and shear stresses are summarized in Table 6. It is noted here that in this case the nodal values based on the finite element calculation are identical to the analytical solution.

Table 6. Nodal values of the internal reactions bending moment (M_y) and shear force (Q_z) at each node

	Element I		Element II	
	Left node	Right node	Left node	Right node
M_y	0	$-\frac{F_0 L}{6}$	$-\frac{F_0 L}{6}$	$-\frac{F_0 L}{4}$
Q_z	$-\frac{F_0}{2}$	$-\frac{F_0}{2}$	$-\frac{F_0}{2}$	$-\frac{F_0}{2}$

From Table 6 it can be concluded that the maximum normal stress in each element is reached at the right node and that the shear stress is constant in each element. Thus, the critical stresses in each element can be calculated as follows:

$$\sigma_{x,\mathrm{I}} = -\frac{F_0 L}{a\, b_{\mathrm{I}}{}^2}, \quad (583)$$

$$\sigma_{x,\mathrm{II}} = -\frac{3F_0 L}{2a\, b_{\mathrm{II}}{}^2}, \quad (584)$$

or for the shear stresses:

$$\tau_{xz,\mathrm{I}} = -\frac{3F_0}{4a\, b_{\mathrm{I}}}, \quad (585)$$

$$\tau_{xz,\mathrm{II}} = -\frac{3F_0}{4a\, b_{\mathrm{II}}}. \quad (586)$$

The objective function of the optimization problem, i.e. the mass of the stepped beam, can be represented as a function of the two design variables $X_1 = b_{\mathrm{I}}$ and $X_2 = b_{\mathrm{II}}$ as follows:

$$F(X_1, X_2) = \varrho L_{\mathrm{I}} a X_1 + \varrho L_{\mathrm{II}} a X_2 = \varrho L a \left(\frac{X_1}{3} + \frac{X_2}{6} \right), \tag{587}$$

where the minimization is to be carried out under consideration of the following seven inequality constraints:

$$g_1(X_1, X_2) = + \frac{F_0 L^3 (19 X_1^3 + 8 X_2^3)}{108 E a X_1^3 X_2^3 r_1 L} - 1 \leq 0 \quad \text{(max. deflection)}, \tag{588}$$

$$g_2(X_1, X_2) = \frac{F_0 L}{a X_1^2} - R_{\mathrm{p}0.2} \leq 0 \quad \text{(normal stress in I)}, \tag{589}$$

$$g_3(X_1, X_2) = \frac{3 F_0 L}{2 a X_2^2} - R_{\mathrm{p}0.2} \leq 0 \quad \text{(normal stress in II)}, \tag{590}$$

$$g_4(X_1, X_2) = \frac{3 F_0}{4 a X_1} - \frac{R_{\mathrm{p}0.2}}{2} \leq 0 \quad \text{(shear stress in I)}, \tag{591}$$

$$g_5(X_1, X_2) = \frac{3 F_0}{4 a X_2} - \frac{R_{\mathrm{p}0.2}}{2} \leq 0 \quad \text{(shear stress in II)}, \tag{592}$$

$$g_6(X_1, X_2) = X_1 - 20a \leq 0 \quad \text{(height-to-width ratio in I)}, \tag{593}$$

$$g_7(X_1, X_2) = X_2 - 20a \leq 0 \quad \text{(height-to-width ratio in II)}. \tag{594}$$

The graphical representation of the objective function and the inequality constraints (g_1, g_2, g_3) in the X_1–X_2 design space is shown in Fig. 72.

From Fig. 72 it can be concluded that considering g_1, g_2, and g_3 can be sufficient if a reasonable range of design variables is chosen. Furthermore, from Fig. 72 it can be concluded that the minimum is achieved for the case that g_1 is tangent to the objective function F. In order to derive an analytical solution for the minimum, one must take into account that the objective function F, in the representation $X_2(X_1) = \frac{6c}{\varrho L a} - 2X_1$, represents straight lines with a slope of -2. Thus, the condition for the minimum is that the curve g_1 assumes a slope of -2. Based on Eq. (588), the expression for g_1 can be rewritten:

$$X_2 = \left(- \frac{19 F_0 L^3 X_1^3}{8 F_0 L^3 - 108 E a r_1 L X_1^3} \right)^{\frac{1}{3}}, \tag{595}$$

which, after calculation, yields the expression of the first-order derivative as follows:

$$\left. \frac{\mathrm{d}X_2}{\mathrm{d}X_1} \right|_{g_1} = \frac{219^{\frac{1}{3}} F^{\frac{4}{3}} L^{\frac{8}{3}}}{(8 F L^2 - 108 E a r_1 {X_1}^3)^{\frac{1}{3}} (27 E a r_1 {X_1}^3 - 2 F L^2)} \overset{!}{=} -2. \tag{596}$$

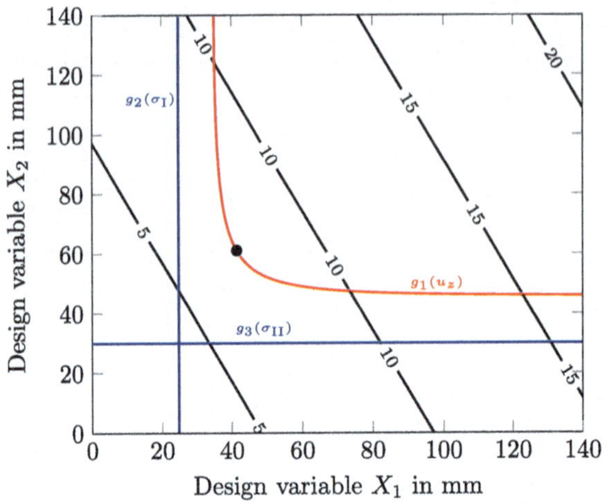

Fig. 72. Optimal design of a simply supported beam with three sections (consideration of symmetry): $L = 2540\,\text{mm}$, $a = 45.16\,\text{mm}$, $r_1 = 0.01$, $F_0 = 2667\,\text{N}$, $E = 68{,}948\,\text{MPa}$, $R_{p0.2} = 247\,\text{MPa}$, $\varrho = 2.691 \times 10^{-6}\,\text{kg/mm}^3$, (original set of equations (587)–(594)). Exact solution for the minimum: $X_{1,\text{extr}} = 41.437$ mm, $X_{2,\text{extr}} = 61.173$ mm (indicated by the • marker)

The last equation can be rearranged so that one side has a root, and Newton's method can be used to determine the root. It is important that the root is sought in a reasonable interval, e.g. $40 \leq X_{1,\text{ extr}} \leq 60$.

References

1. Altenbach, H.: Holzmann/Meyer/Schumpich Technische Mechanik Festigkeitslehre. Springer Vieweg, Wiesbaden (2016)
2. Ashby, M.F., Jones, D.R.H.: Engineering Materials 1: An Introduction to Properties. Applications and Design. Elsevier, Amsterdam (2005)
3. Ashby, M.F.: Materials Selection in Mechanical Design. Butterworth-Heinemann, Burlington (2011)
4. Boresi, A.P., Schmidt, R.J.: Advanced Mechanics of Materials. Wiley, New York (2003)
5. Bruhn, E.F.: Analysis and Design of Flight Vehicle Structures. Jacobs Publishing, Indianapolis (1973)
6. IATA: Jet fuel price monitor. http://www.iata.org/publications/economics/fuel-monitor/Pages/index.aspx. Accessed 15 April 2017
7. Javanbakht, Z., Öchsner, A.: Computational Statics Revision Course. Springer, Cham (2018)
8. Kim, H.-S.: New extruded multi-cell aluminum profile for maximum crash energy absorption and weight Efficiency. Thin Wall. Struct. **40**(4), 311–327 (2002). https://doi.org/10.1016/S0263-8231(01)00069-6
9. Klein, B.: Leichtbau-Konstruktion: Berechnungsgrundlagen und Gestaltung. Vieweg+Teubner, Wiesbaden (2009)

10. Lufthansa Group: Balance issue 2016. https://www.lufthansagroup.com/fileadmin/downloads/en/LH-sustainability-report-2016.pdf. Accessed 25 April 2017
11. Öchsner, A.: Elasto-Plasticity of Frame Structure Elements: Modeling and Simulation of Rods and Beams. Springer, Berlin (2014)
12. Öchsner, A.: A Project-Based Introduction to Computational Statics. Springer, Cham (2018)
13. Öchsner, A., Makvandi, R.: Numerical Engineering Optimization: Application of the Computer Algebra System Maxima. Springer Nature, Cham (2020)
14. Öchsner, A.: Stoff- und Formleichtbau: Leichter Einstieg mit eindimensionalen Strukturen. Springer Vieweg, Wiesbaden (2022)
15. Öchsner, A.: Computational Statics and Dynamics: An Introduction Based on the Finite Element Method. Springer, Singapore (2023)
16. Öchsner, A.: Aufgaben und Lösungen zum Stoff- und Formleichtbau. Springer Vieweg, Wiesbaden (2023)
17. Ohrn, K.E.: Aircraft energy use. In: Capehart, B.L. (ed.) Encyclopedia of Energy Engineering and Technology, vol. 1, pp. 24–30. CRC Press, Boca Raton (2007)
18. Rezvani, M.J., Jahan, A.: Effect of initiator, design, and material on crashworthiness performance of thin-walled cylindrical tubes: a primary multi-criteria analysis in lightweight design. Thin Wall. Struct. 96, 169–182 (2015). https://doi.org/10.1016/j.tws.2015.07.026
19. Rothwell, A.: Optimization Methods in Structural Design. Springer, Cham (2017)
20. Schumacher, A.: Optimierung mechanischer Strukturen: Grundlagen und industrielle Anwendungen. Springer Vieweg, Berlin (2013)
21. Spillers, W.R., MacBain, K.M.: Structural Optimization. Springer, New York (2009)
22. Timoshenko, S.P., Gere, J.M.: Theory of Elastic Stability. McGraw-Hill, Singapore (1963)
23. Vanderplaats, G.N.: Numerical Optimization Techniques for Engineering Design. Vanderplaats Research & Development, Colorado Springs (1999)
24. Warrior, N.A., Turner, T.A., Robitaille, F., Rudd, C.D.: The effect of interlaminar toughening strategies on the energy absorption of composite tubes. Compos. Part A-Appl. Sci. Manuf. **35**(4), 431–437 (2004). https://doi.org/10.1016/j.compositesa.2003.11.001

The manufacturer's authorised representative in the EU is Springer Nature Customer Service Centre GmbH, Europaplatz 3, 69115 Heidelberg, Germany. If you have any concerns regarding our products, please contact ProductSafety@springernature.com

Printed and bound by CPI Group (UK) Ltd, Croydon, CR0 4YY
01/10/2026
02225901-0002